From Clone to Bone
The Synergy of Morphological and Molecular Tools
in Palaeobiology

Since the 1980s, a renewed understanding of molecular development has afforded an unprecedented level of knowledge of the mechanisms by which phenotype in animals and plants has evolved. In this volume, top scientists in these fields provide perspectives on how molecular data in biology help to elucidate key questions in estimating palaeontological divergence and in understanding the mechanisms behind phenotypic evolution. Palaeobiological questions such as genome size, digit homologies, genetic control cascades behind phenotype, estimates of vertebrate divergence dates, and rates of morphological evolution are addressed, with a special emphasis on how molecular biology can inform palaeontology, directly and indirectly, to better understand life's past. Highlighting a significant shift towards interdisciplinary collaboration, this is a valuable resource for students and researchers interested in the integration of organismal and molecular biology.

Robert J. Asher is a Lecturer and Curator of Vertebrates in the University Museum of Zoology, Cambridge, UK. He is a vertebrate palaeontologist, specializing in mammals, with interests in phylogenetics and development.

Johannes Müller is Professor of Palaeozoology at the Natural History Museum, Humboldt University, Berlin, Germany. He is a palaeobiologist, focusing on the evolutionary diversification of fossil and recent reptiles.

Cambridge Studies in Morphology and Molecules: New Paradigms in Evolutionary Biology

This new Cambridge series addresses the interface between morphological and molecular studies in living and extinct organisms. Areas of coverage include evolutionary development, systematic biology, evolutionary patterns and diversity, molecular systematics, evolutionary genetics, rates of evolution, new approaches in vertebrate palaeontology, invertebrate palaeontology, palaeobotany, and studies of evolutionary functional morphology. The series invites proposals demonstrating innovative evolutionary approaches to the study of extant and extinct organisms that include some aspect of both morphological and molecular information. In recent years the conflict between "molecules vs. morphology" has given way to more open consideration of both sources of data from each side, making this series especially timely.

From Clone to Bone

The Synergy of Morphological and Molecular Tools in Palaeobiology

EDITED BY

Robert J. Asher

Museum of Zoology,
University of Cambridge, UK

and

Johannes Müller

Museum für Naturkunde/Humboldt
University of Berlin, Germany

CAMBRIDGE
UNIVERSITY PRESS

CAMBRIDGE
UNIVERSITY PRESS

University Printing House, Cambridge CB2 8BS, United Kingdom

One Liberty Plaza, 20th Floor, New York, NY 10006, USA

477 Williamstown Road, Port Melbourne, VIC 3207, Australia

314-321, 3rd Floor, Plot 3, Splendor Forum, Jasola District Centre, New Delhi - 110025, India

103 Penang Road, #05-06/07, Visioncrest Commercial, Singapore 238467

Cambridge University Press is part of the University of Cambridge.

It furthers the University's mission by disseminating knowledge in the pursuit of
education, learning and research at the highest international levels of excellence.

www.cambridge.org
Information on this title: www.cambridge.org/9781107003262

© Cambridge University Press 2012

First published 2012

A catalogue record for this publication is available from the British Library

Library of Congress Cataloging in Publication data
From clone to bone: the synergy of morphological and molecular tools in palaeobiology /
edited by Robert J. Asher, Johannes Müller.
    pages   cm. – (Cambridge studies in morphology and molecules; 4)
    Includes bibliographical references and index.
    ISBN 978-1-107-00326-2 (Hardback) – ISBN 978-0-521-17676-7 (Paperback)
1. Evolutionary paleobiology.   2. Paleobiology–Methodology.   3. Morphology (Animals)
4. Morphogenesis.   5. Molecular biology.   I. Asher, Robert J.   II. Müller, Johannes, 1973–
    QE721.2.E85F76 2012
    560–dc23   2012014611

ISBN 978-1-107-00326-2 Hardback
ISBN 978-0-521-17676-7 Paperback

Contents

The colour plates are situated between pages 182 and 183.

Contributors

Neal Anthwal, Department of Craniofacial Development, King's College London, UK

Robert J. Asher, Department of Zoology, University of Cambridge, UK

Robin Beck, Department of Mammalogy, American Museum of Natural History, New York, USA

Olaf R. P. Bininda-Emonds, Institute for Biology and Environmental Sciences, Carl von Ossietzky University Oldenburg, Oldenburg, Germany

Emily A. Buchholtz, Department of Biological Sciences, Wellesley College, MA, USA

Xiaoyi Cao, Center for Biophysics and Computational Biology, University of Illinois, Urbana, IL, USA

Merijn A. G. de Bakker, Institute of Biology, Leiden University, The Netherlands

T. Alexander Dececchi, Redpath Museum, McGill University, Montreal, Canada

Carolyn K. Doroba, Department of Animal Biology, University of Illinois, Urbana, IL, USA

Luke B. Harrison, Redpath Museum, McGill University, Montreal, Canada

Rafael Jiménez, Departamento de Genética, University of Granada, Spain

Zerina Johanson, Department of Palaeontology, Natural History Museum, London, UK

Peter Kondrashov, Anatomy Department, Kirksville College of Osteopathic Medicine, A.T. Still University of Health Sciences, Kirksville, MO, USA

Shigeru Kuratani, Laboratory for Evolutionary Morphology, RIKEN Center for Developmental Biology, Kobe, Japan

Hans C. E. Larsson, Redpath Museum, McGill University, Montreal, Canada

Ross D. E. MacPhee, Department of Mammalogy, American Museum of Natural History, New York, USA

José Ezequiel Martín, Instituto de Parasitología y Biomedicina López-Neyra, CSIC, Granada, Spain

Christian Mitgutsch, RIKEN Center for Developmental Biology, Laboratory for Evolutionary Morphology, Kobe, Japan

Johannes Müller, Museum für Naturkunde, Leibniz-Institut für Evolutions- und Biodiversitätsforschung an der Humboldt Universität zu Berlin, Germany

Hiroshi Nagashima, Department of Regenerative and Transplant Medicine, Niigata University, Niigata, Japan

Chris Organ, Department of Organismic and Evolutionary Biology, Harvard University, Cambridge, MA, USA

Michael K. Richardson, Institute of Biology, Leiden University, The Netherlands

Marcelo R. Sánchez-Villagra, Paläontologisches Institut und Museum, University of Zürich, Switzerland

Leonhard Schmid, Paläontologisches Institut und Museum, University of Zürich, Switzerland

Karen E. Sears, Department of Animal Biology and Institute for Genomic Biology, University of Illinois, Urbana, IL, USA

Carl Simpson, Museum für Naturkunde, Leibniz-Institut für Evolutions- und Biodiversitätsforschung, Humboldt University of Berlin, Germany

Moya Smith, Department of Croniofacial and Stem Cell Biology, Dental Institute, King's College London, UK

Abigail S. Tucker, Department of Craniofacial and Stem Cell Biology, Dental Institute, King's College London, UK

Dan Xie, Department of Bioengineering, University of Illinois, Urbana, IL, USA

Sheng Zhong, Department of Animal Biology, Institute for Genomic Biology, and Department of Statistics, University of Illinois, USA

1

Molecular tools in palaeobiology: divergence and mechanisms

ROBERT J. ASHER AND JOHANNES MÜLLER

In 1987, Cambridge University Press published a volume entitled *Molecules and Morphology in Evolution: Conflict or Compromise?* edited by the esteemed British palaeobiologist Colin Patterson. Since the 1980s, we have witnessed a great deal of incorporation of the tools and data of molecular biology into palaeontological hypothesis building and testing. The degree of integration is substantial enough so as to rule out the rather pejorative subtitle of the 1987 volume, 'conflict or compromise'. We believe a new designation is appropriate: 'synergy'. Stated differently, our ability to address major questions in biological history requires the integration of molecular methods and data into the palaeobiologist's toolkit. The antagonism implicit in the notion of 'conflict or compromise' is more an artefact of disciplinary boundaries and analytical traditions, and is not firmly rooted in the data of biology. Palaeobiologists today routinely consider data from molecular biology in their research on the shape and antiquity of the tree of life ('divergence'), and in understanding the genetic and developmental mechanisms behind morphological change ('mechanisms').

This book documents aspects of this synergy, focusing on these two general categories: divergence and mechanisms. It derives from the symposium 'molecular tools in palaeobiology' that took place during the 2009 meetings of the Society of Vertebrate Paleontology in Bristol, UK. In retrospect, we realize that the 'vertebrate' orientation of that conference has resulted in a level of taxonomic focus in this book that excludes many important contributions regarding evolutionary divergence and mechanisms. Nevertheless, this is no small taxonomic category, and there has been much to say about it since 1987.

The unifying theme of that symposium, as in this edited volume, is general: how have molecular methods become critical in particular subfields of palaeobiology, and how does this constitute interdisciplinary synergy to understand life history?

From Clone to Bone: The Synergy of Morphological and Molecular Tools in Palaeobiology, ed. Robert J. Asher and Johannes Müller. Published by Cambridge University Press.
© Cambridge University Press 2012.

Divergence

Many aspects of the synergy we would like to emphasize in this book were evident to the authors of Patterson (1987). For example, on its cover, Patterson's volume depicted Ernst Haeckel's artistically rendered Tree of Life, published in German in 1874. From our perspective in 2011, careful inspection of that tree reveals the extraordinary extent to which the comparative anatomists of the nineteenth century 'got it right' in terms of the overall structure of animal, particularly vertebrate, phylogeny (Figure 1.1). Although biologists of the 1980s were unable to completely tease apart many issues in systematics – such as relations near the base of chordates, between birds and mammals, and among mammalian orders – for the species they had in common, Goodman et al. (1987) and Bishop and Friday (1987) proposed trees that were not far off from those outlined over a century before.

Concordance of vertebrate trees generated by molecular data today with those based on comparative anatomy of the nineteenth century is remarkable: not much has changed compared with the basic structure deciphered by naturalists two centuries ago. Haeckel (1874), Gill (1872) and others of their time considered dozens of major groups and hundreds of species, recognizing the nested relationships of vertebrates, gnathostomes, bony and cartilaginous fish, ray- and lobe-finned fish, tetrapods, amphibians and amniotes, diapsids and synapsids. Since then, there have been a few major questions regarding this understanding of vertebrate phylogeny, such as the possibility of a bird–mammal clade (Huxley 1868; Hedges et al. 1990; Gardiner 1993), the placement of coelacanths with ray-finned fish (Arnason et al. 2004), or the paraphyly of rodents (D'Erchia et al. 1996). However, subsequent analysis of more comprehensive data sets, often by the same investigators, has resolved these issues beyond any reasonable doubt (Hedges 1994; Murphy et al. 2001; Hallström and Janke 2009) and has left the Tree of Haeckel and Gill relatively unscathed. Changes within groups have occurred (particularly mammals), but perhaps the only substantial change among major vertebrate groups has been the recognition that tunicates are closer to vertebrates than Branchiostoma (Delsuc et al. 2008), switching two branches that Haeckel had already placed immediately adjacent to one another (Figure 1.1).

It's worth comparing, for example, the 88-taxa study by Hugall et al. (2007) with Haeckel's tree as displayed on the cover of Patterson (1987). The former (Hugall et al. 2007: figs. 1, 2) places hagfish and lamprey at the base of vertebrates, followed by a monophyletic frog–salamander–caecilian clade (they did not sample Chondrichthyes or Actinopterygia), followed by amniotes, which consist in turn of Mammalia as the sister taxon to an archosaur–squamate clade.

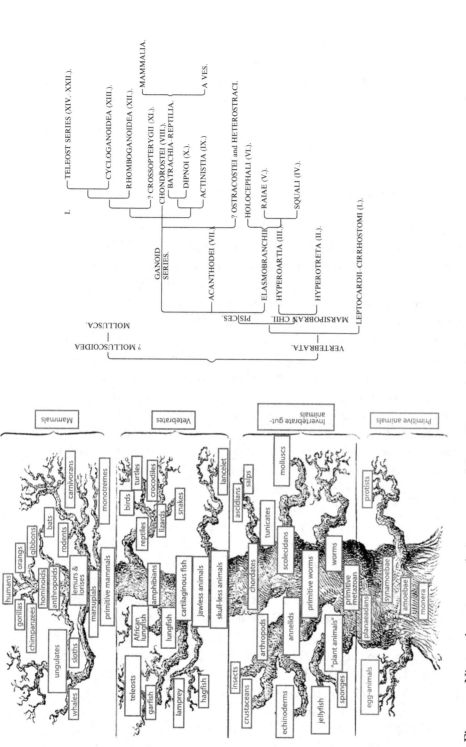

Figure 1.1 Nineteenth-century evolutionary trees of Ernst Haeckel (left) and Theodore Gill (right). Haeckel's image is from the 1897 English version of *Evolution of Man*, repeating his previously articulated views (e.g. Haeckel 1874, p. 513). Gill's phylogeny was published in 1872. Both show a remarkably high level of agreement with current ideas on vertebrate interrelationships, including the monophyly of craniate vertebrates, cyclostomes, gnathostomes, cartilaginous and bony fish, actinopterygians, sarcopterygians, tetrapods, amphibians and amniotes (Hugall *et al.* 2007, Delsuc *et al.* 2008). (See also colour plate.)

The 179-gene data set of Delsuc *et al.* (2008: figs. 2 and 3) across deuterostomes similarly supported vertebrates, gnathostomes, cartilaginous and bony fish, actinopterygians, tetrapods, amphibians and amniotes. This basic structure, as inferred by both Hugall *et al.* (2007) and Delsuc *et al.* (2008), does not differ substantially from that proposed by Haeckel (1874) or Gill (1872), extending even to cyclostomes (i.e. hagfish and lamprey as sister taxa) and turtles within archosaurs (i.e. birds and crocodiles).

To put this in context, there are approximately 2×10^{157} ways in which the 88 taxa sampled by Hugall *et al.* (2007) could be interconnected as rooted, bifurcating trees (Felsenstein 1978). Why is it that out of this extraordinary number of possibilities Hugall *et al.* (2007) honed in on essentially the same pattern as that proposed by Haeckel and Gill in the nineteenth century? The answer of course speaks to the prediction made by Darwin in 1859, that there is a genuine Tree of Life subject to reconstruction using the evidence left behind by the mechanism of descent with modification, evidence that manifests itself on a developmental continuum between genotype and phenotype. Haeckel and Gill knew about the phenotypic side of this continuum. Hugall *et al.* (2007) focused on one small part of genotype: sequences of the nuclear *Rag-1* gene; Delsuc *et al.* (2008) had a much larger genetic data set with fewer sampled vertebrate taxa. That each source of data supports such a specific hypothesis out of such an astronomical number of possibilities is testament to the reality of Darwinian evolution. A largely consistent topological signal across independent data sets is what one would expect if animals actually share genealogical history with one another via the mechanism of descent with modification (Penny *et al.* 1982).

Phylogenetic consensus has been more elusive for parts of the Tree of Life other than vertebrates that (1) contain much more ancient branching points, (2) have not been as thoroughly documented genomically and anatomically, (3) are more prone to phenomena such as lateral gene transfer, and/or (4) have a much more limited fossil record (Woese 1987; Doolittle and Bapteste 2007). Nevertheless, evolutionary biologists who are focused on the most difficult branches on the Tree of Life should not understate the consensus elsewhere that has proven robust. Because it is natural for investigators to focus on areas of controversy and disagreement, uncertainty at one level (e.g. do metazoans share a single common ancestor?) has, on occasion, been inaccurately portrayed as an irreconcilable stumbling block for the entire phylogenetic enterprise. For example, Patterson (1987: p. 4) quoted a particularly pessimistic passage from Ernst Mayr's 1982 book *The Growth of Biological Thought* (p. 218): 'The futile attempts to establish the major phyla of animals induced at least one competent zoologist ... [to call] common descent ... a beautiful myth not established by any factual foundation. ... Honesty compels us to admit that our ignorance concerning these relationships is still great, not to say overwhelming.' The zoologist to which Mayr is referring is Fleischmann (1901), hardly representative of biological thought from Mayr's perspective in 1982. Broadly speaking, this passage takes genuine uncertainty regarding pre-Cambrian divergences and overgeneralizes from this to common descent itself. On the contrary, given the agreement we now have about vertebrate interrelations (for example) such hyperbole is inaccurate and misleading for both students and the general public.

The scrutiny of protein sequences starting in the 1960s did indeed overturn several cherished ideas in systematics, such as an ancient human lineage to the exclusion of other great apes dating to the early Miocene (see Goodman *et al.* 1987; Andrews 1987). Zuckerkandl and Pauling's (1962: table 2) early application of the molecular clock to divergences within great apes yielded a surprisingly prescient result. For the differences they observed in human and gorilla alpha and beta haemoglobin molecules (one and two amino acid substitutions, respectively), they estimated a common ancestor between the two species to have existed approximately 11 million years ago. Such an interpretation is much closer to the view accepted today than to the previous theory that early Miocene apes, such as '*Ramapithecus*', shared ancestry with humans to the exclusion of other great apes (Simons 1972).

Interestingly, the calibration used for Zuckerkandl and Pauling's estimate was 'the common ancestor of man and horse [which] lived in the Cretaceous or possibly in the Jurassic period, say between 100 and 160 millions of years ago' (Zuckerkandl and Pauling 1962: p. 200). In hindsight, it is interesting to note that while some recent molecular clock studies support 'common ancestor of man and horse' (i.e. the divergence between Laurasiatheria and Euarchontoglires) just about 100 million years ago in the Cretaceous (Bininda-Emonds *et al.* 2007, but see Kitazoe *et al.* 2007, Hallström and Janke 2010, or Dos Reis *et al.* 2012), as a calibration this divergence has no palaeontological basis whatsoever (Wible *et al.* 2007; Benton *et al.* 2009). Nevertheless, it yielded a result for hominine divergence which is not far off from that accepted today based on our understanding of both the fossil record and a molecular clock (Lockwood 2007).

To some specialists of the mid/late twentieth century, even one with such a major influence and apparently broad perspective as Ernst Mayr, it seemed that after dethroning the idea of an independent human lineage dating to the early Miocene, all other such cherished ideas were soon to follow. They didn't. Again, and with the important qualification that many surprises have occurred within certain groups (e.g. mammals; see Asher *et al.* 2009), the genomic work of the last decade has confirmed the basic vertebrate topology first recognized in the nineteenth century, not scrambled it (Figure 1.1).

For mammals, at least, there remains today more uncertainty about divergence dates than topology among living clades. Bininda-Emonds *et al.* (this volume) have made substantial contributions towards understanding mammalian divergences. Building upon a supertree meta-analysis of mammals (Bininda-Emonds *et al.* 2007), one which has a far larger taxon sample of mammals than any previous study, they make the case that ordinal divergences within mammals substantially predate the Cretaceous–Tertiary (KT) boundary at 65 million years before present. Palaeontologists have known for some time that the earliest occurrences of undisputed crown-placental mammals do not predate the KT boundary (Kielan-Jaworowska 1978; Asher *et al.* 2005; Wible *et al.* 2009). Given the obvious presence of such clades as rodents and carnivorans in the early Palaeocene, it is reasonable to expect that at least some placental divergences predate the KT boundary. We know that first appearances in the fossil record are not synonymous with actual cladistic divergence (Benton *et al.* 2009). However, a missing record of crown placentals extending to the early Cretaceous, over 120 million years ago (Kumar and Hedges 1998), seems too much in conflict with our growing understanding of the Cretaceous fossil record (Foote *et al.* 1999; Reisz and Müller 2004; Hunter and Janis 2006), particularly given the limitations and pitfalls of the molecular clock (Graur and Martin 2004; Kitazoe *et al.* 2007). Bininda-Emonds *et al.* (this volume) offer perspective on this ongoing debate on the Mesozoic antiquity of crown-placental mammals.

Estimating divergence times from differences in molecular evolution is only one way of measuring rates of evolutionary change. Evolutionary rates can also be assessed in terms of morphological evolution, such as phenotypic or taxonomic change (Polly 2001). Simpson's (1944) classic work *Tempo and Mode in Evolution* indicates how palaeontologists have sought for many years to assess how evolutionary change can be measured and, ideally, quantified from a morphological perspective. The question remains, however, as to how morphological change through time can or should be measured, and if it is possible to use morphological data complementary to molecular methods for assessing rates of evolution. Larsson *et al.* (this volume) address this issue from a novel perspective and present a road map for using morphology not only as a tool to measure evolutionary rates through time, but also to estimate divergence times on a similar basis as molecular approaches.

Recent years have seen tremendous progress in the field of genomics as a result of major advances in DNA extraction and sequencing. At first glance this type of research appears unrelated to palaeobiology because of its strictly genetic nature and necessary focus on living forms. However, not only is it possible to reveal genomic data from extinct organisms (Paäbo 1989; Noonan *et al.* 2005), including some that are truly ancient (Organ *et al.* 2008), but in combination with methods from phylogenetic and comparative biology, fossilized structures can be used in conjunction with genomic data to gain insights into the biology of long extinct taxa (Organ *et al.* 2007; 2009). A review of this issue by Organ (this volume), demonstrates the great potential for 'palaeogenomics' in biology, and how it can reveal insights into both deep time and the recent past. He illustrates this potential with recent work on ancient genome size and sex chromosome reconstruction for species that have been extinct for tens of millions of years.

Estimating rates of speciation and extinction, and thus diversification, is a classical theme in macroevolution and for a long time could only be approached using fossil taxa (Raup 1978; Sepkoski *et al.* 1981). In recent years, major advances in bioinformatics and gene sequencing techniques enabled researchers to also use DNA-derived phylogenetic hypotheses and divergence estimates to calculate diversification rates (see Ricklefs *et al.* 2007). This novel approach did not make the palaeontological method obsolete, however, but instead stimulated and fostered new research on the underlying processes of diversification, resulting in a wealth of new studies from both neontology and palaeontology. The progress in this field also made it possible to reconsider a classical question of evolutionary theory: how does natural selection work, and on how many levels? Simpson and Müller (this volume) address the concept of species selection, an issue with a long tradition in palaeontology and often discussed as a possible evolutionary process. They describe how molecular phylogenies can be used to measure the extent and impact of species selection, and suggest that species selection is ubiquitous, providing deeper insights into the causes of diversification.

Mechanisms

All of the chapters in Patterson's 1987 book concerned the Tree of Life, methods for its reconstruction, and aspects of antiquity as inferred by the fossil record and molecular clocks. The focus of those authors was very much on phylogenetics, and virtually nothing was said regarding a field that is now of considerable importance to palaeontologists: evolutionary development. During the 1980s 'Evo-Devo' was of course a sophisticated discipline (Akam *et al.* 1988; Keynes and Stern 1988), but was generally the domain of those working on model organisms and beyond the practical remit of most palaeontologists.

This is quite different today, and the disciplinary scopes of palaeobiology and evolutionary development have become increasingly intertwined (e.g. Smith and Hall 1990; Hall 2002; Smith 2003; Donoghue and Purnell 2009; Sánchez-Villagra

2010; Schmid and Sánchez-Villagra 2010; Takechi and Kuratani 2010). This coalescence is reflected in the majority of chapters in this book (summarized below) in which concepts of ontogeny and developmental genetics are applied to palaeontological questions, in stark contrast to the virtual absence of developmental subject matter in the chapters of Patterson (1987). While both Patterson's book and this one include too few chapters to comprise an infallible barometer of the course of palaeobiological thought over three decades, it is nevertheless tempting to view this difference as an indication of how the understanding of developmental genetics of model organisms has become a major springboard for hypothesis generation and testing in palaeobiology.

For example, Anthwal and Tucker (this volume) note the diversity of form among mammalian mandibles, in particular the variable presence and size of the condylar, coronoid and angular processes. Certain adult phenotypes such as the small, un-marsupial-like mandibular angle of the adult koala, are late occurrences during development (Sánchez-Villagra and Smith 1997); due to differential growth of other parts of the jaw during ontogeny, the mandibular angle may become more (or less) apparent. The coronoid process of the dentary is similarly variable; the mandibular condyle exhibits some variation across mammals but less than the angular and coronoid processes. Perhaps more importantly, specific phenotypes of the condylar, coronoid and angular processes observed in nature have also been observed within the phenotypic repertoire documented in past studies of transgenic mice and human genetic disorders. Anthwal and Tucker review these morphologies and note the complicated, but tractable, roles of several genetic loci among mouse knockout studies and human pathologies that are likely behind the variation observed in nature.

Buchholtz (this volume) discusses a long history of the study of the vertebrate axial skeleton, focusing on some of the mechanisms by which mammals in particular exhibit phenotypic diversity despite an extraordinary level of conservatism throughout the Order. Relative to other chordates, mammals show relatively little variation in vertebral counts, particularly among more cranial vertebral segments. The identification of skeletal modules, for example those patterned in somitic versus lateral plate mesoderm, has gained some support in the recent literature (Hautier et al. 2010). In the context of recent discoveries regarding the differential expression of *Hox* genes in specific mesodermal tissues (McIntyre et al. 2007), Buchholtz documents phenotypic modules across species and contributes to the understanding of the means by which mammals may deviate from a fundamentally conserved skeletal body plan.

One of the greatest enigmas in vertebrate evolution is the origin of turtles and, relatedly, the turtle shell (Rieppel 2009). Turtles are unique among chordates by possessing a shoulder girdle inside the ribcage. How this change occurred phylogenetically and developmentally has long been a perplexing question. Recent years have seen tremendous progress in both developmental studies on the evolution of the turtle shell (Nagashima *et al.* 2009) and the testudinate fossil record, for example the recent discovery of the Triassic turtle *Odontochelys semitestacea* (Li *et al.* 2008), a taxon with a complete plastron but rudimentary carapace. Kuratani and Nagashima (this volume) provide a thorough overview of the current state of the art and elegantly show how our current knowledge from the turtle fossil record can be combined with innovative studies from developmental biology, providing an explanatory framework for the origin and evolution of the turtle shell. According to Kuratani and Nagashima (this volume), *Odontochelys* resembles an early stage in modern turtle ontogeny, prior to the onset of carapace development, and it is the embryonic carapacial ridge that leads to the reversed positions of scapula and ribs through inward folding of the ventral body wall.

Palaeontology and developmental genetics have been successfully integrated in studies of tetrapod limb evolution, including the 'fin-limb transition', the increase or reduction of digits, and the overall loss of extremities (e.g. Shubin *et al.* 1997; Cohn and Tickle 1999; Shapiro *et al.* 2003). In many cases hypotheses generated from developmental genetics have been used to explain and understand the phenotypic diversity of limbs as revealed by the fossil record, and vice versa (Shubin *et al.* 2009). In the present volume, this area is tackled using two different examples, both focusing on the evolution of the autopodium. In the first one, Mitgutsch *et al.* (this volume) use the 'mole's thumb', a distinctive sesamoid bone found in talpid mammals, to address the longstanding issue of how to determine homology and understand morphological diversification. They emphasize how studies of developmental genetics require morphology to formulate appropriate research questions. In the second contribution, Richardson (this volume) reviews the recent advances in our understanding of the development of the bird wing, a hot topic in vertebrate palaeontology because of its relevance to the interrelationships between modern birds and Mesozoic theropod dinosaurs (e.g. Burke and Feduccia 1997; Wagner and Gauthier 1999; Larsson and Wagner 2002). Richardson shows that if only one single research discipline is considered, a proper understanding of the evolution of the avian wing is impossible, highlighting the need for integration.

Schmid (this volume) critically investigates the basis for inferring molecular mechanisms behind phenotype in long-extinct organisms. Applying basic principles of parsimony (simplicity is optimal), uniformitarianism (modern processes are relevant to past events) and actualism (the past is a manifestation of our own world), Schmid outlines possible genetic mechanisms behind several aspects of the peculiar phenotype of the Triassic actinopterygian fish

Saurichthys, including body regionalization, fin position, jaw elongation, and scale phenotype. Specific mechanisms concerning somitogenesis and the expression of (for example) retinoic acid, *FGF*, *Eda*, and paralogues of *Hox* and *Tbx* are known to influence these phenotypes in modern organisms such as the zebrafish, cichlid fish, chicken and mouse. Close correspondence not only in detailed phenotype of the axial skeleton, fins, jaws and scales, but also correlations among these hard tissues, make a compelling case that the genetic basis of its phenotype is to a signficant extent tractable in a 240-million-year-old vertebrate.

Linking phenotype and genotype is also a focus of Sears *et al.* (this volume), who synthesize a large body of data concerning the embryology and genetics of marsupial forelimb development. They start with the observation that forelimb and hind limb development in marsupials, unlike that observed in other mammalian groups, is relatively uncoupled. At the time of birth, marsupials such as the short-tailed opossum (*Monodelphis domestica*) have no hind limb to speak of, but have just enough of a forelimb and rostrum to enable them to locate and climb from the reproductive tract to the mother's teat. Here, outside of the uterus, the majority of their embryonic and foetal growth occurs. Sears *et al.* (this volume) apply a microarray assay to survey how gene expression is different in forelimb and hind limb, and between mouse and *Monodelphis*. Both comparisons yield a number of interesting differences. Sears *et al.* (this volume) interpret their findings as supportive of a breakdown in the genetic modularity between forelimb and hind limb early in the course of marsupial evolution, enabling the high level of early specialization of the marsupial forelimb given its peculiar reproductive mode. This is potentially why we do not see, either in the fossil record or among living species, marsupials with divergent limb morphology such as hooves, flippers or wings.

Teeth are among the most common and important remains represented in the fossil record. Smith and Johanson (this volume) comprehensively summarize the debate regarding the origin of teeth in vertebrates. Embryologists and comparative anatomists have recognized for some time the common development that teeth share with other ectodermal structures, such as scales and hair. Yet there are important differences. For example, Smith and Johanson note that tooth addition in fossil taxa such as arthrodiran fish (Placodermi) occurs at the end of pre-existing dental rows, and is consistent with the tooth addition pattern observed in contemporary teleosts and lungfish. The capacity for continuous replacement of teeth differs from dermal structures such as scales. They argue that a co-option of genetic networks responsible for purely dermal structures is unlikely to be the primary mechanism by which teeth evolved among vertebrates. Rather, genetic networks regulating pattern of tooth addition and their replacement present in the internal pharyngeal skeleton of a fossil agnathan group is likely to have played a role in the early evolution of teeth.

Conclusions

The chapters in this volume encapsulate some of the most exciting topics in contemporary palaeobiology. It is rewarding to witness, in a relatively short period of time, how the genomics and informatics revolutions of the last 20 years have enriched the study of past life on Earth. Disciplinary boundaries, for example between molecular biology and palaeontology, have become increasingly fuzzy; this has been an unqualified boon to scientists who are keen to test ever-more sophisticated hypotheses concerning Earth history and the evolution of biodiversity. The late twentieth and early twenty-first centuries witnessed an unprecedented level of coalescence among hypotheses regarding the vertebrate Tree of Life. The chapters in this book represent a small but important number of refinements to this understanding, particularly how palaeontologists using the tools of molecular biology now go about inferring the rate and antiquity of vertebrate clades. To an even greater extent, they represent an increased understanding of the genetic basis of phenotype, even for animal groups that have been extinct for millions of years.

REFERENCES

Akam, M., Dawson, I. and Tear, G. (1988). Homeotic genes and the control of segment diversity. *Development*, **104**, 123–33.

Andrews, P. (1987). Aspects of hominoid phylogeny. In *Molecules and Morphology in Evolution: Conflict or Compromise?*, ed. C. Patterson. Cambridge, UK: Cambridge University Press, pp. 23–34.

Arnason, U., Gullberg, A., Janke, A., Joss, J. and Elmerot, C. (2004). Mitogenomic analyses of deep gnathostome divergences: a fish is a fish. *Gene*, **333**, 61–70.

Asher, R. J., Bennett, N. and Lehmann, T. (2009). The new framework for understanding placental mammal evolution. *BioEssays*, **31**, 853–64.

Asher, R. J., Meng, J., Wible, J. R., *et al.* (2005). Stem Lagomorpha and the antiquity of Glires. *Science*, **307**, 1091–4.

Benton, M. J., Donoghue, P. C. J. and Asher, R. J. (2009). Calibrating and constraining molecular clocks. In *The Timetree of Life*, ed. S. B. Hedges and S. Kumar. Oxford, UK: Oxford University Press, pp. 35–86.

Bininda-Emonds, O. R., Cardillo, M., *et al.* (2007). The delayed rise of present-day mammals. *Nature*, **446**, 507–12.

Bishop, M. J. and Friday, A. E. (1987). Tetrapod relationships: the molecular evidence. In *Molecules and Morphology in Evolution: Conflict or Compromise?*, ed. C. Patterson. Cambridge, UK: Cambridge University Press, pp. 123–40.

Burke, A. C. and Feduccia, A. (1997). Developmental patterns and the identification of homologies in the avian hand. *Science*, **278**, 666–8.

Cohn, M. J. and Tickle, C. (1999). Developmental basis of limblessness and axial patterning in snakes. *Nature*, **399**, 474–9.

Delsuc, F., Tsagkogeorga, G., Lartillot, N. and Philippe, H. (2008). Additional molecular support for the new chordate phylogeny. *Genesis*, **46**, 592–604.

D'Erchia, A. M., Gissi, C., Pesole, G., Saccone, C. and Arnason, U. (1996). The guinea-pig is not a rodent. *Nature*, **381**, 597–600.

Donoghue, P. C. J. and Purnell, M. A. (2009). Distinguishing heat from light in debate over controversial fossils. *BioEssays*, **31**, 178–89.

Doolittle, W. F. and Bapteste, E. (2007). Pattern pluralism and the Tree of Life hypothesis. *Proceedings of the National Academy of Sciences of the United States of America*, **104**, 2043–9.

Dos Reis, M., Inoue, J., Hasegawa, M., *et al.* (2012). Phylogenomic datasets provide both precision and accuracy in estimating the timescale of placental mammal phylogeny. *Proceedings of the Royal Society B*, in press.

Felsenstein, J. (1978). The number of evolutionary trees. *Systematic Zoology*, **27**, 27–33.

Fleischmann, A. (1901). *Die Descendenztheorie*. Leipzig, Germany: Arthur Georgi.

Foote, M., Hunter, J. P., Janis, C. M. and Sepkoski, J. J., Jr. (1999). Evolutionary and preservational constraints on origins of biologic groups: divergence times of eutherian mammals. *Science*, **283**, 1310–4.

Gardiner, B. G. (1993). Haematothermia – warm-blooded amniotes. *Cladistics–the International Journal of the Willi Hennig Society*, **9**, 369–95.

Goodman, M., Miyamoto, M. M. and Czelusniak, J. (1987). Pattern and process in vertebrate phylogeny revealed by coevolution of molecules and morphologies. In *Molecules and Morphology in Evolution: Conflict or Compromise?*, ed. C. Patterson. Cambridge, UK: Cambridge University Press, pp. 141–76.

Gill, T. (1872). *Arrangement of the Families of Fishes*. Washington, DC: Smithsonian Institution.

Graur, D. and Martin, W. (2004). Reading the entrails of chickens: molecular timescales of evolution and the illusion of precision. *Trends in Genetics*, **20**, 80–6.

Haeckel, E. (1874). *Natürliche Schöpfungsgeschichte: Gemeinverständliche wissenschaftliche Vorträge über die Entwickelungslehre im Allgemeinen und diejenige von Darwin, Goethe und Lamarck im Besonderen*. Berlin: G. Reimer.

Hall, B. K. (2002). Palaeontology and evolutionary developmental biology: a science of the nineteenth and twenty-first centuries. *Palaeontology*, **45**, 647–69.

Hallström, B. M. and Janke, A. (2009). Gnathostome phylogenomics utilizing lungfish EST sequences. *Molecular Biology and Evolution*, **26**, 463–71.

Hallström, B. M. and Janke, A. (2010). Mammalian evolution may not be strictly bifurcating. *Molecular Biology and Evolution*, **27**, 2804–16.

Hautier, L., Weisbecker, V., Sanchez-Villagra, M. R., Goswami, A. and Asher, R. J. (2010). Skeletal development in sloths and the evolution of mammalian vertebral patterning. *Proceedings of the National Academy of Sciences of the United States of America*, **107**, 18 903–8.

Hedges, S. B. (1994). Molecular evidence for the origin of birds. *Proceedings of the National Academy of Sciences of the United States of America*, **91**, 2621–4.

Hedges, S. B., Moberg, K. D. and Maxson, L. R. (1990). Tetrapod phylogeny inferred from 18S and 28S ribosomal RNA sequences and a review of the evidence for amniote relationships. *Molecular Biology and Evolution*, **7**, 607–33.

Hugall, A. F., Foster, R. and Lee, M. S. (2007). Calibration choice, rate smoothing, and the pattern of tetrapod diversification according to the long nuclear gene RAG-1. *Systematic Biology*, **56**, 543–63.

Hunter, J. P. and Janis, C. (2006). Spiny Norman in the Garden of Eden? Dispersal and early biogeography of Placentalia. *Journal of Mammalian Evolution*, **13**, 89–123.

Huxley, T. H. (1868). On the animals which are most nearly intermediate between birds and reptiles. *Geological Magazine*, **5**, 357–65.

Keynes, R. J. and Stern, C. D. (1988). Mechanisms of vertebrate segmentation. *Development*, **103**, 413–29.

Kielan-Jaworowska, Z. (1978). Evolution of the therian mammals in the Late Cretaceous of Asia. III. Postcranial skeleton in Zalambdalestidae. *Palaeontologia Polonica*, **38**, 3–41.

Kitazoe, Y., Kishino, H., Waddell, P. J., et al. (2007). Robust time estimation reconciles views of the antiquity of placental mammals. *PLoS One*, **2**, e384.

Kumar, S. and Hedges, S. B. (1998). A molecular timescale for vertebrate evolution. *Nature*, **392**, 917–20.

Larsson, H. C. and Wagner, G. P. (2002). Pentadactyl ground state of the avian wing. *Journal of Experimental Zoology*, **294**, 146–51.

Li, C., Wu, X. C., Rieppel, O., Wang, L. T. and Zhao, L. J. (2008). An ancestral turtle from the Late Triassic of southwestern China. *Nature*, **456**, 497–501.

Lockwood, C. A. (2007). *The Human Story*. London: The Natural History Museum.

Mayr, E. (1982). *The Growth of Biological Thought: Diversity, Evolution, and Inheritance*. Cambridge, MA: Belknap Press.

McIntyre, D. C., Rakshit, S., Yallowitz, A. R., et al. (2007). Hox patterning of the vertebrate rib cage. *Development*, **134**, 2981–9.

Murphy, W. J., Eizirik, E., O'Brien, S. J., et al. (2001). Resolution of the early placental mammal radiation using Bayesian phylogenetics. *Science*, **294**, 2348–51.

Nagashima, H., Sugahara, F., Takechi, M., et al. (2009). Evolution of the turtle body plan by the folding and creation of new muscle connections. *Science*, **325**, 193–6.

Noonan, J. P., Hofreiter, M., Smith, D., et al. (2005). Genomic sequencing of Pleistocene cave bears. *Science*, **309**, 597–9.

Organ, C. L., Janes, D. E., Meade, A. and Pagel, M. (2009). Genotypic sex determination enabled adaptive radiations of extinct marine reptiles. *Nature*, **461**, 389–92.

Organ, C. L., Schweitzer, M. H., Zheng, W., et al. (2008). Molecular phylogenetics of mastodon and *Tyrannosaurus rex*. *Science*, **320**, 499.

Organ, C. L., Shedlock, A. M., Meade, A., Pagel, M. and Edwards, S. V. (2007). Origin of avian genome size and structure in non-avian dinosaurs. *Nature*, **446**, 180–4.

Paäbo, S. (1989). Ancient DNA: extraction, characterization, molecular cloning, and enzymatic amplification. *Proceedings of the National Academy of Sciences of the United States of America*, **86**, 1939–43.

Patterson, C. (1987). *Molecules and Morphology in Evolution: Conflict or Compromise?* Cambridge, UK: Cambridge University Press.

Penny, D., Foulds, L. R. and Hendy, M. D. (1982). Testing the theory of evolution by comparing phylogenetic trees constructed from five different protein sequences. *Nature*, **297**, 197–200.

Polly, P. D. (2001). On morphological clocks and paleophylogeography: towards a timescale for *Sorex* hybrid zones. *Genetica*, **112–113**, 339–57.

Raup, D. M. (1978). Cohort analysis of generic survivorship. *Paleobiology*, **4**, 1–15.

Reisz, R. R. and Müller, J. (2004). Molecular timescales and the fossil record: a paleontological perspective. *Trends in Genetics*, **20**, 237–41.

Ricklefs, R. E., Losos, J. B. and Townsend, T. M. (2007). Evolutionary diversification of clades of squamate reptiles. *Journal of Evolutionary Biology*, **20**, 1751–62.

Rieppel, O. (2009). How did the turtle get its shell? *Science*, **325**, 154–5.

Sánchez-Villagra, M. R. (2010). Developmental palaeontology in synapsids: the fossil record of ontogeny in mammals and their closest relatives. *Proceedings of the Royal Society B–Biological Sciences*, **277**, 1139–47.

Sánchez-Villagra, M. R. and Smith, K. K. (1997). Diversity and evolution of the marsupial mandibular angular process. *Journal of Mammalian Evolution*, **4**, 119–44.

Schmid, L. and Sánchez-Villagra, M. R. (2010). Potential genetic bases of morphological evolution in the triassic fish *Saurichthys*. *Journal of Experimental Zoology B–Molecular and Developmental Evolution*, **314B**, 519–26.

Sepkoski, J. J., Bambach, R. K., Raup, D. M. and Valentine, J. W. (1981). Phanerozoic marine diversity and the fossil record. *Nature*, **293**, 435–7.

Shapiro, M. D., Hanken, J. and Rosenthal, N. (2003). Developmental basis of evolutionary digit loss in the Australian lizard *Hemiergis*. *Journal of Experimental Zoology B–Molecular and Developmental Evolution*, **297B**, 48–56.

Shubin, N., Tabin, C. and Carroll, S. (1997). Fossils, genes and the evolution of animal limbs. *Nature*, **388**, 639–48.

Shubin, N., Tabin, C. and Carroll, S. (2009). Deep homology and the origins of evolutionary novelty. *Nature*, **457**, 818–23.

Simons, E. L. (1972). *Primate Evolution: An Introduction to Man's Place in Nature*. New York: Macmillan.

Simpson, G. G. (1944). *Tempo and Mode in Evolution*. New York: Columbia University Press.

Smith, M. M. (2003). Vertebrate dentitions at the origin of jaws: when and how pattern evolved. *Evolution and Development*, **5**, 394–413.

Smith, M. M. and Hall, B. K. (1990). Developmental and evolutionary origins of vertebrate skeletogenic and odontogenic tissues. *Biological Reviews of the Cambridge Philosophical Society*, **65**, 277–374.

Takechi, M. and Kuratani, S. (2010). History of studies on mammalian middle ear evolution: a comparative morphological and developmental biology perspective. *Journal of Experimental Zoology B–Molecular and Developmental Evolution*, **314**, 417–33.

Wagner, G. P. and Gauthier, J. A. (1999). 1,2,3 = 2,3,4: a solution to the problem of the homology of the digits in the avian hand. *Proceedings of the National Academy of Sciences of the United States of America*, **96**, 5111–16.

Wible, J. R., Rougier, G. W., Novacek, M. J. and Asher, R. J. (2007). Cretaceous eutherians and Laurasian origin for placental mammals near the K/T boundary. *Nature*, **447**, 1003–6.

Wible, J. R., Rougier, G. W., Novacek, M. J. and Asher, R. J. (2009). The eutherian mammal *Maelestes gobiensis* from the Late Cretaceous of Mongolia and the phylogeny of Cretaceous Eutheria. *Bulletin of the American Museum of Natural History*, **327**, 1–123.

Woese, C. R. (1987). Macroevolution in the microscopic world. In *Molecules and Morphology in Evolution: Conflict or Compromise?*, ed. C. Patterson. Cambridge, UK: Cambridge University Press, pp. 177–202.

Zuckerkandl, E. and Pauling, L. (1962). Molecular disease, evolution, and genetic heterogeneity. In *Horizons in Biochemistry*, ed. M. Kasha and B. Pullman. New York: Academic Press.

2

Genomics and the lost world: palaeontological insights into genome evolution

CHRIS ORGAN

A consilience of genomics and palaeontology: palaeogenomics

Although genomes are the primary source of heritable information in organisms, over 99% of it has vanished due to pervasive extinction throughout the history of life (Raup 1992). The ability to ask and answer questions (even small ones) about the genomes of extinct organisms opens vast uncharted avenues of research. If one considers palaeontology a subdiscipline of biology, then it is the only biological subdiscipline that uses historical data (fossils) directly for the creation and testing of hypotheses. Palaeontology can bring to genomics direct evidence from the past in the same way it informs ecology and organismal biology about past environments and forms. Palaeontology reveals extinct lineages and morphologies otherwise unknown in extant species, and it can help delineate the limits and scope of form and/or function. It can expose patterns of change over long periods of time, or even lay bare past biological interactions, environmental conditions and biogeography. Palaeogenomic research attempts to integrate these lines of evidence with whole genome data. Although a nascent field, palaeogenomics has already provided insights intractable by looking at data from extant species alone, such as the Neanderthal contribution to the genome of modern humans (Green *et al.* 2010). Palaeogenomics can also help find regions of reduced polymorphism around a gene containing a recently adaptive mutation (the mutation 'sweeps' to fixation carrying linked sequences with it before recombination breaks the region apart – creating what is known as linkage disequilibrium; Figure 2.1).

Many fields of biology, including genomics, test hypotheses about past events, but these tests are limited to comparisons among extant taxa alone to

From Clone to Bone: The Synergy of Morphological and Molecular Tools in Palaeobiology,
ed. Robert J. Asher and Johannes Müller. Published by Cambridge University Press.

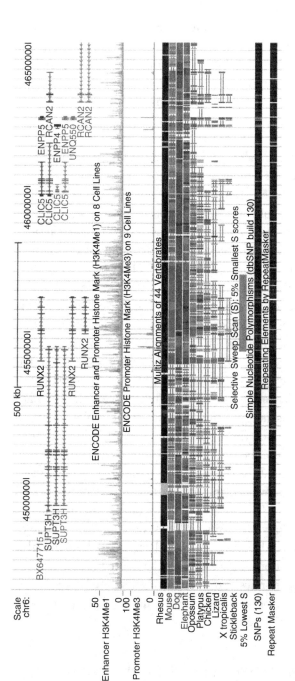

Figure 2.1 Human *Runx2* locus on chromosome six shows evidence of a selective sweep using Neanderthal data (UCSC Genome Browser). The first entry marks the scale of the window, the second entry chromosomal position, the third entry annotated genes within this region. Short vertical bars mark exons and dashed lines mark introns. Note that genes can overlap and that alternate splicing patterns create multiple isoforms for the same gene. *Runx2*, which is a transcription factor involved with bone growth, codes for three isoforms. The next two entries show epigenetic modifications (such as methylation) to histones, which influence gene transcription. The entry listing various taxa reveals varying degrees of conservation over this region. The next entry shows a region in which positive selection may have occurred in *Homo sapiens* based on a strong signal for a depletion of Neanderthal-derived alleles (the third entry up from the bottom marked '5% Lowest S'). The penultimate entry shows single nucleotide polymorphisms (SNPs) while the last shows the location of repeats. Note the abundance of elements (darkness of the entries forms a nearly black line) in the last two entries. (See also colour plate.)

discern past events and processes. This is not to understate the importance of inferences derived from extant species in evolutionary biology (Schmid, this volume). On the contrary, extremely important discoveries have been made using this approach (e.g. the endosymbiotic theory of mitochondria Sagan 1967; Margulis 1975). Confusingly, the term 'palaeogenomics' has been defined as precisely this kind of inquiry – what evolutionary biologists and palaeontologists would call character state reconstruction (Birnbaum *et al.* 2000; Bottjer *et al.* 2006; Muffato 2009; Salse *et al.* 2009). This research, although relevant, does not fall under the umbrella of palaeontological science because fossil data are not used directly to form or test hypotheses.

Genomics research can influence palaeontology as will. Molecular sequence data can be used to construct phylogenies and palaeontological data can be used to constrain divergence times of those trees (e.g. Bininda-Emonds *et al.* and Larsson *et al.*, this volume). Rates of molecular evolution can then be used to estimate divergence times for lineages lacking a robust fossil record and therefore help palaeontological hypothesis formation of taxonomy or aid in directing field work to strata of certain ages. This approach has become important for virtually all fields of biology including medicine (Pybus *et al.* 2007; Rambaut *et al.* 2008). Two online resources that provide divergence time data demonstrate this fact: Time Tree (Hedges *et al.* 2006) provides inferred divergence times for any two taxa within its database; Date-A-Clade (Benton *et al.* 2009) provides fossil evidence for constraining divergence times. Perhaps the most exciting aspect of this synergy is that the field of molecular evolution can make predictions of divergence times testable by palaeontologists when new fossils are discovered and described.

A well-established relationship also exists between palaeontology and evolutionary developmental biology. However, the integration of palaeontology and genomics (in both directions) has been slow, in part because genomic biology does not translate to morphology as easily as does developmental biology. Nevertheless, palaeogenomics holds great promise, because without insights from palaeontological data, evolutionary genomics will be limited to comparisons of extant taxa. Luckily, palaeogenomics and ancient DNA studies have already shown that combining genetics and genomics with palaeontology is tractable across the geological timescale (Figure 2.2).

Palaeogenomic insights from the recent past

Nucleotide sequences can be obtained from recently extinct species, thereby allowing a direct glimpse into genomes of the past. The first sequence-based studies focused not on whole genomes but on using genomic fragments

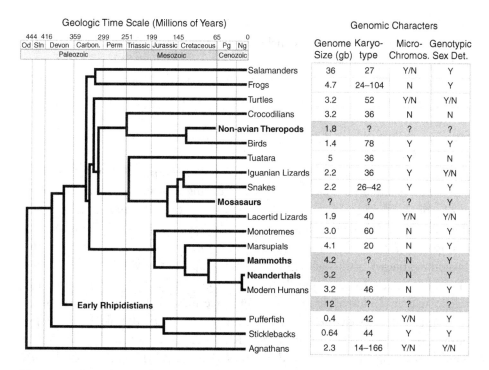

Figure 2.2 Basic genomic parameters for select extant and extinct vertebrates. The left portion of the figure shows a generally accepted phylogeny for vertebrates (see Asher and Müller, this volume). On the right side of the figure, genome size average is measured in gigabases, karyotype in the number of chromosomes, microchromosomes as present or absent, and genotypic sex determination as present/absent. Absence of data is indicated with a question mark, polymorphisms with 'Y/N'.

to elucidate phylogenetic relationships. For example, the extinct quagga (*Equus quagga*) was shown to be more closely related to zebras than to horses (Higuchi *et al.* 1984). Ancient DNA research also helped to determine colonization patterns in prehistoric Pacific Islanders (Hagelberg and Clegg 1993), as well as parasite–host coevolution in relation to disease (Salo *et al.* 1994), and intestinal symbiotic interactions (DeSalle *et al.* 1992; Cano *et al.* 1994; Rhodes *et al.* 1998).

This work was based on sequence fragments obtained from fossil material, but whole mitochondrial genome sequences have also been obtained for a variety of extinct species. Mitochondria inhabit eukaryotic cells, producing most of a cell's energy through the electron transport chain. The circular genomes of mitochondria are small (about 16 kb) and contain roughly 37 genes. These are highly conserved across eukaryotes, yet use a slightly altered genetic code from that used in the nuclear genome (Boore 1999). Although the order of genes in the mitochondrial genome is also highly conserved, rearrangements have occurred in various lineages, including different clades of animals (Kumazawa and Nishida 1995; Janke and Arnason 1997; Rest *et al.* 2003; Macey *et al.* 2004; Kumazawa 2007). Conservation of the mitochondrial genome, its abundance in cells (which can number in the thousands, in myocytes, for example), and its maternally bound inheritance system have made mitochondrial DNA a focus of molecular phylogenetics and evolutionary genetic studies (Ballard and Whitlock 2004).

Sequencing the mitochondrial genome has allowed scientists to reconstruct past events, such as population bottlenecks during human evolution (Gherman *et al.* 2007). Because of its utility in phylogenetics and population genetics, the mitochondrial genome has also played a crucial role in palaeogenomics (Cooper *et al.* 2001; Paabo *et al.* 2004; Noonan *et al.* 2005; Green *et al.* 2006). Insights garnered from such studies include improved phylogenetic resolution of extant species: for example, Pleistocene fossils helped resolve the origin of the polar bear (Lindqvist *et al.* 2010). Yet, discerning phylogenetic relationships for some recent species may remain difficult even with the aid of palaeogenomics, as is the case for rhinoceroses (Rhinocerotidae) where the addition of whole mitochondrial sequence data from the woolly rhinoceros (*Coelodonta antiquitatis*) did not clarify within-clade relationships. Whole mitochondrial genome data have also played an important role in phylogeography studies. For example, mitochondrial genome sequence comparison of extinct moas with ratites suggests a Late Cretaceous vicariant speciation of ratites in Gondwana (Cooper *et al.* 2001). Similarly, comparisons of 160 woolly mammoth (*Mammuthus primigenius*) sequences have helped to reveal the contribution of New World populations to global diversity within the genus (Debruyne *et al.* 2008).

Next generation sequencing (Margulies *et al.* 2005) has improved our ability to sequence not just the mitochondrial genome, but also the nuclear genome from frozen, mummified or otherwise preserved specimens. Hair is an excellent source of relatively high quality DNA for sequencing whole genomes, as was shown for prehistoric humans (Rasmussen *et al.* 2010) and the mammoth (Miller *et al.* 2008). Unlike other sources of DNA, such as bones or skin, it appears that the keratinized casing of the hair shaft protects DNA from environmental degradation, including direct damage and contamination from bacteria and fungi. This may mean that obtaining sequences from extinct mammals will be easier than for other clades, though investigations using similar structures in other species, such as feathers or scales, are lacking. Taphonomy plays a role as well. For example, DNA is fairly stable when dry, though frozen specimens and those preserved in amber are potentially the best sources for obtaining samples that can be sequenced (Cano *et al.* 1993;

Kelman and Moran 1996). Sequencing advances have also increased our ability to screen for errors caused by C->U post-mortem miscoding lesions and control for contamination using bioinformatics and other protocols (Brotherton *et al.* 2007; Burbano *et al.* 2010). For example, a method called direct multiplex sequencing uses high-throughput sequencing of sample barcodes coupled with multiplex PCR (Stiller *et al.* 2009). With this approach, 96% of the *Ursus spelaeus* (cave bear) mitochondrial genome was sequenced in only two runs of a 454 Life Sciences sequencer. Exciting 'third generation' sequencing technologies may further improve our ability to obtain reliable sequences from fossils. Single molecule sequencing is one such approach that identifies base pairs while passing through a nanopore (Clarke *et al.* 2009). It does not require fluorescent labelling and is low cost, accurate and very fast. If implemented in a massively parallel system, single molecule sequencing might help improve sequencing from recent fossil tissues.

Advances in contamination control and next generation sequencing technologies improve the potential to uncover genetic adaptations in extinct species to environmental change over geologic time. Genetic adaptations can be as simple as single nucleotide polymorphisms (SNPs) within the coding portions of genes. Genes are, of course, the primary information unit of the genome. Although the definition of a gene has changed through time, most genes have portions that code for proteins called exons and non-coding elements called introns that are spliced out during processing into proteins (but note that not all genes code for proteins: some code for RNA products, like ribosomal RNA). A given gene can make proteins from different combinations of its exons by splicing (Brett *et al.* 2002; Jaffe 2003), but this is not the only source of information variation. Gene expression is controlled by complex networks of regulation. The primary sequence of DNA may be methylated, a common mechanism of epigenetic gene regulation (Zemach *et al.* 2010). Genes also regulate one another through networks of expression, which may respond to environmental cues, or they may be controlled by small strands of RNA that bind to and prevent messenger RNA from being translated into protein (Mattick 2004; Chen and Rajewsky 2007).

Some of this genetic information is lost during fossilization, even if DNA can be sequenced. Methylation patterns and the location of genes on chromosomes, or even the number of chromosomes within a genome, may not be elucidated using next generation sequencing technologies. However, finding changes among genes in extinct species on a genome-wide scale is feasible in recently extinct species. The sequence of the Neanderthal genome (Green *et al.* 2010) enabled scientists to discern 78 protein-altering nucleotide substitutions in genes that became fixed in modern humans after the *Homo sapiens* – *H. neanderthalensis* split a few hundred thousand years ago. Of these, several have multiple protein-altering mutations, including *SPAG17*, a gene that codes for a protein involved in the machinery that powers sperm flagella. These data suggest that intrasexual selection (sperm competition) may have been important during the evolution of *H. sapiens*. Adaptations to climate change have also been suggested for several unique genetic mutations in the woolly mammoth, though more work is needed to test this hypothesis (Miller *et al.* 2008).

Like small sequence fragments and whole mitochondrial genomes, whole nuclear genome sequences help resolve questions about molecular evolution and phylogenetics. For instance, the woolly mammoth genome has helped constrain rates of molecular evolution within elephants (Miller *et al.* 2008) and analysis of the Neanderthal genome has revealed that non-African modern humans likely interbred with Neanderthals during our migration out of Africa (Green *et al.* 2010). Remains of a 4000-year-old Greenland native have also revealed that the Saqqaq people of Greenland migrated from Siberia independently from the migration that gave rise to modern Inuit and Native Americans (Rasmussen *et al.* 2010). Sequence data from fossils also have the potential to mitigate problems arising from long braches in molecular phylogenies (Bergsten 2005) and to refine estimates of heterotachy in molecular evolution.

Despite the excitement and promise of ancient DNA research (Paabo 1989), nucleic acids cannot be sequenced from specimens older than about 500 000 years using current technology (Paabo *et al.* 2004; Hebsgaard *et al.* 2005). As noted above, degradation, post-mortem miscoding lesions and contamination pose challenging problems to sequencing-based palaeogenomic studies. In addition, low coverage (the number of times a stretch of DNA is repeatedly sequenced) remains a serious hurdle for establishing confidence in sequences obtained from fossils. These limitations are being addressed by optimizing sequencing reactions (King *et al.* 2009; Heyn *et al.* 2010) and informatics that account for or detect damage in ancient DNA (Mateiu and Rannala 2008; Green *et al.* 2009; Pruefer *et al.* 2010), but hurdles will likely remain that challenge the field.

Palaeogenomic insights from deep time

Palaeontology offers a direct glimpse of life from the depths of Earth's history. For all the promise of sequence-based palaeogenomics, it is likely that sequences can only be obtained for recently extinct species. However, evidence suggests that some types of proteins are preserved for millions of years (Asara *et al.* 2007; Organ *et al.* 2008b). Using mass spectrometry, sequence fragments of collagen α1 and α2 were obtained from the fossilized bones of a 160 000- to 600 000-year-old mastodon (*Mammut americanum*) and a 68-million-year-old *Tyrannosaurus rex* (Asara and Schweitzer 2008). These fragments were shown to contain phylogenetic information (Organ *et al.* 2008b), which highlights their potential use in phylogenetic inference. From a proteomic perspective, these biomolecules were small fragments of one protein family – not a whole protein or a whole proteome (i.e. all the proteins produced or capable of being produced in an organism). These reports were also controversial (Asara and Schweitzer 2008; Buckley *et al.* 2008; Pevzner *et al.* 2008), though reanalysis by independent researchers supported the original study (Bern *et al.* 2009). If collagen molecules remain partially intact over long periods of time, other structural proteins, such as keratin, may also be preserved, though the preservation needed to retrieve amino acid sequences from fossils is likely to be rare.

Another way to study the genomes of long extinct species is by using comparative phylogenetic methods that directly integrate a genomic correlate that fossilizes. By analysing the correlate one can say something about the genome (Figure 2.3). This approach has long been used by palaeontologists and in the 1990s was formalized in cladistic terms as the extant phylogenetic bracket (EPB) (Bryant and Russell 1992; Witmer 1995). The EPB approach uses sister groups to make cladistically informed inferences about character states. However, this approach works only at the categorical level, it lacks statistical rigour (testing), and the absence of branch length information in cladograms renders the EPB weak compared with phylogenies whose branch lengths represent the amount of overall character change. Ideally, we want an approach that uses all the information in a phylogeny, can accommodate statistical hypothesis testing, employs models of character evolution, can be used for both continuous and discrete data types, and can account for phylogenetic uncertainty.

Phylogenetic comparative methods, developed to control for the expected lack of independence of species data (Felsenstein 1985), are a family of such tools. Recent advances in comparative methods use Bayesian modelling to account for uncertainty (Huelsenbeck *et al.* 2000, 2003; Pagel and Lutzoni 2002; Pagel *et al.* 2004; Pagel and Meade 2006). More importantly for palaeogenomics, such tools can make predictions using correlate data (Garland and Ives 2000; Organ *et al.* 2007, 2009). These predictions are not ancestral character state reconstructions, but inferences of character state values for terminal taxa using a regression model (for continuous characters), a fossilized correlate (an independent data point) and phylogenetic information. They might be better termed 'retrodictions' or predictions of past events (Scriven 1959). Phylogenetically informed predictions and comparative phylogenetic character

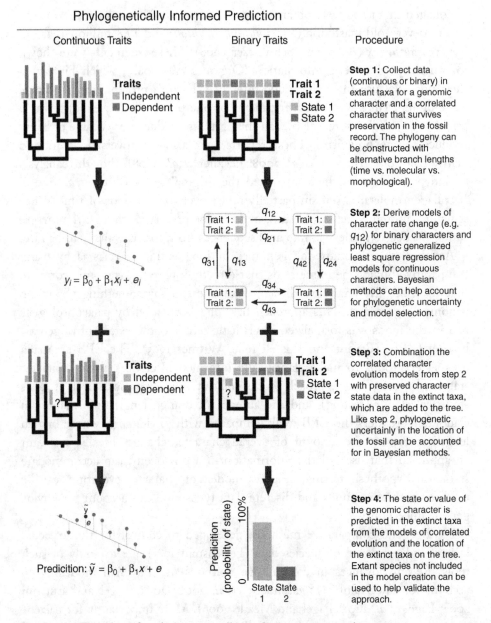

Figure 2.3 Flowchart for phylogenetically predicting genomic character states in extinct taxa. The rationale and general approach for making phylogenetically informed predictions in extinct taxa is the same for continuous and binary characters.

analysis allow palaeogenomic and macroevolutionary hypotheses to be evaluated rigorously (Organ *et al.* 2007, 2009).

The best example of the phylogenetic comparative approach in palaeogenomics is in the study of genome size. A simple relationship among genome size, nucleus size and cell size does not exist (Neumann and Nurse 2007). A robust correlation between cell size and nucleus size (genome size), however, has been recognized in a wide array of cell types in plants and animals (Gregory 2005), presumably because a bigger nucleus is required to encase a bigger genome. In eukaryotes, genome size is important because it is primarily determined by the density of repetitive elements (i.e. mobile genetic elements containing a reverse transcriptase gene and simple repeats called microsatellites). To a lesser extent, the number and size of introns also play a role. Genome size varies across species and is measured either by weight (reported in picograms, pg) or by the number of nucleotide base pairs in thousands (kilobases, kb), millions (megabases, mb), or billions (gigabases, gb). In vertebrates, for example, birds have small genomes that generally range between 1 and 2 pg, equivalent to roughly 1–2 gb (Tiersch and Wachtel 1991; Gregory 2002), and that are sparsely populated with repetitive elements (Shedlock 2006; Shedlock *et al.* 2007). The larger genomes of reptiles and mammals, which have ranges of 2–3 pg and 3–4 pg, respectively (Gregory 2010), have more repetitive elements.

Although inferring genome size from bone-cell lacunae size data in extinct organisms was proposed in the 1960s (Vialli and Sacchi Vialli 1969), it took several more years before an analysis was conducted in extinct lungfish and amphibians (Thomson 1972; Thomson and Muraszko 1978). These studies found that, remarkably, although dipnoans (lungfish) have descended in relative morphological stasis, their genome gradually expanded through time. Differential tempo and mode of evolution across the genotype and phenotype is just one exciting insight afforded by these studies. Later research used conodont epithelial cells (Conway Morris and Harper 1988) and angiosperm stomata (Masterson 1994) as genomic correlates. However, these studies lacked a phylogenetic framework and it is known that analysing character data without accounting for phylogeny leads to inflated error rates (Rohlf 2006; Freckleton 2009).

Our analysis of genome size in dinosaurs was the first application of phylogenetically informed predictions in palaeogenomics. We used osteocyte lacunae size as a correlate for genome size (Organ *et al.* 2007). The results showed that non-avian theropod dinosaurs possessed genomes as small as those of modern birds – that the small, streamlined genomes of birds arose in their saurischian ancestors. This finding is counter to the hypothesis that the metabolic demands of flight selected for smaller genome size (Hughes and Hughes 1995), but important questions remain. For example, have genomes been expanding or contracting across lineages since the common amniote ancestor? Using extant comparisons, the ancestor is inferred to have had a small (i.e. bird/reptile-sized) genome with expansion of genome size occurring in the mammal lineage. Alternatively, the ancestor may have had a medium (mammal-sized) genome with contraction along the reptile line (Shedlock et al. 2007). This question is important because the evolution of genome size is related to the underlying evolutionary dynamics of repetitive content. Palaeogenomic sampling of 14 extinct tetrapod genera from the Palaeozoic and early Mesozoic Eras suggests that the genome size of the ancestral amniote was mammal-like, with contractions occurring along the diapsid lineage, and no directional change along the line to mammals (Organ et al. 2011).

Because repetitive elements determine genome size in amniotes to a large extent, it is also likely that the repeat-poor genome of extant birds was inherited from their dinosaur (saurischian) ancestors (Organ et al. 2007; Organ and Brusatte 2009). Repetitive elements can play important functional roles in genome evolution (Kazazian 2004). For example, roughly 45% of the human genome is composed of LINEs and SINEs (long and short interspersed nuclear elements), including the Alu-element group of repeats (Deininger and Batzer 1993). As their names suggest, LINEs and SINEs become interspersed throughout the genome as they proliferate, as opposed to juxtaposing themselves. LINEs proliferate autonomously by copying and pasting themselves throughout the genome, leaving behind a parental copy. SINEs proliferate using the reverse transcriptase (a polymerase enzyme that transcribes single-stranded RNA into double-stranded DNA) of LINE elements (Ohshima et al. 1996). Unlike the high retroelement content of the human genome noted above, which is typical for a eutherian mammal, only 15–20% of bird genomes are composed of repetitive elements (Epplen et al. 1978; Schmid 1996; Hillier et al. 2004). Intriguingly, the dominant group of retroelements in reptiles, the CR1 family (Shedlock 2006), appears to be going extinct in the chicken and zebra finch genomes (Wicker et al. 2004) because fully intact CR1s without disruptive mutations are exceedingly rare (Hillier et al. 2004; Warren et al. 2010). Palaeogenomic work suggests that this trend began prior to the evolution of Aves (Organ et al. 2007; Organ and Brusatte 2009).

Phylogenetic comparative analysis has also been used to study the evolution of the number, size and type of chromosomes – the karyotype (Olmo and Signorino 2005; Organ et al. 2008a) as well as the palaeogenomics of sex-determining mechanisms (Organ et al. 2009). Chromosomes are, by convention, numbered according to their size, so that chromosome 2 is bigger than chromosome 4. In adult diploid animals,

chromosomes are composed of two pairs of chromatids, with one pair pater-
nally derived and the other maternally derived. The number, sizes and shapes
of chromosomes within a cell vary widely across clades. For example,
mammals have a relatively small number of large, so-called
'macrochromosomes', while many birds and non-avian reptiles have a small
number of macrochromosomes and a large number of microchromosomes
(Olmo 2005). Although nearly all birds have microchromosomes (some
raptors are the exception), the closest living relatives of birds, crocodilians,
lack them (Iwabe et al. 2005). Avian microchromosomes appear to be structur-
ally different from macrochromosomes. They have higher G+C content (Burt
2002; Axelsson et al. 2004, 2005), higher gene density (Hillier et al. 2004;
Fowler et al. 2008; Griffin et al. 2008) and recombination rate (Rodionov et al.
1992a, 1992b; Edwards and Dillon 2004; Backstrom et al. 2006). However,
these differences are based on comparisons of only two bird genomes, chicken
and zebra finch (Gallus and Taeniopygia), with those of mammals.

Chromosome structure is evolutionarily dynamic (Robinson and Ruiz-Herrera
2008). Rearrangements, such as translocations and inversions, can move entire
chunks of a chromosome and all the genes it contains to a new chromosome.
This process can break old linkages between genes and form new ones. Loci on
chromosomes may confer beneficial functions in one sex to the detriment of the
other sex (Arnqvist and Rowe 2005; Charlesworth et al. 2005). Over time such
sexually antagonistic loci may result in the evolution of sex chromosomes from
autosomes (Graves 2006) – it is assumed that incubation temperature determined
the sex of offspring ancestrally. Important changes in chromosome biology then
occur, such as reduced recombination, which leads to the accumulation of
deleterious mutations in one of the emerging sex chromosomes. One of the sex
chromosome pairs contains either male (XY system) or female (ZW system)
specific genes.

Within reptiles, and possibly other groups of vertebrates, different sex-
determining mechanisms (temperature or genotype) evolve repeatedly (Organ
and Janes 2008). The primary adaptive benefit of genotypic sex determination
may be that it helps maintain balanced sex ratios in populations regardless
of fluctuating environmental temperatures (Bull 1983). Palaeogenomic work
has shown that genotypic sex determination was important for opening evolutionary pathways to viviparity in amniotes (Organ et al. 2009). There is also
evidence linking the evolution of genotypic sex determination with the evolution of sexual dimorphism (Mank 2009). Using phylogenetically informed
predictions of binary characters (the presence or absence of genotypic sex
determination) and viviparity as a correlate, several clades of extinct marine
reptiles (mosasaurs, sauropterygians and ichthyosaurs) were shown to have

likely possessed genotypic sex determination (Organ *et al.* 2009). This finding links a genomic character with marine radiations in amniotes, an association difficult to test using the extant biodiversity of obligate marine amniotes.

Despite the success and promise of palaeogenomics using comparative methods with correlate fossil data, the approach has its own limitations. The biggest drawback is that, regardless of the statistical rigour, accuracy of phylogenies and fit of models, the conclusions are based on inferences, not direct evidence or experiment. That said, errors in tree topology and branch lengths have been shown generally to have little impact on phylogenetic regression, so long as the errors are not deep within the tree (Martins *et al.* 2002; Stone 2011). Another problem with inferences from comparative methods is the adequacy of correlations in extant species. For example, bone cell and lacunae size is variable across the skeleton (de Ricqlès *et al.* 1991), but is more uniform in compact bone in the diaphysis of long bones (Canè *et al.* 1982). Analyses should attempt to minimize intra-skeletal variation by sampling lacunae from homologous bones, such as tibiae. Luckily, intra-skeletal variation has been shown to have minimal impact on inferred genome size, so long as adequate phylogenetic comparative methods are used (Montanari *et al.* 2011). This variability clearly adds noise to inferences of genome size, which is why phylogenetic signal and multiple samples within a clade are important in such studies. Moreover, it is likely that only broad aspects of the genome will be open to study using this approach.

Conclusion and future directions

Many aspects of genome architecture have not been studied from a palaeogenomics perspective, let alone outside mammals or model organisms. For example, the macroevolutionary trends of whole genome duplications are not well understood, nor are the dynamics of chromosome evolution, including sex chromosomes evolution and degradation. Have G-C isochores arisen convergently in endothermic amniotes or have they been lost convergently in ectothermic species from a common ancestral genome that was rich in isochores? How many conserved non-coding elements existed in the genomes of extinct lineages and what might this reveal concerning their function and evolution? What about gene families? Genes periodically, and randomly, duplicate during replication (Figure 2.4). Thereafter, duplicated genes accumulate differences compared with the ancestral copy, taking on specialized or new roles within the organism (Ohno 1970; Nowak *et al.* 1997; Hittinger and Carroll 2007; Osada and Innan 2008; Organ *et al.* 2010). This is a potent source of variation for adaptive evolution, yet

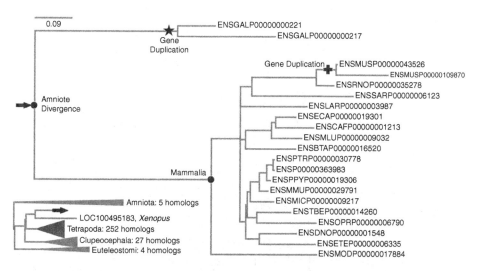

Figure 2.4 A gene tree for the small *MHC- class II beta chain* subfamily. The large tree is a subcomponent of the cladogram at bottom left (based on 329 genes and 149 duplication nodes), as indicated by the black arrow. The dark triangle in the lower left (Tetrapoda) denotes a paralogue clade. Within the subtree, the *MHC- class II beta chain* gene duplicated sometime after the divergence of amniotes within the chicken lineage (denoted by a star; ENSGALP codes for chicken (*Gallus*) protein) and after the mouse lineage split from the rat lineage (denoted by a plus; ENSMUSP codes for mouse (*Mus*) protein and ENSRNOP codes for rat (*Rattus*) protein). The tip names are designated as Ensembl proteins and branch lengths are in units of molecular change per site. Note that unlike trees of organisms, a taxon may appear multiple times in a gene tree and that the entire gene tree for *MHC- class II beta chain* is complex.

the tempo and mode of gene families over macroevolutionary timescales is poorly understood. Whole genome comparisons are now allowing researchers to estimate when duplications occurred. In the future we may be able to elucidate how gene duplication has helped species adapt to environmental changes over geologic timescales (Figure 2.5).

Palaeogenomics is an even younger science than genomics, but one that has already made important contributions for understanding genome biology and evolution. And though limited by the vagaries of preservation, technological hurdles and the ability to address certain questions, palaeogenomic research is critical for placing into proper context genomics of extant organisms.

Summary

Palaeogenomics is concerned with the structure, function and evolution of genomes in extinct organisms. Palaeontological data can complement

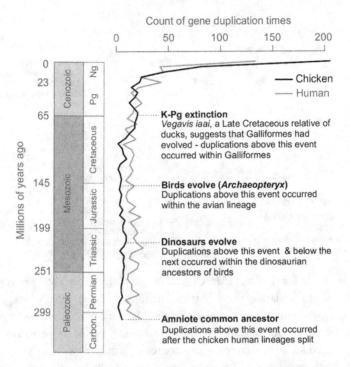

Figure 2.5 The history of gene duplications along the chicken lineage (modified from Organ *et al*. 2010), showing the timing of duplications (divergence of paralogues) in the *Gallus gallus* genome after the evolution of amniotes. The timing of duplications on the line to humans is provided as a comparison. Duplication times were estimated using a molecular clock with *Gallus/Homo* comparisons. Pivotal events during the evolution of this lineage are noted on the figure. Pg indicates Palaeogene and Ng indicates Neogene.

genomic data and expand insights into genomic biology including the tempo and mode of molecular evolution, the reconstruction of ancestral genes and genomes, or facilitate hypothesis testing of genotype–phenotype–environment interactions. Because palaeontology is largely concerned with extinct organisms and large-scale patterns of evolution, comparative genomics offers palaeontologists new insights into the mechanisms behind the divergence of higher level taxa and the genetics responsible for the evolution of novel phenotypes within lineages. Preservation and its associated hurdles constrain the ability to adequately address some palaeogenomic hypotheses. Nevertheless, by leveraging fossil data, palaeogenomics allows a better glimpse of the genomic past and renders tractable certain evolutionary genomic hypotheses that would otherwise remain untestable.

Acknowledgements

I wish to thank Nicole Hobbs, Dan Janes and two anonymous reviewers for comments on this chapter. Robert Asher and Johannes Müller, the editors of this volume, also deserve many thanks for their enthusiam, vision, and editorial guidance.

REFERENCES

Arnqvist, G. and Rowe, L. (2005). *Sexual Conflict*. Princeton, NJ: Princeton University Press.

Asara, J. M. and Schweitzer, M. H. (2008). Response to comment on 'protein sequences from mastodon and *Tyrannosaurus rex* revealed by mass spectrometry'. *Science*, **319**(5859), 33d.

Asara, J. M., Schweitzer, M. H., Freimark, L. M., Phillips, M. and Cantley, L. C. (2007). Protein sequences from mastodon and *Tyrannosaurus rex* revealed by mass spectrometry. *Science*, **316**(5822), 280–5.

Axelsson, E., Smith, N. G. C., Sundstrom, H., Berlin, S. and Ellegren, H. (2004). Male-biased mutation rate and divergence in autosomal, Z-linked and W-linked introns of chicken and turkey. *Molecular Biology and Evolution*, **21**(8), 1538–47.

Axelsson, E., Webster, M. T., Smith, N. G. C., Burt, D. W. and Ellegren, H. (2005). Comparison of the chicken and turkey genomes reveals a higher rate of nucleotide divergence on microchromosomes than macrochromosomes. *Genome Research*, **15**(1), 120–5.

Backstrom, N., Brandstrom, M., Gustafsson, L., *et al.* (2006). Genetic mapping in a natural population of collared flycatchers (*Ficedula albicollis*): conserved synteny but gene order rearrangements on the avian Z chromosome. *Genetics*, **174**(1), 377–86.

Ballard, J. W. O. and Whitlock, M. C. (2004). The incomplete natural history of mitochondria. *Molecular Ecology*, **13**(4), 729–44.

Benton, M. J., Donoghue, P. C. J. and Asher, R. J. (2009). Calibrating and constraining molecular clocks. In *The Timetree of Life*, ed. S. B. Hedges and S. Kumar, pp. 35–86. Oxford, UK: Oxford University Press.

Bergsten, J. (2005). A review of long-branch attraction. *Cladistics*, **21**(2), 163–93.

Bern, M., Phinney, B. S. and Goldberg, D. (2009). Reanalysis of *Tyrannosaurus rex* mass spectra. *Journal of Proteome Research*, **8**(9), 4328–32.

Birnbaum, D., Coulier, F., Pébusque, M.-J. and Pontarotti, P. (2000). 'Paleogenomics': looking in the past to the future. *Journal of Experimental Zoology*, **288**, 21–2.

Boore, J. L. (1999). Animal mitochondrial genomes. *Nucleic Acids Research*, **27**(8), 1767–80.

Bottjer, D. J., Davidson, E. H., Peterson, K. J. and Cameron, R. A. (2006). Paleogenomics of echinoderms. *Science*, **314**, 956–9.

Brett, D., Pospisil, H., Valcartel, J., Reich, J. and Bork, P. (2002). Alternative splicing and genome complexity. *Nature Genetics*, **30**, 29–30.

Brotherton, P., Endicott, P., and Sanchez, J. J. (2007). Novel high-resolution characterization of ancient DNA reveals C > U-type base modification events as the sole cause of post mortem miscoding lesions. *Nucleic Acids Research*, **35**(17), 5717–28.

Bryant, H. N. and Russell, A. P. (1992). The role of phylogenetic analysis in the inference of unpreserved attributes of extinct taxa. *Philosophical Transactions of the Royal Society of London B*, **337**, 405–18.

Buckley, M., Walker, A., Ho, S. Y. W., *et al.* (2008). Comment on 'Protein sequences from mastodon and *Tyrannosaurus rex* revealed by mass spectrometry'. *Science*, **319** (5859), 33c.

Bull, J. J. (1983). *Evolution of Sex Determining Mechanisms*. Menlo Park, CA: Benjamin/ Cummings.

Burbano, H. A., Hodges, E., Green, R. E., *et al.* (2010). Targeted investigation of the neandertal genome by array-based sequence capture. *Science*, **328**(5979), 723–5.

Burt, D. W. (2002). Origin and evolution of avian microchromosomes. *Cytogenetic and Genome Research*, **96**(1–4), 97–112.

Canè, V., Marotti, G., Volpi, G. (1982). Size and density of osteocyte lacunae in different regions of long bones. *Calcified Tissue International*, **34**, 558–63.

Cano, R. J., Poinar, H. N., Pieniazek, N. J., Acra, A. and Poinar, G. O., Jr. (1993). Amplification and sequencing of DNA from a 120–135-million-year-old weevil. *Nature*, **363**(6429), 536–8.

Cano, R. J., Borucki, M. K., Higby-Schweitzer, M., *et al.* (1994). Bacillus DNA in fossil bees: an ancient symbiosis? *Applied Environmental Microbiology*, **60**(6), 2164–7.

Charlesworth, D., Charlesworth, B., and Marais, G. (2005). Steps in the evolution of heteromorphic sex chromosomes. *Heredity*, **95**(2), 118–28.

Chen, K. and Rajewsky, N. (2007). The evolution of gene regulation by transcription factors and microRNAs. *Nature Reviews Genetics*, **8**, 93–103.

Clarke, J., Wu, H.-C. and Jayasinghe, L. (2009). Continuous base identification for single-molecule nanopore DNA sequencing. *Nature Nanotechnology*, **4**(4), 265–70.

Conway Morris, S. and Harper, E. (1988). Genome size in conodonts (Chordata): inferred variations during 270 million years. *Science*, **241**, 1230–2.

Cooper, A., Lalueza-Fox, C., Anderson, S., *et al.* (2001). Complete mitochondrial genome sequences of two extinct moas clarify ratite evolution. *Nature*, **409**, 704–7.

de Ricqlès, A., Meunier, F. J., Castanet, J. and Francillon-Vieillot, H. (1991). Comparative microstructure of bone. In *Bone*, Vol. 3, ed. B. K. Hall, pp. 1–78. London: CRC Press.

Debruyne, R., Chu, G., King, C. E., *et al.* (2008). Out of America: ancient DNA evidence for a New World origin of Late Quaternary woolly mammoths. *Current Biology*, **18**(17), 1320–6.

Deininger, P. L. and Batzer, M. A. (1993). Evolution of retroposons. *Evolutionary Biology*, **27**, 157–96.

DeSalle, R., Gatesy, J., Wheeler, W. and Grimaldi, D. (1992). DNA sequences from a fossil termite in Oligo-Miocene amber and their phylogenetic implications. *Science*, **257**, 1933–6.

Edwards, S. V. and Dillon, M. (2004). Hitchhiking and recombination in birds: evidence from Mhc-linked and unlinked loci in red-winged blackbirds (*Agelaius phoeniceus*). *Genetical Research*, **84**, 175–92.

Epplen, J. T., Leipoldt, M., Engel, W. and Schmidtke, J. (1978). DNA sequence organization in avian genomes. *Chromosoma*, **69**, 307–21.

Felsenstein, J. (1985). Phylogenies and the comparative method. *American Naturalist*, **125**(1), 1–15.

Fowler, K. E., Skinner, B. M., Robertson, L. B. W., *et al.* (2008). Molecular cytogenetic maps of turkey, duck and zebra finch and their implications for genome evolution. *Chromosome Research*, **16**(7), 1043–4.

Freckleton, R. P. (2009). The seven deadly sins of comparative analysis. *Journal of Evolutionary Biology*, **22**(7), 1367–75.

Garland, T., Jr. and Ives, A. R. (2000). Using the past to predict the present: confidence intervals for regression equations in phylogenetic comparative methods. *American Naturalist*, **155**, 346–64.

Gherman, A., Chen, P. E., Teslovich, T. M., *et al.* (2007). Population bottlenecks as a potential major shaping force of human genome architecture. *PLoS Genetics*, **3**(7), e119.

Graves, J. A. M. (2006). Sex chromosome specialization and degeneration in mammals. *Cell*, **124**(5), 901–14.

Green, R. E., Krause, J., Ptak, S. E., *et al.* (2006). Analysis of one million base pairs of Neanderthal DNA. *Nature*, **444**(7117), 330–6.

Green, R. E., Briggs, A. W., Krause, J., *et al.* (2009). The Neandertal genome and ancient DNA authenticity. *EMBO Journal*, **28**(17), 2494–502.

Green, R. E., Krause, J., Briggs, A. W., *et al.* (2010). A draft sequence of the Neandertal genome. *Science*, **328**(5979), 710–22.

Gregory, T. R. (2002). A bird's-eye view of the C-value enigma: genome size, cell size, and metabolic rate in the class Aves. *Evolution*, **56**(1), 121–30.

Gregory, T. R. (2005). Genome size evolution in animals. In *The Evolution of the Genome*, ed. T. R. Gregory, pp. 4–71. Boston, MA: Elsevier Academic Press.

Gregory, T. R. (2010). *Animal Genome Size Database*. Online: www.genomesize.com/.

Griffin, D. K., Robertson, L. B., Tempest, H. G., *et al.* (2008). Whole genome comparative studies between chicken and turkey and their implications for avian genome evolution. *BMC Genomics*, **9**.

Hagelberg, E. and Clegg, J. B. (1993). Genetic polymorphisms in prehistoric Pacific islanders determined by analysis of ancient bone DNA. *Proceedings of the Royal Society B*, **252**(1334), 163–70.

Hebsgaard, M. B., Phillips, M. J. and Willerslev, E. (2005). Geologically ancient DNA: fact or artifact? *Trends in Microbiology*, **13**(5), 212–20.

Hedges, S. B., Dudley, J. and Kumar, S. (2006). *TimeTree: A Public Knowledge-base*. Pennsylvania and Arizona State Universities, USA Online: www.timetree.org.

Heyn, P., Stenzel, U., Briggs, A. W., *et al.* (2010). Road blocks on paleogenomes – polymerase extension profiling reveals the frequency of blocking lesions in ancient DNA. *Nucleic Acids Research*, **38**(16), e161.

Higuchi, R., Bowman, B., Freiberger, M., Ryder, O. A. and Wilson, A. C. (1984). DNA sequences from the quagga, an extinct member of the horse family. *Nature*, **312**, 282–4.

Hillier, L. W., Miller, W. and Birney, E. (2004). Sequence and comparative analysis of the chicken genome provide unique perspectives on vertebrate evolution. *Nature*, **432**(7018), 695–716.

Hittinger, C. T. and Carroll, S. B. (2007). Gene duplication and the adaptive evolution of a classic genetic switch. *Nature*, **449**, 677–81.

Huelsenbeck, J. P., Nielsen, R. and Bollback, J. P. (2003). Stochastic mapping of morphological characters. *Systematic Biology*, **52**, 131–58.

Huelsenbeck, J. P., Rannala, B. and Masly, J. P. (2000). Accommodating phylogenetic uncertainty in evolutionary studies. *Science*, **288**(5475), 2349–50.

Hughes, A. L. and Hughes, M. K. (1995). Small genomes for better flyers. *Nature*, **377** (6548), 391.

Iwabe, N., Hara, Y., Kumazawa, Y., et al. (2005). Sister group relationship of turtles to the bird-crocodilian clade revealed by nuclear DNA-coded proteins. *Molecular Biology and Evolution*, **22**(4), 810–13.

Jaffe, S. (2003). Alternative splicing goes mainstream. *The Scientist*, **December 15**, 28–30.

Janke, A. and Arnason, U. (1997). The complete mitochondrial genome of *Alligator mississippiensis* and the separation between recent Archosauria (birds and crocodiles). *Molecular Biology and Evolution*, **14**(12), 1266–72.

Kazazian, H. H. J. (2004). Mobile elements: drivers of genome evolution. *Science*, **303**, 1626–32.

Kelman, Z. and Moran, L. (1996). Degradation of ancient DNA. *Current Biology*, **6**(3), 223.

King, E. C., Debruyne, R., Kuch, M., Schwarz, C. and Poinar, H. N. (2009). A quantitative approach to detect and overcome PCR inhibition in ancient DNA extracts. *BioTechniques*, **47**, 941–9.

Kumazawa, Y. (2007). Mitochondrial genomes from major lizard families suggest their phylogenetic relationships and ancient radiations. *Gene*, **388**(1–2), 19–26.

Kumazawa, Y. and Nishida, M. (1995). Variations in mitochondrial transfer-RNA gene organization of reptiles as phylogenetic markers. *Molecular Biology and Evolution*, **12**(5), 759–72.

Lindqvist, C., Schuster, S. C., Sun, Y., et al. (2010). Complete mitochondrial genome of a Pleistocene jawbone unveils the origin of polar bear. *Proceedings of the National Academy of Sciences of the United States of America*, **107**(11), 5053–7.

Macey, J. R., Papenfuss, T. J., Kuehl, J. V., Fourcade, H. M. and Boore, J. L. (2004). Phylogenetic relationships among amphisbaenian reptiles based on complete mitochondrial genomic sequences. *Molecular Phylogenetics and Evolution*, **33**(1), 22–31.

Mank, J. E. (2009). Sex chromosomes and the evolution of sexual dimorphism: lessons from the genome. *American Naturalist*, **173**(2), 141–50.

Margulies, M., Egholm, M., Altman, W. E., et al. (2005). Genome sequencing in microfabricated high-density picolitre reactors. *Nature*, **437**(7057), 376–80.

Margulis, L. (1975). Symbiotic theory of the origin of eukaryotic organelles; criteria for proof. *Symposia of the Society for Experimental Biology*, **29**, 21–38.

Martins, E. P., Diniz-Filho, J. A. F., and Housworth, E. A. (2002). Adaptive constraints and the phylogenetic comparative method: a computer simulation test. *Evolution*, **56**(1), 1–13.

Masterson, J. (1994). Stomatal size in fossil plants: evidence for polyploidy in majority of angiosperms. *Science*, **264**, 421–3.

Mateiu, L. M. and Rannala, B. H. (2008). Bayesian inference of errors in ancient DNA caused by postmortem degradation. *Molecular Biology and Evolution*, **25**(7), 1503–11.

Mattick, J. (2004). RNA regulation: a new genetics? *Nature Reviews Genetics*, **5**, 316–23.

Miller, W., Drautz, D. I., Ratan, A., *et al.* (2008). Sequencing the nuclear genome of the extinct woolly mammoth. *Nature*, **456**(7220), 387–90.

Montanari, S., Brusatte, S. L., De Wolf, W. and Norell, M. A. (2011). Variation of osteocyte lacunae size within the tetrapod skeleton: implications for palaeogenomics. *Biology Letters*, **7**, 751–4.

Muffato, M. C. and Crollius, H. R. (2009). Paleogenomics in vertebrates, or the recovery of lost genomes from the mist of time. *BioEssays*, **30**, 122–34.

Neumann, F. R. and Nurse, P. (2007). Nuclear size control in fission yeast. *Journal of Cell Biology*, **179**(4), 593–600.

Noonan, J. P., Hofreiter, M., Smith, D., *et al.* (2005). Genomic sequencing of Pleistocene cave bears. *Science*, **309**(5734), 597–9.

Nowak, M. A., Boerlijst, M. C., Cooke, J. and Maynard Smith, J. (1997). Evolution of genetic redundancy. *Nature*, **388**(6638), 167–71.

Ohno, S. (1970). *Evolution by Gene Duplication*. Heidelberg, Germany: Springer-Verlag.

Ohshima, K., Hamada, M., Terai, Y. and Okada, L. (1996). The 3′ ends of short interspersed repetitive elements are derived from the 3′ ends of long interspersed repetitive elements. *Molecular and Cellular Biology*, **16**, 3756–64.

Olmo, E. (2005). Rate of chromosome changes and speciation in reptiles. *Genetica*, **125**(2–3), 185–203.

Olmo, E. and Signorino, G. (2005). Chromorep: a reptile chromosomes database. Online: http://ginux.univpm.it/scienze/chromorep/ (retrieved April 1, 2008).

Organ, C. L. and Brusatte, S. (2009). Sauropod dinosaurs evolved moderately sized genomes unrelated to body size. *Proceedings of the Royal Society of London B*, **276**(1677), 4303–8.

Organ, C. L. and Janes, D. E. (2008). Evolution of sex chromosomes in Sauropsida. *Integrative and Comparative Biology*, **48**(4), 512–19.

Organ, C. L., Shedlock, A. M., Meade, A., Pagel, M. and Edwards, S. V. (2007). Origin of avian genome size and structure in non-avian dinosaurs. *Nature*, **446**(7132), 180–4.

Organ, C. L., Moreno, R. G., and Edwards, S. V. (2008a). Three tiers of genome evolution in reptiles. *Integrative and Comparative Biology*, **48**(4), 494–504.

Organ, C. L., Schweitzer, M. H., Zheng, W. X., *et al.* (2008b). Molecular phylogenetics of mastodon and *Tyrannosaurus rex*. *Science*, **320**(5875), 499.

Organ, C. L., Janes, D. E., Meade, A. and Pagel, M. (2009). Genotypic sex determination enabled adaptive radiations of extinct marine reptiles. *Nature*, **461**, 389–92.

Organ, C. L., Rasmussen, M., Baldwin, M. W., Kellis, M. and Edwards, S. V. (2010). A phylogenomic approach to the evolutionary dynamics of gene duplication in birds. In *Evolution After Gene Duplication*, ed. K. Dittmar and D. Liberles, pp. 253–68. Hoboken, NJ: Wiley & Sons.

Organ, C. L., Canoville, A., Reisz, R. R. and Laurin, M. (2011). Paleogenomic data suggest mammal-like genome size in the ancestral amniote and derived large genome size in amphibians. *Journal of Evolutionary Biology*, **24**(2), 372–80.

Osada, N. and Innan, H. (2008). Duplication and gene conversion in the *Drosophila melanogaster* genome. *PLoS Genetics*, **4**(12), e1000305.

Paabo, S. (1989). Ancient DNA: extraction, characterization, molecular cloning, and enzymatic amplification. *Proceedings of the National Academy of Sciences of the United States of America*, **86**(6), 1939–43.

Paabo, S., Poinar, H., Serre, D., *et al.* (2004). Genetic analyses from ancient DNA. *Annual Review of Genetics*, **38**, 645–79.

Pagel, M. and Lutzoni, F. (2002). Accounting for phylogenetic uncertainty in comparative studies of evolution and adaptation. In *Biological Evolution and Statistical Physics*, ed. M. Lässig and A. Valleriani, pp. 148–61. Berlin: Springer-Verlag.

Pagel, M. and Meade, A. (2006). Bayesian analysis of correlated evolution of discrete characters by reversible-jump Markov chain Monte Carlo. *American Naturalist*, **167**, 808–25.

Pagel, M., Meade, A., and Barker, D. (2004). Bayesian estimation of ancestral character states on phylogenies. *Systematic Biology*, **53**(5), 673–84.

Pevzner, P. A., Kim, S. and Ng, J. (2008). Comment on 'Protein sequences from mastodon and *Tyrannosaurus rex* revealed by mass spectrometry'. *Science*, **321** (5892), 1040b.

Pruefer, K., Stenzel, U., Hofreiter, M., *et al.* (2010). Computational challenges in the analysis of ancient DNA. *Genome Biology*, **11**(5), R48.

Pybus, O. G., Rambaut, A., Belshaw, R., *et al.* (2007). Phylogenetic evidence for deleterious mutation load in RNA viruses and its contribution to viral evolution. *Molecular Biology and Evolution*, **24**(3), 845–52.

Rambaut, A., Pybus, O. G., Nelson, M. I., *et al.* (2008). The genomic and epidemiological dynamics of human influenza A virus. *Nature*, **453**, 615–19.

Rasmussen, M., Li, Y., Lindgreen, S., *et al.* (2010). Ancient human genome sequence of an extinct Palaeo-Eskimo. *Nature*, **463**(7282), 757–62.

Raup, D. M. (1992). *Extinction: Bad Genes or Bad Luck?*. New York: W.W. Norton.

Rest, J. S., Ast, J. C., Austin, C. C., *et al.* (2003). Molecular systematics of primary reptilian lineages and the tuatara mitochondrial genome. *Molecular Phylogenetics and Evolution*, **29**(2), 289–97.

Rhodes, A. N., Urbance, J. W., Youga, H., *et al.* (1998). Identification of bacterial isolates obtained from intestinal contents associated with 12,000-year-old mastodon remains. *Applied Environmental Microbiology*, **64**(2), 651–8.

Robinson, T. J. and Ruiz-Herrera, A. (2008) Defining the ancestral eutherian karyotype: a cladistic interpretation of chromosome painting and genome sequence assembly data. *Chromosome Research*, **16**, 1133–41,

Rodionov, A. V., Chelysheva, L. A., Solovei, I. V. and Myakoshina, Y. A. (1992a). Chiasma distribution in the lampbrush chromosomes of the chicken, *Gallus gallus domesticus* – hot spots of recombination and their feasible role in proper disjunction of homologous chromosomes at the 1st meiotic division. *Genetika*, **28**(7), 151–60.

Rodionov, A. V., Myakoshina, Y. A., Chelysheva, L. A., Solovei, I. V. and Gaginskaya, E. R. (1992b). Chiasmata in the lampbrush chromosomes of *Gallus gallus domesticus* – the cytogenetic study of recombination frequency and linkage map lengths. *Genetika*, **28**(4), 53–63.

Rohlf, F. J. (2006). A comment on phylogenetic correction. *Evolution*, **60**(7), 1509–15.

Sagan, J. (1967). On the origin of mitosing cells. *Journal of Theoretical Biology*, **14**(3),255–74.

Salo, W. L., Aufderheide, A. C., Buikstra, J. and Holcomb, T. A. (1994). Identification of *Mycobacterium tuberculosis* DNA in a pre-Columbian Peruvian mummy. *Proceedings of the National Academy of Sciences of the United States of America*, **91**(6), 2091–4.

Salse, J., Abrouk, M., Bolot, S., *et al.* (2009). Reconstruction of monocotelydoneous proto-chromosomes reveals faster evolution in plants than in animals. *Proceedings of the National Academy of Sciences of the United States of America*, **106**(35), 14 908–13.

Schmid, C. (1996). Alu: structure, origin, evolution, significance and function of one-tenth of human DNA. *Progress in Nucleic Acid Research and Molecular Biology*, **53**, 283–319.

Scriven, M. (1959). Explanation and prediction in evolutionary theory: satisfactory explanation of the past is possible even when prediction of the future is impossible. *Science*, **130**(3374), 477–82.

Shedlock, A. M. (2006). Phylogenomic investigation of CR1 LINE diversity in reptiles. *Systematic Biology*, **55**(6), 902–11.

Shedlock, A. M., Botka, C. W., Zhao, S., *et al.* (2007). Phylogenomics of nonavian reptiles and the structure of the ancestral amniote genorne. *Proceedings of the National Academy of Sciences of the United States of America*, **104**(8), 2767–72.

Stiller, M., Knapp, M., Stenzel, U., Hofreiter, M. and Meyer, M. (2009). Direct multiplex sequencing (DMPS) – a novel method for targeted high-throughput sequencing of ancient and highly degraded DNA. *Genome Research*, **19**(10), 1843–8.

Stone, E. A. (2011). Why the phylogenetic regression appears robust to tree misspecification. *Systematic Biology*, **60**(3), 245–60.

Thomson, K. S. (1972). An attempt to reconstruct evolutionary changes in the cellular DNA content of lungfish. *Journal of Experimental Zoology*, **180**, 363–72.

Thomson, K. S. and Muraszko, K. (1978). Estimation of cell size and DNA content in fossil fishes and amphibians. *Journal of Experimental Zoology*, **205**, 315–20.

Tiersch, T. R. and Wachtel, S. S. (1991). On the evolution of genome size of birds. *Journal of Heredity*, **82**(5), 363–8.

Vialli, M. and Sacchi Vialli, G. (1969). Morfometria delle lacune ossee di vertebrati attuali e fossili alla luce delle conoscenze di biologia cellulare. *Rendiconti Istituto Lombardo Scienze e Lettere, Sezione B*, **103**, 234–54.

Warren, W. C., Clayton, D. F., Ellegren, H., *et al.* (2010). The genome of a songbird. *Nature*, **464**(7289), 757–62.

Wicker, T., Robertson, J. S., Schulze, S. R., *et al.* (2004). The repetitive landscape of the chicken genome. *Genome Research*, **15**, 126–36.

Witmer, L. M. (1995). The extant phylogenetic bracket and the importance of reconstructing soft tissues in fossils. In *Functional Morphology in Vertebrate Paleontology*, ed. J. J. Thomason, pp. 19–33. New York: Cambridge University Press.

Zemach, A., I. McDaniel, E., Silva, P. and Zilberman, D. (2010). Genome-wide evolutionary analysis of eukaryotic DNA methylation. *Science*, **328**(5980), 916–19.

3

Rocking clocks and clocking rocks: a critical look at divergence time estimation in mammals

OLAF R. P. BININDA-EMONDS, ROBIN M. D. BECK AND ROSS D. E. MACPHEE

Introduction

Much has been written about the molecular revolution in phylogenetics and the ongoing conflict between molecules and morphology (Hillis 1987; Patterson 1987; Springer *et al.* 2004). With reference to therian mammals at least, the supposed conflict has been largely overblown: there is in fact general agreement between the two data sources, something unfortunately overshadowed by a handful of persistent 'problem children'. The taxonomic content of most mammalian orders and other traditional higher-level taxa originally proposed purely on the basis of morphology has remained unscathed by the application of molecular sequence analysis. Even within these taxa, conflicts between molecular and morphological hypotheses of relationships are comparatively rare and usually relatively minor. For instance, a comparative study within Carnivora (Bininda-Emonds, 2000) revealed that most data sources and methods of analysis pointed at the same general solution, a few admittedly problematic taxa (e.g. Felidae) notwithstanding. In the end, the frequency and nature of disagreements over tree topology is arguably of the same order of magnitude within the separate spheres of molecular and morphological systematics as it is between them (Patterson *et al.* 1993). In many ways, the situation in mammals parallels that in vertebrates, where a fairly robust tree including gnathostomes, actinopterygians, sarcopterygians, tetrapods, amniotes and diapsids (among many other groups) has been supported by comparative anatomy since the 1800s (Asher and Müller, this volume).

Instead, many of the more celebrated conflicts in mammals tend to represent a lack of information, especially on the morphological side. The evolutionary

From Clone to Bone: The Synergy of Morphological and Molecular Tools in Palaeobiology, ed. Robert J. Asher and Johannes Müller. Published by Cambridge University Press.

tree of eutherian mammals presented by Novacek (1992), which exemplified the state-of-the-art morphological opinion at the time, is conspicuous today not for being very wrong (although some clades within it have been overturned by molecular information), but for its lack of resolution. Insectivora was long recognized to be a taxonomic wastebasket for any small brown mammal with sharp teeth that wasn't a rodent and couldn't fly. With time, the application of ever more detailed morphological data to the problem resulted in the exclusion of Macroscelidea, Scandentia and a host of early Cenozoic clades better placed elsewhere, but ultimately it just wasn't possible to tease more out of the data. Molecular data, however, revealed the non-monophyly of the remaining, 'lipotyphlan' insectivores, now allocated to two different mammalian super-orders (see Springer *et al.* 2004).

In many cases, new information or improved analyses resolved apparent conflicts. For instance, Cetartiodactyla was first proposed based on molecular data (Graur and Higgins 1994), but subsequently received strong morphological support with the discovery of fossils of early, terrestrial whales that preserved features previously thought diagnostic of non-cetacean artiodactyls (e.g. *Pakicetus*; Geisler and Uhen 2003; Gingerich *et al.* 2001; Thewissen *et al.* 2001). On the flip side, the morphologically strongly supported grouping Glires, which unites rodents and lagomorphs and was rejected by several early molecular analyses (e.g. Graur *et al.* 1996; Misawa and Janke 2003), is now consistently recovered by most molecular studies with sufficient taxon sampling (see Springer *et al.* 2004). More recently, the morphological study of Wible *et al.* (2007) recovered a monophyletic Euarchontoglires (a placental superorder first recognized on the basis of molecular data; Murphy *et al.* 2001) and clades that are similar (although not identical) in taxonomic content to the molecularly supported placental superorders Laurasiatheria (Waddell *et al.* 1999) and Afrotheria (Stanhope *et al.* 1998).

At higher taxonomic levels at least, arguably the only major phylogenetic conflict between molecules and morphology is the monophyly of Afrotheria, which unites paenungulates, aardvark, elephant shrews and afrosoricidan 'insectivores'. Despite extensive research (Asher 1999, 2001, 2007; Whidden 2002; Asher *et al.* 2003; Sánchez-Villagra *et al.* 2007; Seiffert 2007b; Wible *et al.* 2007; Asher and Lehmann 2008; Seiffert 2010), strong morphological support (at least with afrosoricidans included) for this placental superorder is still scarce. The failure of morphology to recognize the monophyly of Afrotheria (and the non-monophyly of Lipotyphla) means that many of the groups that are accepted today, like Laurasiatheria, were not strictly recovered in the past, thereby giving the appearance of conflict. However, a more historically accurate view reveals that the content of the higher-level taxa, albeit not identical, is often very similar between morphological and molecular studies.

Instead of conflict, therefore, the combination of molecules and morphology is achieving a robust consensus with respect to mammalian higher-level relationships, with a recent study (Lee and Camens 2009) showing that hidden support for relationships originally founded on molecular grounds exists within morphological data sets. The same general lack of conflict, however, cannot be said to be true for estimates of divergence times within Mammalia and for the origin and initial radiations of the ordinal crown groups in particular. Here the conflict is real and continuing, and plays in the expected direction: molecular-based estimates are consistently older than fossil-based ones (although not universally; see Douzery *et al.* 2003; Kitazoe *et al.* 2007). What makes this conflict particularly interesting is both the scale of the difference and the implied relative importance of the Cretaceous–Palaeogene (K–Pg; traditionally known as the Cretaceous–Tertiary or KT) mass extinction event for basal divergences within Placentalia. Whereas fossil-based estimates (as exemplified by Wible *et al.* 2007) typically place both the origins and the basal diversifications of the crown groups close to the K–Pg boundary, most molecular-based estimates (as exemplified by Bininda-Emonds *et al.* 2007, 2008) indicate that the crown groups both originated as well as began radiating well within the Late Cretaceous.

In this review, our goal is not to find the answer to the problem of when existing mammalian (primarily therian) crown groups evolved, but instead to determine where the difficulties might lie in finding such an answer. The use of fossil, molecular, or both kinds of data for estimating divergence times is coupled with any number of crucial assumptions, many of which are hardly ever mentioned explicitly. Very often, the 'blame' for any conflict is simply passed to the other side, but it could equally be the case that both data sources are wanting and providing flawed estimates. Here, with particular reference to examples from the mammalian radiation, we elucidate some of the assumptions and potential sources of error underlying each data type and describe their respective strengths and weaknesses. The hope is that the increased awareness of both sides of the discussion will help bring fossil- and molecular-based date estimates closer together with time.

Of rocks: assumptions underlying fossil-based date estimates

The fossil record, whether directly in the form of fossil taxa or indirectly through inferred palaeobiogeographic or stratigraphic information, ultimately represents the single (and only!) data source against which all our divergence time estimates are calibrated. It is accepted that individual fossils can, at best, be as old as their associated nodes on a phylogeny, and that they will often moderately to severely underestimate nodal age. This discrepancy often derives simply from sampling issues, with fossils of the oldest members of particular lineages not yet having been discovered (or possibly not even preserved in the first place). Thus, whether any given fossil has any real relevance for dating actual divergence times is often questionable when there is no basis for estimating how well fossil members of the focal lineages have been sampled. Such factors undoubtedly play a role in debates about the validity of molecular-based versus fossil-based dates for the deeper mammalian divergences, including those of the ordinal crown groups. Indeed, the incompleteness of the fossil record arguably represents the default explanation in any cases of severe conflict.

In the case of placental mammals, it remains possible that the oldest members of the extant superorders and orders might have been present in regions that are as yet poorly sampled for particular intervals (e.g. some parts of Gondwana). However, there are other, less appreciated factors involving the interpretation of fossil data (e.g. our ability to recognize basal members of the mammalian superorders and orders in the fossil record) that may be equally if not more important and whose examination might lead to new insights concerning the interpretation of morphological data in a phylogenetic context. In the following sections we examine a wide range of issues associated with interpreting fossil evidence, including the epistemological nature of characters and character definition, the significance of gaps in the fossil record, how the phylogenetic affinities of fossils are established, and the contingent relevance of biogeography in interpreting evolutionary history.

Defining and using morphological characters in phylogenetics: the funnel of induction and the ratchet

Even without limitations incurred by the incompleteness of the fossil record, our interpretation of fossil material and its taxonomic affinities relies on a fundamental aspect of modern systematics, the definition and analysis of discrete morphological characters. Certainly, the process of character description has intensified greatly over the past 35–40 years, largely as a result of the search for characters useful for making phylogenetic inferences. In the case of mammalian systematics, this activity has produced a number of important insights into morphological evolution, such as the likely primitive therian premolar number (Giallombardo and AToL Mammal Morphology Team 2010; Kielan-Jaworowska et al. 2004) and mammalian cochlear form and mechanics (Luo et al. 2011a).

In working with morphological characters, we need to be aware of the distinction drawn by Hennig (1966), who recognized that semaphoronts – the 'holders' of characters, whether individuals or groups, adults or juveniles – are empirically real; whereas character phylogenies are abstractions based on characters detected on semaphoronts and interpreted in the context of a given study. Within the fossil record, semaphoronts are often temporally clumped and are separated (in the simplest case) by intervals in which nothing very much like them may occur. Coeval semaphoronts can often be defined in terms of characters apparently unique to them, as well as in terms of characters that link them monophyletically with other groups up or down the tree. The end result is, metaphorically, a series of beads on a string, each bead more or less different from, but still related to, those above or below; what keeps them all together in the mind of the systematist are the successively derived character states of features they purportedly hold in common.

Several points concerning this procedure need to be examined. First, and most obviously, is that the entire exercise, from drafting definitions of characters to interpreting a final tree, proceeds entirely inductively (Bryant 1989, 1991). Semaphoronts do not advertise their diagnostic characters; these need to be selected by an informed mind in which prior inductive knowledge plays a substantial role. The teeth of fossil mammals are usually heavily sampled for characters, both because they are among the most readily fossilized elements of the mammalian skeleton and because the dentition is often a very good (if imperfect) guide to relationships. Another issue concerns how morphological character states are defined, counted, or measured; these decisions operate with few constraints, especially compared with the restricted state space for molecular data (i.e. nucleotide bases or amino acids, or perceived gaps between them). The choice of descriptor language used in character definition may differ sharply between systematists, even when describing the same thing, with consequences for how characters may be scored across taxa and, possibly, the form of the final tree (Cartmill 1981; MacPhee 1994). This is not a minor problem. The current drive toward comparatively large morphological character sets and more control over descriptors arising from community efforts like - MorphoBank (www.morphobank.org, O'Leary 2011; also Novacek and AToL Mammal Morphology Team 2008; Vogt et al. 2010) is welcome at one level (e.g. access to information, ability to recognize poor-quality characters), but because the characters are still generated in much the same way as before, nothing has changed procedurally.

Another serious problem is the general tendency, in most real records, for the quality of the information extractable from semaphoronts to decline as one goes backward in time (Benton and Donoghue 2007; Donoghue and Purnell 2009; Sansom et al. 2010). This is not only because the fossils themselves may become less complete or adequately preserved, as already noted, but because candidate derived characters, based on prior experience and used to operationally define a given group of interest, are either not represented in the preserved anatomy or are no longer interpretable as part of the same character-state complex. This circumstance – the 'character funnel' – is commonly observed in many kinds of inductive processes in which some set of initial propositions about a phenomenon is serially depleted (in this case, across time) until the phenomenon can no longer be stipulated on the basis of the propositions that remain, reducing that its likelihood can be characterized as either true or false, or perhaps present versus absent (Hacking 2001; Vickers 2010). Incidentally, the possibility that morphological characters ever disappear in a formal sense need not detain us; certainly, the recognition criteria for such characters do, which ultimately terminates any chain of induction about character 'states'.

A good example of the funnel in action is the result achieved when only characters present in existing members of the crown group are utilized to determine group membership for fossil candidates. If at some point character states antecedent to the crown group's defining synapomorphies cannot be recognized, the funnel empties and the chain of induction terminates with the last member in which the derived character(s) can be recognized. Of course, in any real situation there will be more characters in play and very probably considerable variation as to when individual characters drop out. However, the end result is the same: analysis stops with the earliest definable member of a group, and its age provides the (minimum) constraint on divergence time.

As we note later on, a danger when relying solely on diagnostic characters for the placement of fossils is that the process can easily stray into the realm of typology rather than phylogenetics. It is curious that, in securing ingroup placement for a fossil, we often demand that the diagnostic features of the clade be identifiable even when such features may not be universally present in the extant members due to loss or additional character modification. For example, the presence of limbs is diagnostic for (fossil) tetrapods, although they are lost or highly modified in a variety of extant forms. By contrast, the use of a phylogenetic perspective in combination with fossil evidence can often overturn or amend hypotheses based on extant taxa only. If extant crown-group morphology were the sole basis for investigating the systematics of mammals, our understanding of their evolution would be meagre, if not flawed, indeed. For example, careful study of fossils reveals that epipubic bones, which are diagnostic for extant marsupials, are actually plesiomorphous at a much higher level (Mammaliaformes) and therefore of little cladistic utility for defining groups at lower hierarchical levels (Novacek et al. 1997).

In practice, fossils can play an important role in determining the content of monophyletic groups, particularly for the purpose of defining stem membership. Specifically, character evidence from fossils provides basic phylogenetic structure that would be impossible to induce merely from extant crown-group morphology. It acts in much the same way that a ratchet provides control over an object while work is performed on it, thus helping to ensure that the job will be completed satisfactorily. Perhaps the most compelling example of the induction ratchet at work – in which something other than mere data reshuffling occurred – is the paper by Luo *et al.* (2001a) in which critical features of tribosphenic mammals from the Mesozoic of northern and southern continents were compared (cf. Rich *et al.* 1997; Flynn *et al.* 1999; Rauhut *et al.* 2002; Rougier *et al.* 2007). Earlier, the fact that all such taxa had superficially similar eutherian-style molars was taken as evidence that they had to be related as eutherians (Rich *et al.* 1997). Luo *et al.* (2001a) instead argued that the southern forms are better regarded as highly derived monotreme relatives, grouped as australosphenidans, in which a descriptively tribosphenic cusp pattern was independently evolved. As a result of this analysis, much more was learned about the groups concerned and new phylogenetic relationships suggested. Our point here is not whether the australosphenidan/boreosphenidan dichotomization is correct – something that can only be justified with still more inductively based propositions about molar evolution – but rather that palaeontological evidence should not be thought of as a source of evolutionary knowledge that is somehow less privileged compared with neontological evidence. The two are, in fact, properly conceived as reciprocally illuminating the understanding of systematists as they set about trying to interpret the fragments of evolutionary history on which they work.

Of course, the implied promise here is that more description of more fossils will always yield better phylogenies. Actual cases reveal that 'better' (that is, more resolved or more compatible) phylogenies often do result when fossils are included (Donoghue *et al.* 1989), but such improvements to systematic knowledge are not necessarily helpful for determining divergence dates. Thus, the recent large increase in the number of named taxa of plesiadapiforms has greatly improved our understanding of the morphology, adaptations and within-group relationships of these stem primates (e.g. Bloch *et al.* 2007). However, perhaps because of the countervailing effect of the character funnel (combined with the difficulty in scoring potentially diagnostic characters in available fossils; see above), this effort has not, so far, resulted in any consensus regarding how such taxa are related to the extant crown group beyond the mere fact that they reside somewhere near it on the primate tree trunk.

If even the best-researched groups suffer from such uncertainties, definitive statements about poorly known groups are clearly premature. The widely cited Explosive Model of placental evolution (Archibald and Deutschmann 2001) in its strictest form denies the existence of Cretaceous members of the extant modern orders, either because they had not evolved to that time point or, less restrictively, because none are yet known. But how representative are the 40-odd recognized genera of Cretaceous eutherians of the diversity that existed on the planet as a whole 65 to 100 million years (Ma) ago? The majority of these taxa are known solely from highly incomplete dentitions, and all but three are from northern continents. If the loss of information through the funnel were commonly offset by the information gained via the ratchet, it is possible that the status of possible Cretaceous antecedents of crown-group orders might need to be re-evaluated. It is relevant to note that the problem is not unique to the placental/eutherian distinction, but instead occurs at all higher-level positions along the hierarchy. A good example is the paucity of characters that appear to distinguish reliably the molars of basal eutherians and metatherians (Rougier *et al.* 1998; Luo *et al.* 2002, 2003; Averianov *et al.*, 2003; Wible *et al.* 2005). On the whole, therefore, although additional fossil evidence bearing on these problems would be very welcome, it may never be sufficient to stipulate divergence dates with any greater certainty than at present.

Gaps in the fossil record and their interpretation

Gaps in the fossil record can occur if the relevant taxa are restricted to biogeographical regions that are poorly sampled (or completely unsampled) for fossils, for which fossil-bearing strata of the appropriate age are unavailable, or for which the original conditions for fossil preservation were poor. Recent discoveries have partially 'filled in' some gaps in the fossil record of therian mammals, but it is clear that major lacunae remain. How we interpret these gaps can impinge greatly on our views of the likely time and place of origin of the modern mammalian orders and superorders.

Based on the results of their comprehensive morphological analysis of Cretaceous eutherian mammals and representatives from many placental crown groups, Wible *et al.* (2007) favoured a Laurasian origin for all placental higher taxa at or near the K–Pg boundary, with explosive divergence and extension of clade ranges onto southern terranes occurring thereafter. Certainly, this Laurasia-first interpretation finds better support in the fossil record than does any other interpretation: the oldest generally accepted stem and crown members of the extant placental orders are almost exclusively from northern continents. However, the overall accuracy of this view depends on the likelihood that the evidentiary gaps are meaningful and accurate, and that true placentals have not been missed due to insufficient prospecting or, possibly, misinterpretation of existing fossils (see below).

Of particular interest at the moment are the allegedly euarchontan-like *Deccanolestes* and the possible 'condylarth' *Kharmerungulatum*, both from the Maastrichtian of India. Regardless of the precise affinities of either fossil (which are disputed), both minimally demonstrate the presence of eutherians during the Late Cretaceous in this as yet relatively poorly sampled region. However, if either or both taxa are indeed placentals (as has been argued for *Deccanolestes*; cf. Boyer *et al.* 2010; Prasad *et al.* 2010; Seiffert 2010, but see Goswami *et al.* 2011), at least some divergences within Placentalia must have occurred before the K–Pg boundary, a possibility in fact allowed by Wible *et al.* (2007). In addition, accepting this conclusion would require that one either reject exclusively Laurasian origins for the modern placental orders (see below) or invoke palaeo-geographically unlikely early contacts between northward-moving India and either Africa or Asia (see Krause 1986; Rose *et al.* 2008, 2009; Boyer *et al.* 2010; Prasad *et al.* 2010; Smith *et al.* 2010). It would also support the hypothesis that India (and possibly other poorly sampled southern landmasses such as Africa and Antarctica) may have played a 'Garden of Eden' role (*sensu* Foote *et al.* 1999) for placental mammals during the Cretaceous (see also Krause and Maas 1990).

Even if both taxa are stem eutherians rather than placentals, their presence in the Late Cretaceous of India is still biogeographically surprising given that India was an island continent by about 80 Ma ago and experienced its maximum degree of isolation during the latest Cretaceous (Ali and Aitchison 2008). This raises questions as to how and when eutherians reached India, and whether eutherians were distributed more widely in southern continents during the Late Cretaceous. Providing confident answers to these questions is prob-ably impossible given the paucity of the Late Cretaceous and early Palaeogene fossil record for most southern continents. For example, the Cretaceous fossil record for Africa lacks any verified, published examples of eutherians, but the late Mesozoic terrestrial record for that continent is generally so poor that it can be called into question whether the observed absence of evidence is likely to be real evidence of absence. Similarly, there is no Late Cretaceous fossil record whatsoever of mammals in Australasia: there is only a single site containing mammals (the early Eocene Tingamarra Local Fauna; Godthelp *et al.* 1992) for the time period spanning ~110 to 25 Ma ago. However, the presence of a probable terrestrial eutherian at Tingamarra (*Tingamarra porterorum*; Godthelp *et al.* 1992) hints at a wider distribution of Eutheria within Gondwana than is perhaps generally assumed.

Even the South American fossil record, which is the best of all the major southern landmasses, is puzzling in many ways. Pre- and post-K–Pg mammal records for this continent are staggeringly different, implying that a major biogeographical shift had to have occurred there at the close of the Mesozoic

(Pascual and Ortiz-Jaureguizar 1992, 2007). The therian groups characteristic of South America during the Cenozoic (metatherians, eutherian 'condylarths', 'ungulates' and xenarthrans) were seemingly absent during the Late Cretaceous, with the continent populated instead by a variety of non-therian groups ('symmetrodonts', dryolestoids, monotremes and gondwanatherians), most or all of which had disappeared completely by the end of the Palaeocene. At about this time (i.e. middle to late Palaeocene), therians make their first appearance in the South American fossil record (de Muizon 1991; de Muizon et al. 1998; de Muizon and Cifelli 2000; Marshall and de Muizon 1988). In the case of the metatherians and 'condylarths', their arrival in South America was almost certainly the result of dispersal from North America, possibly via a proto-Antillean connector inferred to have existed for a brief period around the end of the Cretaceous (Gayet 2001; Iturralde-Vinent 2006). The origins of xenarthrans are far more puzzling. The earliest record of Xenarthra is from the late Palaeocene or early Eocene Itaborai fauna in Brazil (Bergqvist et al. 2004; Gelfo et al. 2009). The Itaboraian xenarthrans are relatively derived and include probable crown-group forms (cingulates; Bergqvist et al. 2004), suggesting a considerable period of unsampled prior evolution in South America or elsewhere. However, fossils of Xenarthra, which should be easily identifiable based on the presence of osteoderms (a distinctive xenarthran apomorphy rarely found in any other therian clade), are as yet unknown from older sites in South America or indeed from any other continent (MacPhee and Reguero 2010). The same is true for putative xenarthran relatives, which are either lacking entirely from Late Cretaceous (and later) fossil records of other Gondwanan fragments (e.g. Africa and Australia) or do not display expected diagnostic traits such as xenarthry (e.g. the Laurasian palaeanodonts).

The nature of gaps, however, is that they can be filled by new material, and often substantially so. If correct, the suggestion of Seiffert et al. (2007) and Seiffert (2010) that *Widanelfarasia bowni* and *Dilambdogale gheerbranti* from late Eocene localities in the Fayum area of northern Egypt are possible stem members of a tenrec–golden mole clade pushes the first split within that clade to a minimum of 37 Ma ago. Previously, the oldest known fossils relevant to this divergence were the tenrecoids *Protenrec*, *Erythrozootes* and *Parageogale* and the chrysochlorid *Prochrysochloris* from the early Miocene of East Africa (Butler 1984), some 17 Ma younger. Moreover, the latter fossils are at least zalambdodont in aspect, a distinctive dental morphology of extant tenrecs and golden moles, which the Fayum fossils are not. This new minimum is, in any case, still at least 25–30 Ma younger than the presumed divergence predicted by most molecular studies (e.g. Douady and Douzery 2003; Poux et al. 2008).

Some first appearance records, already reasonably ancient, have been pushed back even more, but only modestly. For example, recent discoveries have pushed evidence for both Proboscidea (*Eritherium azzouzorum* from Morocco; Gheerbrant 2009) and Cetartiodactyla (*Ganungulatum xincunliense* from the Nongshanian of China; Ting *et al.* 2007; Clyde *et al.* 2010) into the middle Palaeocene (both ~60 Ma ago). (The cetartiodactylan affinities of *Ganungulatum* admittedly are not certain, however; see below.) Although only slightly older, these two new discoveries require us to re-evaluate the evolutionary scenarios attached to them given that both orders are well nested within their respective superorders (Afrotheria and Laurasiatheria): their earlier, middle Palaeocene appearance leaves only 5 Ma for the deeper, superordinal splits to have occurred under the Explosive Model of placental origins (see Wible *et al.* 2007).

Whether future finds will actually push the inferred origins of placental orders over the K–Pg boundary, as predicted by most molecular studies, remains to be seen. A recent discovery bearing on this issue is the recovery of a single upper premolar referred to *Protungulatum*, an archaic 'ungulate' sometimes regarded as an early placental (e.g. Spaulding *et al.* 2009), in rocks of unquestionably latest Cretaceous age in Montana (Archibald *et al.* 2011). This is perhaps not that surprising because this taxon is already well known from the earliest Palaeocene and the new fossil is thought to be only a few hundred thousand years earlier than the K–Pg boundary. Although Archibald *et al.* (2011; also Wible *et al.* 2007) discount a placental affinity for *Protungulatum*, this discovery remains important because it raises two critical questions. First, what else is being missed and why? Second, in addition to problems associated with apomorphy lag (see below) and the character funnel, can we place the often highly fragmentary Cretaceous and Palaeocene mammal fossils within a phylogenetic framework robustly enough to be confident of their relationships to the extant taxa? In short, can we ever hope to recognize the earliest putative members of a group for what they are?

Establishing the phylogenetic affinities of fossils

Whether fossil data are used to infer divergence times in isolation or as calibration points for molecular-based analyses, there are two crucial prerequisites for their use: these data must be (1) associated accurately with particular nodes on the phylogenetic tree for which we want to derive divergence time information and (2) appropriate to act as estimators for the divergence date of the node in question *even if* they are properly placed. These issues are intertwined, logically as well as empirically, and become increasingly important and problematic the further back in time we go, when both character information and taxonomic sampling become increasingly limited and limiting. Using current concepts and analyses, we shall concentrate on some of the associated epistemological and empirical issues that complicate the search for evolutionary origins and diversifications within mammals.

Associating a fossil taxon with a particular node on a phylogenetic tree requires both that (1) the taxon possesses one or a combination of candidate apomorphies enabling it to be plausibly referred to one of the lineages descending from that node (preferably demonstrated via formal phylogenetic analysis) and (2) its available fossil remains actually do preserve at least one *diagnostic* apomorphy that can be recognized for what it is. The first condition is largely definitional under a cladistic framework. Without the possession of attendant apomorphies, no matter how weakly buttressed, we have no basis for associating any taxon (fossil or extant) with a particular clade except at the most uninformative level (e.g. 'cf. Mammalia'). Instead, the second condition is the more problematic given that missing information can prevent the accurate placement of fossil material when the first condition is fulfilled.

In this regard, data completeness for fossil material is frequently a problem. In mammals, the majority of fossil taxa are known only from partial dental remains (usually molars of adult individuals). Thus, information from the fossil specimens alone is often insufficient to confidently resolve their phylogenetic affinities. For example, *Eomaia* and *Sinodelphys* from the 125-Ma-old Yixian Formation of China represent among the oldest known generally accepted members of Eutheria and Metatheria, respectively, and therefore represent critical data points for inferring the age of the placental–marsupial split. However, the determination of their taxonomic affinities was based partly on morphological features that are only identifiable because of the exceptional preservation of the Yixian specimens (Ji *et al.* 2002; Luo *et al.* 2003). If both taxa were known only from the more typical fragmentary dental remains, it is questionable whether either could be confidently distinguished from tribosphenic stem therians (e.g. both retain the plesiomorphic condition of eight upper post-canines, rather than seven as in crown-group placentals and marsupials).

An example of the more typical difficulty associated with assigning mammalian taxa known only from fragmentary dental remains is provided by the Middle Jurassic (Bathonian) Malagasy fossil *Ambondro mahabo* (Flynn *et al.* 1999), known only from three lower teeth in a jaw fragment. On the basis of this relatively meagre evidence, it has been serially associated with monotremes (Luo *et al.* 2001b), eutherians (Woodburne *et al.* 2003) or therians as a whole (Rowe *et al.* 2008) in robust cladistic analyses. Such differences in opinion are, in fact, typically encountered in the interpretative history of many Mesozoic mammal fossils. A comparison of phylogenies by Luo and co-workers (e.g. Ji *et al.* 2002;

Luo *et al.* 2002, 2003, 2007; Kielan-Jaworowska *et al.* 2004; Luo and Wible 2005) with those of Woodburne *et al.* (2003), Rougier *et al.* (2007) and Rowe *et al.* (2008) reveals major differences in how these authors portray the relationships of key taxa such as monotremes, Mesozoic tribosphenic forms from Gondwana (including *Ambondro*), allotherians (multituberculates, haramiyidans and gondwanatherians) and the various groups of triconodonts and symmetrodonts. In turn, these differences in opinion will have a major impact on the choice of fossil taxa suitable for dating the monotreme–therian split and thus the root of crown-group Mammalia. For instance, the choice of *Ambondro* by Bininda-Emonds *et al.* (2007) for this purpose might turn out to be unduly conservative, particularly if haramiyidans, the first record of which is from the Late Triassic, are indeed crown-group mammals (Luo *et al.* 2002, 2007; Luo and Wible 2005; Rowe *et al.* 2008). This would mean that the deepest divergences within living mammals might be even older than estimated by Bininda-Emonds *et al.* (2007, 2008). The recent discovery of the putative eutherian *Juramaia sinensis* (contemporaneous with *Ambondro* at ~160 Ma; Luo *et al.*, 2011b) would strongly hint that this is indeed the case.

More generally, the crucial structures needed to place a given fossil are often not preserved in the fossil record. For instance, arguably the most distinctive morphological apomorphy of Cetartiodactyla is their apparently uniquely derived 'double-pulleyed' astragalus (Schaeffer 1947; Rose 1996; Luckett and Hong 1998). However, determining the presence or absence of this feature in fossil taxa requires the discovery of post-cranial material, which is comparatively much rarer than dental material. Thus, *Ganungulatum xincunliense* from the middle Palaeocene (Nongshanian) of China might be the oldest known cetartiodactyl (as suggested by Ting *et al.* 2007), but, in the absence of potentially definitive post-cranial evidence, this identification rests on relatively minor dental features. Instead, the oldest cetartiodactyl known to have a double-pulleyed astragalus, the early Eocene *Diacodexis*, is at least 4 Ma younger. Similarly, the early Eocene Australian marsupial *Djarthia murgonensis* was originally described based on dental material that was insufficient to determine its higher-level relationships (Godthelp *et al.* 1999); it was only after the subsequent referral of tarsal remains preserving diagnostic apomorphies that *Djarthia* could be identified as the oldest unequivocal member of the crown-group marsupial clade Australidelphia (Beck *et al.* 2008).

A similar problem arises when the potentially diagnostic morphological apomorphies occur early in ontogeny. In such cases, adult material, even when abundantly available, can be uninformative with regard to the question at hand. Unfortunately, well-preserved fossils of juvenile mammals are comparatively rare (but see Rougier *et al.* 1998; Shoshani *et al.* 2006) because their hard parts are usually much less robust than those of adults. Thus, unequivocally documenting the presence of a petrosal-derived bulla (a widely accepted diagnostic apomorphy of crown-group primates; MacPhee 1981) in a given case requires young specimens of the taxon to show that no suture ever intervenes between the bulla and promontorium. When this criterion is applied to Plesiadapiformes, none could currently qualify as crown-group primates simply because there are no known taxa in which bullar composition can be fully resolved (Boyer 2009; but for a contrary opinion regarding *Ignacius graybullianus*, see Kay *et al.* 2001). Similarly, determining the presence or absence of the highly derived, potentially diagnostic pattern of tooth replacement seen in living marsupials (in which replacement is restricted to the last premolar; Luo *et al.* 2004) in fossils requires adequate juvenile material that is only rarely available (Cifelli and de Muizon 1998).

Finally, correctly placing fossils in a phylogeny can also prove difficult when there are few or no genuinely diagnostic apomorphies for the group of interest. The latter is the case for all placental superorders except for the morphologically highly derived xenarthrans. For example, putative morphological synapomorphies of Afrotheria, such as an increased number of thoracolumbar vertebrae (Sánchez-Villagra *et al.* 2007) or delayed eruption of the permanent dentition (Asher and Lehmann 2008), show considerable homoplasy both within Afrotheria and between afrotherians and other placentals (see Asher and Lehmann 2008: their fig. 3). Thus, confidently assigning fossils to Afrotheria might be expected to be extremely difficult; indeed, the 'pseudoextinction' analyses of Springer *et al.* (2007) argue that morphological data alone are often insufficient to correctly place extant placentals in their appropriate superorder (but see Asher and Hofreiter 2006; Asher *et al.* 2008; Lee and Camens 2009).

This situation raises the possibility that the earliest members of each of the placental superorders were morphologically little different from stem eutherians. Potential support for this hypothesis is provided by a series of recent papers (Hooker 2001; Boyer *et al.* 2010; Prasad *et al.* 2010; Seiffert 2010; Smith *et al.* 2010; but see Goswami *et al.* 2011) that highlight close dental and post-cranial similarities between *Deccanolestes*, the adapisoriculids *Adapisoriculus*, *Bustylus* and *Remiculus* from the Palaeocene of Europe, and *Afrodon*, a taxon originally known from dental specimens from the latest Palaeocene and earliest Eocene of North Africa but subsequently also identified in the early Palaeocene of Europe. Adapisoriculids have been identified as plesiadapiforms (Storch 2008; Smith *et al.* 2010) and hence members of the placental supraordinal clade Euarchonta. *Deccanolestes* may be a euarchontan (Hooker 2001; Boyer *et al.* 2010; Smith *et al.* 2010) or a stem eutherian (Wible *et al.* 2007). Notably, while Goswami *et al.* (2011) argued against crown placental affinities for *Deccanolestes* and *Afrodon*, their placement as stem euarchontans could not be significantly rejected using Templeton tests. Finally, *Afrodon* may be an adapisoriculid (and hence a possible euarchontan; Smith *et al.* 2010) or an afrotherian (Seiffert 2010). It seems probable that at least one of these taxa is indeed a crown-group placental, yet lacks diagnostic apomorphies that would unequivocally identify it as such; the position of *Purgatorius* (widely thought to be a euarchontan) outside Placentalia in the analysis of Wible *et al.* (2007) may represent another example of this (see Boyer *et al.* 2010). If *Deccanolestes* is a stem eutherian (Wible *et al.* 2007), adapisoriculids are euarchontans (Boyer *et al.* 2010; Smith *et al.* 2010; but see Goswami *et al.* 2011) and *Afrodon* is an afrotherian (Seiffert 2010), then their close morphological similarities are presumably plesiomorphic retentions. This, in turn, raises the possibility that the earliest members of the four superorders may have been 'adapisoriculid-like' in dental, and possibly also post-cranial, morphology (Seiffert 2010).

Fossils as divergence date estimators

A traditional palaeontological view is that a lineage begins with the 'first' or earliest taxon that can be assigned to that lineage on the basis of shared derived characters. In deference to evolutionary theory, an indeterminate (but usually small) amount of time is often allowed for 'prior evolution' to take care of the problem that a fossil taxon that already possesses the earmark apomorphies of a monophyletic group must itself descend from an ancestor that lacked those apomorphies (or expressed them differently), yet nevertheless also belonged to that group. Looked at in this way, the independent history of the modern placental orders has usually been assumed by palaeontologists to begin within a few million years of the K–Pg boundary for most groups because this is as far back as the 'first taxa + prior evolution' estimate allows one to push the available data without interpolating the lengthy ghost lineages (e.g. of up to 20–30 Ma, if not more) many molecular-based results would require.

However, the true period of time between a lineage's origin and the acquisition of its first diagnostic apomorphies is probably unknowable, and there seems no a-priori reason why it could not be quite lengthy. In the absence of compelling evidence for a 'morphological clock' (see Larsson *et al.*, this volume), it seems probable that it varies considerably both between different clades and between lineages within the same clade. If this period is long, then even a perfect fossil record will considerably underestimate the time of origin because the oldest fossil taxon with diagnostic apomorphies will be (much) younger than its parent node. Verifying the existence of such 'apomorphy lag' is difficult, because its most obvious manifestation will be a large discrepancy between (older) molecular and (younger) fossil estimates of the age of the node in question, for which several other explanations are possible (as discussed elsewhere in this chapter), and because, by definition, it cannot be identified by phylogenetic analyses of morphological data. As such, it represents a convenient, but possibly inherently untestable, ad hoc explanation that can be invoked whenever molecular and fossil divergence dates are incongruent.

Reconciliation with palaeobiogeography

Phylogenetic trees that include information on nodal divergence dates – such as the ones discussed in this chapter – have implications. Palaeobiogeographical interpretations of the distributional history of Late Cretaceous and earliest Cenozoic eutherians, for example, can be strongly influenced by one's preferred reconstruction of their phylogeny. Thus, a clear implication of the Explosive Model (like that favoured by Wible *et al.* 2007) is that post-divergence distributions must have been largely accomplished by over-water transport of propagules rather than dispersal across terrestrial portals (e.g. pre-rift conjugate terranes or land bridges).

Recent plate-tectonic and palaeogeographic reconstructions for the interval 150–50 Ma ago (e.g. Aitchison *et al.* 2007; Eagles 2007, 2010; Whittaker *et al.* 2007; Ali and Huber 2010; Ali and Krause 2011) indicate that, with the exception of southern South America/West Antarctica (via the Antarctic Peninsula) and probably also eastern Antarctica/Australia (via the South Tasman Rise), the major terranes comprising Gondwana did not remain in proximate contact after the close of the Mesozoic. Dispersal of the ancestors of australidelphian marsupials into Australia before the final, apparently gradual separation of the latter from East Antarctica (Woodburne and Case 1996; Whittaker *et al.* 2007; Beck 2008) could have involved an all-terrestrial route any time from the late Mesozoic until the earliest Eocene. Evans *et al.* (2008) make a case for several terrestrial vertebrate taxa (but no eutherians) dispersing across Antarctica via a dry-land route joining India/Madagascar to southern South America.

These points are of interest here because the lack of portals among Gondwanan fragments by the early Cenozoic should have acted as a very strong constraint on mammalian movements. Yet this factor is either ignored or side-stepped by studies advocating an Explosive Model. The necessary corollary to this argument is that nothing except non-placental eutherians (and other kinds of mammaliaforms) could have occupied Gondwanan daughter terranes until the start of the Cenozoic, when the modern orders came into existence. This is effectively a Palaeogene version of the Sherwin–Williams Effect (Hershkovitz 1968; Clemens 1986) that divergences of major mammalian lineages are always assumed to have occurred on northern landmasses, with movement to the south only taking place significantly later. The alternative is that initial differentiation of Placentalia not only began well within the Cretaceous, but also occurred among clades already occupying Gondwanan terranes whose fossils are as yet unknown. Is this plausible? Fossil taxa such as the Maastrichtian *Deccanolestes* and *Kharmerungulatum* from India hint that it might be (see also Krause and Maas 1990), although the generally poor Cretaceous fossil record for Africa, India and Australasia (see above) prevents a definite assessment of this hypothesis.

And clocks: assumptions underlying molecular-based date estimates

The molecular-clock hypothesis

All molecular-based methods of divergence time estimates rely crucially on two factors: (1) calibration data in the form of fossil and/or biogeographic events and (2) the assumption that molecular data evolve in a more-or-less clock-like fashion (the molecular-clock hypothesis). We have already dealt with the data and inference issues underlying the first factor and will only add here that the fact that poor calibration data can cause problems should be obvious (although the scale of any errors is often unknown). Indeed, in cases of conflict, molecular phylogeneticists typically question the quality of the calibration data in the first instance (e.g. Near and Sanderson 2004; Inoue *et al.* 2010; Pyron 2010). But what of the other side of the equation?

The molecular-clock hypothesis dates to the early 1960s when Zuckerkandl and Pauling (1962, 1965) noted that, unlike morphological evolution (but see Polly 2001), amino-acid changes accumulated at a relatively constant rate over the long term in haemoglobin. This hypothesis, in turn, formed an important cornerstone of the neutral theory of evolution first proposed by Kimura (1983). In many ways, however, the molecular-clock hypothesis has been misinterpreted to imply an absolutely constant rate of molecular evolution or a strict molecular clock. Even from the earliest days, however, it was realized that there was no universal clock. In generalizing their observations, Zuckerkandl and Pauling (1962, 1965) were careful to make it protein specific. Today, we know that different proteins run according to different clocks in line with the degree of inferred functional constraint they are under (more constrained genes tend to evolve more slowly; see data in Nei 1987; Ohta 1995) and differences between the faster mitochondrial and slower nuclear genomes are also apparent. In addition, it was noted that the clock could fluctuate, often greatly, over the short term. It was only over evolutionary time spans that a clock-like behaviour became apparent.

It is well accepted today that the rate of molecular evolution also varies between organismal groups (Britten 1986; Drake *et al.* 1998) and often dramatically so (e.g. viruses, the extreme speedsters, juxtaposed against the more stately mammals) because of differences in the accuracy of the DNA replication machinery and/or the inverse relationship between substitution rate and body size, whether directly or indirectly through generation time. This rate variation can also occur on a more taxonomically restricted scale and also within the same gene (heterotachy). Among mammals, for instance, rodents have a comparatively high rate of molecular evolution, whereas the clock is appreciably slower in the great apes (the 'hominid slowdown') and cetaceans; other rate differences across mammals are also apparent (see Bininda-Emonds 2007).

Although the idea of a global clock has largely been discredited for the reasons provided above, it is unfortunately still commonly applied. For instance, a substitution rate of 2% per million years for mitochondrial DNA has been applied for primates, birds and arthropods, among other groups (see Brown *et al.* 1979; Brower 1994; Ho 2007)! In addition to the improbability of such diverse groups all possessing the same rate of evolution, the use of such a value also ignores the documented rate variation among mtDNA genes (e.g. estimated as 19.2× across genes for mammals in Bininda-Emonds 2007).

Given that the molecular clock is not so much of a Rolex as it is a Timex, what hope is there then for molecular-based date estimates? Fortunately, clock-like evolution does appear to occur on more restricted timescales (and thus also taxonomic scales) before heterotachy acts to change the rate of evolution significantly. In fact, there might be a surprising amount of (local) clock-like activity. For instance, the comprehensive study of Bininda-Emonds (2007) revealed very few mammalian clades (42 of 1282) or individual branches in the mammalian tree (74 of 3332) where the rate of evolution differed significantly from the overall mammalian average. Significant changes in rate compared with an ancestral node or branch were even more rare (38 and 20, respectively). Mapping of the cytochrome b (*MT-CYB*) data set from Bininda-Emonds *et al.* (2007) onto the topology of the mammalian supertree (Bininda-Emonds *et al.* 2007) reveals more surprises. Using ModelTEST v3.7 (Posada and Crandall 1998), PAUP* v4.0b10 (Swofford 2002), and a likelihood ratio test to examine each clade in turn, a clock-like rate of evolution could not be rejected at the 0.05 level (corrected for multiple comparisons using a Holm–Bonferroni correction; Holm 1979) for 237 of the 561 clades in total (= 42.2%) (red + green lineages in Figure 3.1)! As would be expected with the idea of a local clock, the clock-like clades were both significantly younger (16.5 ± 13.7 versus 22.1 ± 20.2 Ma (mean ± SD); Mann–Whitney $U = 3.348 \times 10^4$; $z = -2.593$, $P = 0.009518$) and smaller in size (5.6 ± 3.6 versus 25.9 ± 34.6 species; $U = 1.674 \times 10^4$; $z = -11.5$, $P = 1259 \times 10^{-30}$) than the non-clock-like clades. However, the oldest clock-like clade, which links the three representatives in the tree for Afrosoricida and

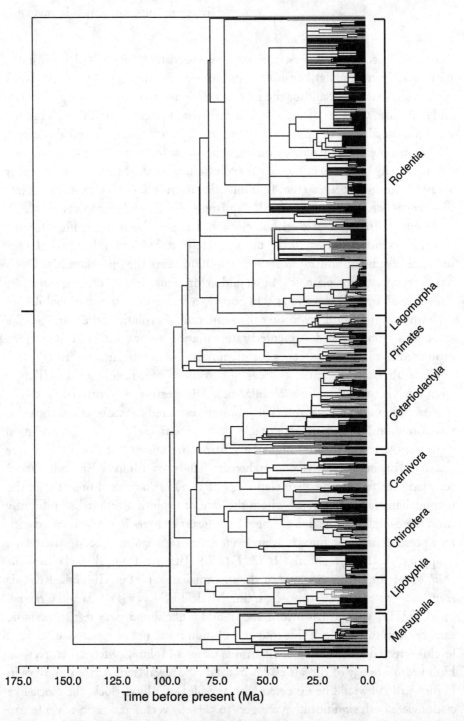

Figure 3.1 The mammal supertree of Bininda-Emonds *et al.* (2007, 2008) restricted to those species for which *MT-CYB* data were available. Clades that evolve in a clock-like fashion according to a likelihood ratio test with a nominal *P*-value of 0.05 are highlighted in either green (clade and all its descendant clades are clock-like) or red (only the focal clade is clock-like). (See also colour plate.)

Macroscelidea, is some 88.3 Ma old, a much longer timescale for clock-like behaviour than perhaps expected by many. (The largest clock-like clade comprised 22 rodent species, spanning from Echimydae plus *Capromys pilorides*, and was 21 Ma old.)

Admittedly, the previous comparisons run into a statistical problem of non-independence given the hierarchical structure of the tree, meaning that clock-like clades can contain other clock-like clades, therefore overestimating their frequency. However, this need not be the case, and there are several instances where a clock-like status emerged despite not being present in the subclades. Moreover, the non-independence problem only occurs because we have examined all the constituent clades in turn: the clock-like clades would retain this characteristic if they were examined in isolation. If, however, we apply the more conservative criterion that all subclades in a clade must evolve in a clock-like fashion for the clade to be recognized as such, then 96 such clades still remain (green lineages in Figure 3.1), with no appreciable difference in average size (5.0 ± 2.8 species; maximum = 17 species within the shrew genus *Sorex*) or age (17.9 ± 1.53 Ma; maximum = 88.3 Ma for the Afrosoricida + Macroscelidea clade) than before.

Thus, although the idea of a global clock is largely untenable, local clocks do appear to exist and might, in fact, be more widespread than is commonly assumed. In fact, local clocks play a crucial role in molecular dating studies given that they underlie most modern methods in some form. True local clock methods such as non-parametric rate smoothing (NPRS; Sanderson 1997), PATHd8 (Britton *et al.* 2007), or relDate (Purvis 1995; Bininda-Emonds *et al.* 2007) allow for different rates of molecular evolution among different lineages, but assume the rate to be constant within each lineage. Relaxed clock methods such as penalized likelihood (PL; Sanderson 2002) and those implemented in PhyBayes (Aris-Brosou and Yang 2002) and MULTIDIVTIME (Thorne and Kishino 2002) model rates as being autocorrelated along branches, essentially assuming that the rate of molecular evolution itself is a heritable character that can evolve over time. Yet another solution is provided by BEAST (Drummond and Rambaut 2007), which, in addition to a strict molecular clock, also implements uncorrelated relaxed clock models in which the rates of molecular evolution are drawn from an underlying rate distribution (e.g. exponential or log normal distributions). These two categories of methods represent the extremes of a continuum: as the lineages for local clock methods become increasingly restricted taxonomically, the methods will increasingly resemble relaxed clock methods (albeit without any model for the rate of molecular evolution). Even a global clock is simply an extreme version of a local clock method, with only a single lineage being specified.

Revisiting the use of fossils as calibration points in molecular dating analyses

In molecular dating analyses, fossils (and biogeographic or stratigraphic data) not only act as calibration points to indicate how fast the molecular clock is ticking, but are often also used to constrain the range of divergence dates for the same nodes as well. Because fossils can, at best, only be as old as their associated nodes, they have typically been used to provide minimum age constraints for that node. A frequent criticism of molecular studies then was that although many nodes possessed lower bounds, comparatively few possessed upper bounds, potentially contributing to the inflation of molecular divergence time estimates compared with strictly fossil based ones. This criticism could apply to the mammal supertree. As noted above, only the root was fixed at 166.2 Ma (based on the Middle Jurassic Malagasy fossil *Ambondro mahabo*); all other fossil calibrations were minimal age estimates and upper bounds were not specified. Thus, although the maximum age of the entire tree was capped, no such restriction existed for nodes within the tree. That being said, there is no evidence for the divergence times bunching up against the 166.2 Ma ceiling, which would indicate potential age inflation among the internal nodes.

The use of maximal age constraints is becoming increasingly common (see Benton *et al.* 2009; Phillips *et al.* 2009) and is usually associated with range constraints. Indeed, the case could be made within extant mammals that at least some of the fossil calibrations for the ordinal crown groups could approximate maximal age constraints for particularly well sampled and morphologically distinct taxa. For example, the teeth of crown-group rodents are highly diagnostic (Meng and Wyss 2005) and, of the hard anatomy, teeth are the most commonly preserved mammalian fossils. Yet, no single rodent cheek tooth has ever been identified from Cretaceous strata, suggesting strongly that the crown-group rodents are restricted to the Cenozoic. A similar case (albeit not based on tooth characters exclusively) could be made for crown-group primates with their well-researched fossil record (Bloch *et al.* 2007, but see Tavaré *et al.* 2002; Martin *et al.* 2007). Any alternative explanations (e.g. preservation bias) seem dubious, as they would have to explain why there is no trace of the Cretaceous members of these two clades that are otherwise well represented in the Cenozoic fossil record. In the case of rodents, the presence in Cretaceous sediments of probable ecological equivalents, namely multituberculates (Krause 1986), also requires that such hypothetical preservation biases would be driven by phylogeny rather than by ecology, which seems unlikely.

Such upper bounds, however, remain inherently problematic in that they can easily be overturned by the old saw of a single new discovery. Unlike lower bounds, reasonable upper bounds are also more difficult to specify robustly, the primate and rodent examples above being the possible exceptions. Methods including the application of phylogenetic bracketing (Reisz and Müller 2004; Hug and Roger 2007) and stochastic modelling (Tavaré *et al.* 2002) do exist for this purpose, but have yet to find wide use and also have specific shortcomings (see Ho and Phillips 2009). Moreover, a molecular overestimate may indicate that a given calibration is too young and that new fossil material is waiting to be discovered. An excellent example here is the recent discovery of crown-group strepsirhine primates from the middle Eocene of Egypt (Seiffert *et al.* 2003; Seiffert 2007a). These fossils are both relatively congruent with prior molecular studies that indicated a 50–62 Ma date for the first splits within crown-Strepsirhini and also supported the prevailing contention among palaeontologists that the previous earliest fossil record for the group (some 20 Ma younger from the early Miocene) was too young.

More recently, an important development in molecular dating methods has been the modelling of uncertainty in calibration data via the use of soft bounds and/or a variety of probabilistic distributions (Drummond *et al.* 2006; Yang and Rannala 2006), thereby causing calibration data to become prior probabilities instead of point estimates. These same methods can be used to specify (probabilistic) upper constraints for any calibration point as well. As important as these advances have been, it still must be realized that the validity of the (many) available models remains to be established, as do reasonable parameters for them. Most represent standard statistical models (e.g. normal, log-normal, exponential or gamma), but this does not automatically guarantee that they represent an accurate portrayal of modelling the uncertainty in calibration data. For instance, the discoveries of *Murtoilestes abramovi* (Averianov and Skutschas 2001) at ~120–128 Ma and *Eomaia scansoria* (Ji *et al.* 2002) at ~125 Ma pushed back the evidence for the origin of eutherian mammals by some 13% from previous estimates (Cifelli 1999), probably well outside the upper bounds many users would specify for the statistical models. More recently, the description of *Juramaia sinensis* (Luo *et al.* 2011b) at ~160 Ma adds another astonishing 28% to these two estimates. Actual research to quantify exactly how uncertain fossil calibrations might be is rare (but see Tavaré *et al.* 2002). In any case, the use of fuzzy calibrations/constraints does seem to improve and/or focus molecular date estimates (Inoue *et al.* 2010) and the use of some vaguely realistic bounds has to be preferable to the open-ended scenario that was in play before. The changes brought about by using bounded estimates can sometimes be dramatic, as shown by Ho and Phillips (2009). In their study, the use of bounded (including distributional) calibrations shifted the inferred origin of Neoaves across the K–Pg boundary into the Cretaceous, in contrast to the use of point estimates, which instead inferred a Palaeogene origin. Underlining this difference is that the 95% HPD (highest posterior density) intervals for the respective estimates did not overlap at all.

In the future, additional improvements to model accuracy could be achieved by the modelling of clade diversification and variation in fossil recovery potential in time and space (Inoue *et al.* 2010; Marjanović and Laurin 2008). Such parameters are not incorporated currently in existing molecular-dating software.

Modelling the rate of molecular evolution

Modern molecular dating methods are invariably model driven, usually in a likelihood framework. Minimally, this involves likelihood estimates of the amount of molecular evolution, but can also include modelling how these data translate into actual divergence times for the relaxed clock methods. The use of models can help us to represent the evolutionary process more precisely, thereby improving our estimates, but they are simultaneously problematic in that they are simple, tractable implementations of what are probably extremely complicated processes. Both the simplification itself and choosing which simplification is the right one can have important consequences.

Likelihood-derived branch lengths are actually rates measured in average number of substitutions per site per unit time, therefore confounding the amount of evolutionary change with the time in which it has taken place. Thus, whereas a long branch means that a lot of evolution has occurred, it does not reveal whether it has occurred over a long time span at a slow rate, over a short time span at a fast rate, or something in between. However, given that modern molecular dating methods necessarily assume some form of relative rate constancy, long branches default to meaning more evolution over longer times. Thus, without the appropriate calibration data, changes in the rate of molecular evolution will be difficult to detect (although less so for large changes) and, in fact, will tend to be smoothed out by the different methods (but see Kitazoe *et al.* 2007).

Evolutionary scenarios do exist, however, where a dramatic fluctuation in the rate of molecular evolution can be imagined, if not likely, to occur. One example is adaptive radiations, where a high rate of phenotypic evolution is coupled with a burst of speciation in a short time span. It seems likely that the rate of molecular evolution, which ultimately underlies the other two phenomena, would also spike during these intervals (e.g. Kitazoe *et al.* 2007), even in the apparent absence of a long-term relationship between the rates of phenotypic and molecular evolution (Davies and Savolainen 2006; Bromham *et al.* 2002, but see Seligmann 2010). Exactly such an event is inferred by

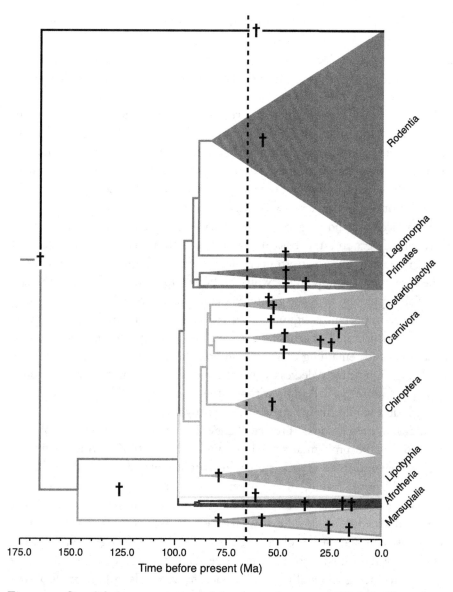

Figure 3.2 Simplified representation of the mammal supertree (Bininda-Emonds *et al.* 2007, 2008) showing the approximate temporal and phylogenetic position of the 30 fossil calibration points used (marked with †). The K–Pg boundary is indicated by the dashed vertical line. (See also colour plate.)

palaeontologists to have occurred immediately after the K–Pg boundary with the demise of the non-avian dinosaurs opening up new niches for the surviving forms (including mammals). However, given the paucity of suitable fossils before this time to recalibrate the clock (see Figure 3.2), such an adaptive radiation for mammals could be easily missed by molecular methods. Curiously, some molecular studies do detect an adaptive radiation of the placental mammal orders, but place it well in advance of the K–Pg boundary at 80 Ma or older (Springer *et al.* 2003; Bininda-Emonds *et al.* 2007). Could this be the missing post-K–Pg radiation of mammals?

The effect on the rate of molecular evolution of the restricted time span of an adaptive radiation can be modelled by constraining the basal divergences for all ordinal crown groups of the mammal supertree to occur between 62.5 and 65.5 Ma (which is the extreme case and admittedly unrealistic for more recent crown groups such as Hyracoidea) and redating the rest of the tree. Doing so results in no significant difference across all node ages compared with the original ages, but differences for those ages associated with each of the nodes linking the ordinal crown groups, the basal-most nodes within the crown groups, and the remainder of the crown-group nodes (Table 3.1; Figure 3.3). The change for the basal crown-group nodes, which all bunch up under the K–Pg boundary, is especially apparent, as is the greater variation for the remaining crown-group nodes. More importantly, significant differences in the inferred rates of evolution occur, with branch specific rates of evolution being significantly increased for the basal and remaining crown-group branches and the degree of local rate shifts being significantly decreased for the branches linking the crown groups (Table 3.1; Figure 3.4). Again, without the necessary calibration data, such increased rates would likely be smoothed out by most programs.

In addition, even gradual, directed changes in the rate of molecular evolution could prove to be problematic for most molecular methods, which generally model rate evolution as a stochastic and/or autocorrelated process (see Drummond *et al.* 2006). Again, in the absence of appropriate calibration data, it might not be possible to account for concerted changes in the clock (e.g. either a continual speedup or slowdown with time). It is known, for example, that Cretaceous mammals were, on average, smaller than Palaeogene and Recent forms (Alroy 1999) and also that body size tends to cluster among mammals (e.g. rodents being generally small, cetaceans being large). It is also well established that the rate of molecular evolution in mammals correlates strongly and inversely with body size (Martin and Palumbi 1993; Bromham *et al.* 1996; Lanfear *et al.* 2010), meaning that there might have been a gradual, possibly lineage-specific, overall slowdown in the mammalian clock over time. Correcting for this artefact by incorporating body size or generation time information into the models, however, would have the effect of drawing the deeper mammalian divergences even farther back in time given that the inferred amounts of molecular evolution would have actually occurred in a shorter timeframe than has been reconstructed currently. Indirect evidence here is provided by the

Table 3.1 Statistical comparison of branch-specific rates of evolution and shifts in the rate of evolution compared with an ancestral node for the mammal supertree when the basal divergences of all ordinal crown groups are constrained to occur between 62.5 and 65.5 Ma, thereby mimicking the Explosive Model of placental evolution. All comparisons used a Wilcoxon signed rank test; *P*-values with an asterisk indicate significant differences at a nominal *P*-value of 0.05 corrected for multiple comparisons (Holm 1979). Values were obtained using the Perl script moleRat.pl v1.0 following the procedure in Bininda-Emonds (2007). Taxonomic partitions were defined as follows: stem – all branches basal to the ordinal crown groups; basal – basal node of the ordinal crown groups and all immediate daughter and granddaughter nodes; crown group – all remaining nodes in the crown group. Only nodes that were older than 35 Ma according to Bininda-Emonds *et al.* (2008) were included.

Variable	Partition	n	Median of original value	Median value under Explosive Model	Trend relative to original values	W	z	P
Node age	All	119	54.9	62.5	+	3861	0.7717	0.4403
	Stem	24	86.3	86.2	–	280	3.714	0.0002*
	Basal	49	56.5	62.5	+	1091	4.76	1.94×10^{-6}*
	Crown	46	46.2	50.9	+	958	4.561	5.08×10^{-6}*
Branch-specific rate of evolution (normalized)	All	118	−3.90	−2.24	+	4131	2.807	0.0050*
	Stem	24	−3.16	−1.98	+	195	1.286	0.1985
	Basal	49	−5.39	−0.37	+	1053	4.382	1.18×10^{-5}*
	Crown	45	−2.68	−2.59	+	813.5	3.341	0.0008*
Rate shift relative to an ancestral node (normalized)	All	133	−0.10	−0.23	–	4805	0.7849	0.4325
	Stem	23	0.45	−0.08	–	237	3.011	0.0026*
	Basal	50	−0.46	3.08	+	732	0.9122	0.3620×10^{-1}
	Crown	60	−0.11	−1.56	–	1106	1.406	0.1597

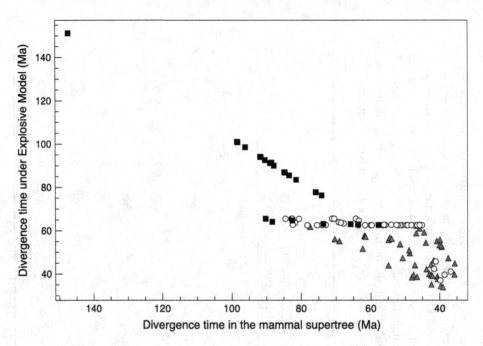

Figure 3.3 Graphical comparison of node ages from the mammal supertree
(Bininda-Emonds *et al.* 2007, 2008) and a redating that models the Explosive Model of
placental evolution. Nodes are partitioned taxonomically as (1) all branches linking
the ordinal crown groups (black squares), (2) basal node of the ordinal crown groups
and all immediate daughter and granddaughter nodes (white circles), and (3) all
remaining nodes in the crown group (grey triangles). Only nodes that were older
than 35 Ma according to Bininda-Emonds *et al.* (2008) were included.

mammal supertree (Bininda-Emonds *et al.* 2007, 2008), where the molecular
date estimates for the origin of the Cetacea, a taxon characterized by large
body sizes and slow generation times (and therefore a slow rate of molecular
evolution), severely underestimate the fossil calibration (22.9 versus 52.2 Ma,
respectively). By contrast, when Springer *et al.* (2003) restricted their analyses
to species that appear to have maintained (small) body sizes similar to those
of Cretaceous mammals, divergence time estimates compared with the full
analysis were largely unchanged, if not slightly younger.

Even in the absence of such scenarios as above, it remains that the application of a single, simple model of evolution often will not reflect biological
reality. Most of the models being used currently can account for rate variation
between sites in a given sequence (rate heterogeneity as modelled using gamma
distribution), but not between clades in a tree despite there being good evidence
for heterotachy in many groups. Indeed, it has been shown that accounting for

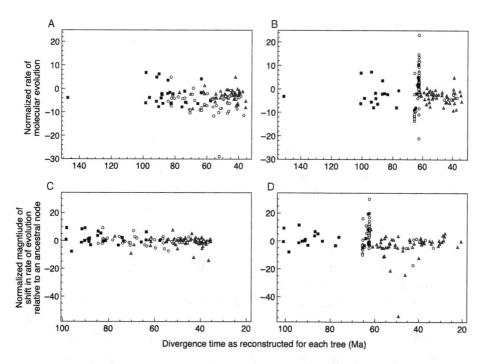

Figure 3.4 Graphical comparison of rates of molecular evolution inferred from the mammal supertree (A, C) and a redating that models the Explosive Model of placental evolution (B, D). Comparisons are made for both normalized branch-specific rates of evolution (A, B) and normalized relative shifts in the rate of evolution compared with an ancestral node (C, D) (following Bininda-Emonds 2007). Taxonomic partitions are as in Figure 3.3.

the latter might dramatically impact our molecular-based divergence time estimates. In using a model to correct for heterotachy within placental mammals (and particularly for speedups in the lineages leading to Euarchontoglires and Laurasiatheria and a subsequent slowdown within the latter), Kitazoe *et al.* (2007) reconstructed divergence time estimates for the ordinal crown groups that more closely agreed with those from the palaeontological Explosive Model (although a number of lineages still crossed into the Cretaceous). In so doing, however, it should be noted that their model of rate variation favours relatively recent radiations. They also did not use any calibration data older than 65 Ma, which might otherwise have pulled the divergence times older, and also used some maximal age constraints. Thus, although they constrained the divergence between the rabbit and the pika to 36–55 Ma, their molecular estimate for this node at ~55 Ma (i.e. at the upper limit of their constraint) suggests that this fossil calibration may be an underestimate (see Arnason *et al.* 2008). An outstanding problem, however, is recognizing heterotachy a-priori or during the course of the analysis. Here the application of relative rates tests (Sarich and Wilson 1973; Tajima 1993), which appear to have fallen somewhat out of favour, might be a viable solution.

In a related vein, choosing the right model of evolution and the right parameters for it are also important considerations. There has been much work into the modelling of mtDNA data, with several studies showing that the use of purine-pyrimidine (RY) coding for the third positions often improves the accuracy of phylogenetic estimates from these data (Phillips and Penny 2003; Phillips et al. 2004). Phillips (2009) has extended this research to show that conventional coding for mtDNA leads to proportionately longer branches deep in the vertebrate tree compared with RY-coding when using deep calibration points. This, in turn, leads to proportionately older divergence time estimates, which again might help to explain the discrepancy between molecular- and fossil-based dates for the placental orders.

Complicating this story is the question of whether only a single model across the tree (as opposed to the sequence data) should be used. Analysis of the *MT-CYB* data for the mammal supertree also reveals that the use of a single substitution model across the tree is unrealistic. Of the 561 nodes for which an optimal model could be estimated using ModelTEST, 6 of the 14 base models in ModelTEST were indicated (HKY, TrN, K81uf, TIM, TVM, GTR) and 23 of the 56 models incorporating correction for invariant sites (+I) and rate heterogeneity (+G). Thus, analogous to the relationship between heterotachy and gamma-corrected rate heterogeneity, a tree-based counterpart to partition-specific substitution models is arguably needed as well.

In estimating (molecular) divergence times, we are chasing an unknown, or worse, compounding two unknowns: the topology of the phylogenetic tree and its branch lengths. However, unlike the case with the pattern of relationships, we often have no good a-priori idea of what a reasonable divergence time estimate should be. That living primates form a single clade is a reasonable expectation, as borne out by application of phylogenetic methods. But, we have little to no intuitive idea precisely how old crown-group primates should be beyond a rough estimate of between 55 (Bloch et al. 2007) and, at the outside, 80–100 Ma (Tavaré et al. 2002; Martin et al. 2007).

Thus, in the absence of a largely complete fossil record, it is often difficult to assess the accuracy of our estimates. Simulation studies can always be undercut by the argument that the parameters used are not realistic. In the end, we are left with a congruence-based approach, both with fossil data as well as with other molecular-based estimates. But, analogous to the conflict with fossil data, even different molecular-based estimates often differ widely. Some of the earliest molecular studies argue for a 'Short-fuse Model' (*sensu* Archibald and Deutschmann 2001) of placental evolution, whereby both the origins and basal diversifications of many of the ordinal crown groups appear well within the Cretaceous (Springer 1997; Kumar and Hedges 1998; Penny *et al.* 1999). More recent studies favour a 'Long-fuse' Model (*sensu* Archibald and Deutschmann 2001), where only the origins occur in the Cretaceous, but the radiations of the crown groups primarily occur in the Cenozoic in agreement with conventional interpretations of the fossil evidence (Springer *et al.* 2003). Still other results argue for an intermediate scenario between these extremes. So, although there is agreement among most studies that placental mammals are older than their first appearances in the fossil record, there is little consensus on just how much older this might be.

In many cases, the different estimates arise because of the different assumptions and models being used by the molecular methods. A cogent example is provided by Welch *et al.* (2005), who examined why analysis of the same data set by two different methods could give wildly divergent estimates for the timing of the divergence between protostomes and deuterostomes. Whereas an earlier local clock-based study (Bromham *et al.* 1998) placed this split at no earlier than 680 Ma, a relaxed-clock study (Aris-Brosou and Yang 2003) indicated a more recent date of 582 ± 112 Ma in line with the Cambrian explosion hypothesis. The difference, Welch *et al.* (2005) concluded, was due in part to the statistically tractable, but probably unrealistic, assumptions being made in the relaxed-clock analysis: a constant net speciation rate through time and a random sampling of the sequences at the tips. Even between comparable methods like the two relaxed-clock methods MCMCTREE (Yang and Rannala 2006; Rannala and Yang 2007) and MULTIDIVTIME, numerous, often subtly different assumptions are being made (see Inoue *et al.* 2010), with no clear indication of which are the more realistic. Molecular dating methods then are arguably even more 'black box' than are phylogenetic inference methods, where different implementations tend to differ more with respect to the search algorithms used rather than in the underlying assumptions being made.

Issues of data quality

As with any other scientific exercise, the quality of the results hangs upon the quality of the data and the methods used to analyse them. Interestingly, issues of data quality in molecular dating studies typically tend to focus on the quality of the calibration points. This is not without justification given that the molecular divergence time estimates directly depend on these data. A badly chosen calibration point, therefore, can have far-reaching effects.

As a result, there has been a reasonable amount of attention paid to the calibration data, including how to counteract any potential problems, to account for uncertainty in them (see above), and/or to assess their influence on the results. For instance, a general, common-sense guideline is that as many calibration points that are spread out across the phylogeny should be used as far as is possible. This strategy both minimizes the potential negative impact of any single questionable calibration point as well as regularly resets the clock to prevent having to interpolate rates of evolution from distant calibration points. Indeed, this point arguably represents a potential weakness of the dating for the mammal supertree in that the large majority of the calibration data used were necessarily from Cenozoic fossils (see Figure 3.2), meaning that all divergence events up to and including the K–Pg boundary hang primarily on a few, very old calibration points.

In addition, it is naturally important that the calibration data be as robust as possible and associated with the proper node on the tree (Reisz and Müller 2004). As discussed earlier, only fossils where the fossil record is reasonably extensive should be used (or where there is otherwise no concern about substantially underestimating the oldest known member of a given group) and fossils should be placed within their focal taxon using robust, explicit criteria (e.g. possessing at least one synapomorphy in common with the taxon; cf. Bininda-Emonds et al. 2007). For several mammalian clades (notably bats, which appear to have particularly low preservation potential; Teeling et al. 2005), however, the first of these desiderata may be unattainable. Finally, there have been several suggestions regarding sensitivity analysis involving the calibration data. These range from the informal usage of several different calibration dates for any given node to assess the overall stability of the molecular dates (e.g. Springer et al. 2003) to more formal methods such as fossil cross-validation (Near and Sanderson 2004) or the likelihood method of Pyron (2010) that can help to identify calibration data that disagree strongly with the remainder of the data set. The principle behind the Near and Sanderson (2004) method is to fix the dates for those nodes with calibration data singly in turn and then to assess how the dates for the remaining nodes agree with their (unused) calibration data. A variant on this procedure (fossil-based model cross-validation) is also used to help shape the parameters of the rate-smoothing model used in PL (Near and Sanderson 2004). The method of Pyron (2010) goes a step beyond this to compare possible placements for fossil taxa where their phylogenetic affinity is ambiguous, thereby potentially providing novel, molecular-based information as to the placement of a given fossil.

What has received surprisingly little attention, however, is the quality (or suitability) of the molecular data being used in the analyses. Although the dating methods and the models behind them are attracting increasing interest, the raw sequence data themselves have largely been ignored. This is surprising precisely because so much attention has been paid to sequence data quality and suitability in the related area of phylogenetic analysis. It is well accepted that genes evolve at different rates and are maximally informative at different phylogenetic levels: conservative genes are preferentially used to infer relationships between higher-level taxa whereas more variable genes are used closer to the species level. Hundreds of papers have also addressed the problems of saturation/multiple hits and its role in long-branch attraction (for a comprehensive review, see Bergsten 2005). Yet in cases where methods indicate a conflict between one or more calibration points and the molecular dates, it is usually the calibration points that are called into question when they could instead actually be informing models of sequence evolution that are being misled by model misspecification or rate heterogeneity (including heterotachy). So why are the sequence data typically taken for granted in molecular-dating analyses?

Perhaps the answer to the contradiction lies with the fact that the same genes are often used to derive the divergence times as were used to obtain the phylogeny in the first place, meaning that the sequence data have already been vetted to some degree. Indeed, programs like BEAST can derive the topology and date estimates for a given data set simultaneously. It remains, however, that the taxonomic and genomic scope of our phylogenetic analyses is increasing rapidly. For instance, the dating of the mammal supertree required a multigene data set comprising 68 genes of different rates to adequately cover all 4554 species and the 160+ Ma time span. The same is true of other, similar large-scale studies, which are becoming increasingly common. However, not all genes in such data sets will be optimally informative throughout the tree and some might be positively misinformative in places (e.g. saturated genes or sites for deep divergences). This problem can be ameliorated to a certain extent by the use of likelihood-based models of sequence evolution that can account for phenomena like saturation, particularly those where the models can be fitted individually to different partitions (e.g. genes or codons). However, as shown by the study of Phillips (2009; see above), the efficacy of this correction depends crucially on using the right model.

Although Bayesian methods can account for uncertainty in the molecular data, they still must use all the molecular data provided, even when parts of the data set appear to be positively misleading or disruptive. An unexplored line of research lies in developing the molecular counterpart of fossil cross-validation, where 'gene cross-validation' would identify and outright eliminate or otherwise downweight outlier genes (or, more generally, partitions) that conflict strongly with the remainder of the data set.

Conclusions

Although our focus here has been largely restricted to the mammalian radiation, it is important to note that the issues we have identified herein go well beyond this taxonomic group. Analogous conflicts between fossil- and molecular-based date estimates have been noted for many other groups, with apparent explosive radiations of the animal phyla (Precambrian versus Cambrian) and birds (K–Pg boundary) being contradicted by molecular data (Cooper and Fortey 1998). Reconciling the disparate divergence time estimates within mammals, as well as for these other groups, will require a critical assessment of the data source and assumptions being made on both sides of the equation and thus the active collaboration of palaeontologists and molecular systematists. Hopefully, this strategy will prove to be an important step in achieving a consensus as to when mammals (and other taxa) evolved, analogous to the growing agreement regarding their relationships to one other.

Summary

A consensus is emerging on phylogenetic relationships within Mammalia, one contributed to, in hindsight, by both morphological and molecular data. This state of general agreement, however, does not extend to divergence time estimates within the group: fossil- and molecular-based dates tend to differ considerably, especially as regards the origins of and initial diversifications within the ordinal crown groups. In this chapter, we take a critical look at both fossil- and molecular-based frameworks for divergence time estimation, with a particular focus on the situation as it affects placental mammals. Our goal was not to determine when these (or any other) mammals evolved, but rather to highlight the assumptions underlying the analysis of each type of data. This exercise provided some insights into how fossil and molecular date estimates can disagree so profoundly as well as suggestions for achieving a better consensus than exists at present.

Acknowledgements

We thank Rob Asher and Johannes Müller for organizing an interesting and diverse symposium at SVP 2009 and for inviting us to present our views there. We thank Andres Giallombardo for discussion of the topics raised in this chapter as well as two anonymous reviewers and Rob Asher for their helpful comments. Funding for Robin Beck was provided by NSF grant DEB-0743039 (in collaboration with Rob Voss at the AMNH) and for the research of Ross MacPhee on Gondwanan mammalian evolution by NSF grants OPP-0636639 and ANT-1142052.

REFERENCES

Aitchison, J. C., Ali, J. R. and Davis, A. M. (2007). When and where did India and Asia collide? *Journal of Geophysical Research*, **112**, 1–19.

Ali, J. R. and Aitchison, J. C. (2008). Gondwana to Asia: plate tectonics, paleogeography and the biological connectivity of the Indian sub-continent from the Middle Jurassic through latest Eocene (166–35 Ma). *Earth-Science Reviews*, **88**, 145–66.

Ali, J. R. and Huber, M. (2010). Mammalian biodiversity on Madagascar controlled by ocean currents. *Nature*, **463**, 653–6.

Ali, J. R. and Krause, D. W. (2011). Late Cretaceous bioconnections between Indo-Madagascar and Antarctica: refutation of the Gunnerus Ridge causeway hypothesis. *Journal of Biogeography*, **38**(10), 1855–72.

Alroy, J. (1999). The fossil record of North American mammals: evidence for a Paleocene evolutionary radiation. *Systematic Biology*, **48**, 107–18.

Archibald, J. D. and Deutschmann, D. H. (2001). Quantitative analysis of the timing of the origin and diversification of extant placental orders. *Journal of Mammalian Evolution*, **8**, 107–24.

Archibald, J. D., Zhang, Y., Harper, T. and Cifelli, R. L. (2011). *Protungulatum*, confirmed Cretaceous occurrence of an otherwise Paleocene eutherian (placental?) mammal. *Journal of Mammalian Evolution*, **18**, 153–61.

Aris-Brosou, S. and Yang, Z. (2002). Effects of models of rate evolution on estimation of divergence dates with special reference to the metazoan 18S ribosomal RNA phylogeny. *Systematic Biology*, **51**, 703–14.

Aris-Brosou, S. and Yang, Z. (2003). Bayesian models of episodic evolution support a late Precambrian explosive diversification of the Metazoa. *Molecular Biology and Evolution*, **20**, 1947–54.

Arnason, U., Adegoke, J. A., Gullberg, A., *et al.* (2008). Mitogenomic relationships of placental mammals and molecular estimates of their divergences. *Gene*, **421**, 37–51.

Asher, R. J. (1999). A morphological basis for assessing the phylogeny of the 'Tenrecoidea' (Mammalia, Lipotyphla). *Cladistics*, **15**, 231–52.

Asher, R. J. (2001). Cranial anatomy in tenrecid insectivorans: character evolution across competing phylogenies. *American Museum Novitates*, **3352**, 1–54.

Asher, R. J. (2007). A web-database of mammalian morphology and a reanalysis of placental phylogeny. *BMC Evolutionary Biology*, **7**, 108.

Asher, R. J. and Hofreiter, M. (2006). Tenrec phylogeny and the noninvasive extraction of nuclear DNA. *Systematic Biology*, **55**, 181–94.

Asher, R. J. and Lehmann, T. (2008). Dental eruption in afrotherian mammals. *BMC Biology*, **6**, 14.

Asher, R. J., Novacek, M. J. and Geisler, J. H. (2003). Relationships of endemic African mammals and their fossil relatives based on morphological and molecular evidence. *Journal of Mammalian Evolution*, **10**, 131–94.

Asher, R. J., Geisler, J. H. and Sanchez-Villagra, M. R. (2008). Morphology, paleontology, and placental mammal phylogeny. *Systematic Biology*, **57**, 311–17.

Averianov, A. O. and Skutschas, P. P. (2001). A new genus of eutherian mammal from the Early Cretaceous of Transbaikalia, Russia. *Acta Palaeontologica Polonica*, **46**, 431–6.

Averianov, A. O., Archibald, J. D. and Martin, T. (2003). Placental nature of the alleged marsupial from the Cretaceous of Madagascar. *Acta Palaeontologica Polonica*, **48**, 149–51.

Beck, R. M. D. (2008). Form, function, phylogeny and biogeography of enigmatic Australian metatherians. PhD dissertation, University of New South Wales, Australia.

Beck, R. M. D., Godthelp, H., Weisbecker, V., Archer, M. and Hand, S. J. (2008). Australia's oldest marsupial fossils and their biogeographical implications. *PLoS One*, **3**, e1858.

Benton, M. J. and Donoghue, P. C. J. (2007). Paleontological evidence to date the tree of life. *Molecular Biology and Evolution*, **24**, 26–53.

Benton, M. J., Donoghue, P. C. J. and Asher, R. J. (2009). Calibrating and constraining molecular clocks. In *The Timetree of Life*, ed. S. B. Hedges and S. Kumar. Oxford, UK: Oxford University Press.

Bergqvist, L. P., Abrantes, E. A. L. and Avilla, L. (2004). The Xenarthra (Mammalia) of São José de Itaboraí Basin (upper Paleocene, Itaboraian), Rio de Janeiro, Brazil. *Geodiversitas*, 323–37.

Bergsten, J. (2005). A review of long-branch attraction. *Cladistics*, **21**, 163–93.

Bininda-Emonds, O. R. P. (2000). Factors influencing phylogenetic inference: a case study using the mammalian carnivores. *Molecular Phylogenetics and Evolution*, **16**, 113–26.

Bininda-Emonds, O. R. P. (2007). Fast genes and slow clades: comparative rates of molecular evolution in mammals. *Evolutionary Bioinformatics Online*, **3**, 59–85.

Bininda-Emonds, O. R. P., Cardillo, M., Jones, K. E., *et al.* (2007). The delayed rise of present-day mammals. *Nature*, **446**, 507–12.

Bininda-Emonds, O. R. P., Cardillo, M., Jones, K. E., *et al.* (2008). Corrigendum: The delayed rise of present-day mammals. *Nature*, **456**, 274.

Bloch, J. I., Silcox, M., Boyer, D. M. and Sargis, E. J. (2007). New Paleocene skeletons and the relationship of plesiadapiforms to crown-clade primates. *Proceedings of the National Academy of Sciences of the United States of America*, **104**, 1159–64.

Boyer, D. M. (2009). New cranial and postcranial remains of late Paleocene Plesiadapidae ('Plesiadapiformes', Mammalia) from North America and Europe: description and evolutionary implications. PhD thesis, Stony Brook University, New York.

Boyer, D. M., Prasad, G. V. R., Krause, D. W., *et al.* (2010). New postcrania of *Deccanolestes* from the Late Cretaceous of India and their bearing on the evolutionary and biogeographic history of euarchontan mammals. *Naturwissenschaften*, **97**, 365–77.

Britten, R. J. (1986). Rates of DNA sequence evolution differ between taxonomic groups. *Science*, **231**, 1393–8.

Britton, T., Anderson, C. L., Jacquet, D., Lundqvist, S. and Bremer, K. (2007). Estimating divergence times in large phylogenetic trees. *Systematic Biology*, **56**, 741–52.

Bromham, L., Rambaut, A. and Harvey, P. H. (1996). Determinants of rate variation in mammalian DNA sequence evolution. *Journal of Molecular Evolution*, **43**, 610–21.

Bromham, L., Rambaut, A., Fortey, R., Cooper, A. and Penny, D. (1998). Testing the Cambrian explosion hypothesis by using a molecular dating technique. *Proceedings of the National Academy of Sciences of the United States of America*, **95**, 12 386–9.

Bromham, L., Woolfit, M., Lee, M. S. and Rambaut, A. (2002). Testing the relationship between morphological and molecular rates of change along phylogenies. *Evolution*, **56**, 1921–30.

Brower, A. V. (1994). Rapid morphological radiation and convergence among races of the butterfly *Heliconius erato* inferred from patterns of mitochondrial DNA evolution. *Proceedings of the National Academy of Sciences of the United States of America*, **91**, 6491–5.

Brown, W. M., George, M., Jr. and Wilson, A. C. (1979). Rapid evolution of animal mitochondrial DNA. *Proceedings of the National Academy of Sciences of the United States of America*, **76**, 1967–71.

Bryant, H. N. (1989). An evaluation of cladistic and character analyses as hypothetico-deductive procedures, and the consequences for character weighting. *Systematic Zoology*, **38**, 214–27.

Bryant, H. N. (1991). The polarization of character transformations in phylogenetic systematics: role of axiomatic and auxiliary assumptions. *Systematic Zoology*, **40**, 433–45.

Butler, P. M. (1984). Macroscelidea, Insectivora and Chiroptera from the Miocene of east Africa. *Palaeovertebrata*, **14**, 117–200.

Cartmill, M. (1981). Hypothesis testing and phylogenetic reconstruction. *Journal of Zoological Systematics and Evolutionary Research*, **19**, 73–96.

Cifelli, R. L. (1999). Tribosphenic mammal from the North American Early Cretaceous. *Nature*, **401**, 363–6.

Cifelli, R. L. and de Muizon, C. (1998). Tooth eruption and replacement pattern in early marsupials. *Comptes Rendus de l'Académie des Sciences, IIA–Earth and Planetary Science*, **326**, 215–20.

Clemens, W. (1986). On Triassic and Jurassic mammals. In *The Beginning of the Age of Dinosaurs: Faunal Change across the Triassic-Jurassic Boundary*, ed. K. Padian. Cambridge, UK: Cambridge University Press.

Clyde, W. C., Ting, S., Snell, K. E., *et al.* (2010). New paleomagnetic and stable-isotope results from the Nanxiong Basin, China: implications for the K/T boundary and the timing of Paleocene mammalian turnover. *Journal of Geology*, **118**, 131–43.

Cooper, A. and Fortey, R. (1998). Evolutionary explosions and the phylogenetic fuse. *Trends in Ecology and Evolution*, **13**, 151–5.

Davies, T. J. and Savolainen, V. (2006). Neutral theory, phylogenies, and the relationship between phenotypic change and evolutionary rates. *Evolution*, **60**, 476–83.

de Muizon, C. (1991). La fauna de mamíferos de Tiupampa (Paleoceno Inferior, Formación Santa Lucía), Bolivia. In *Fósiles y Facies de Bolivia*, Vol. 1 *Vertebrados*, ed. R. Suárez-Soruco. Santa Cruz, Bolivia: Revista Téchnica de Yacimientos Petrolíferos Fiscales Bolivianos.

de Muizon, C. and Cifelli, R. L. (2000). The 'condylarths' (archaic Ungulata, Mammalia) from the early Palaeocene of Tiupampa (Bolivia): implications on the origin of the South American ungulates. *Geodiversitas*, **22**, 47–150.

de Muizon, C., Cifelli, R. L. and Bergqvist, L. P. (1998). Eutherian tarsals from the early Paleocene of Bolivia. *Journal of Vertebrate Paleontology*, **18**, 655–63.

Donoghue, M. J., Doyle, J. A., Gauthier, J., Kluge, A. G. and Rowe, T. (1989). The importance of fossils in phylogeny reconstruction. *Annual Review of Ecology and Systematics*, **20**, 431–60.

Donoghue, P. C. J. and Purnell, M. A. (2009). Distinguishing heat from light in debate over controversial fossils. *BioEssays*, **31**, 178–89.

Douady, C. J. and Douzery, E. J. P. (2003). Molecular estimation of eulipotyphlan divergence times and the evolution of 'Insectivora'. *Molecular Phylogenetics and Evolution*, **28**, 285–96.

Douzery, E. J. P., Delsuc, F., Stanhope, M. J. and Huchon, D. (2003). Local molecular clocks in three nuclear genes: divergence times for rodents and other mammals and incompatibility among fossil calibrations. *Journal of Molecular Evolution*, **57**, S201–13.

Drake, J. W., Charlesworth, B., Charlesworth, D. and Crow, J. F. (1998). Rates of spontaneous mutation. *Genetics*, **148**, 1667–86.

Drummond, A. J. and Rambaut, A. (2007). BEAST: Bayesian evolutionary analysis by sampling trees. *BMC Evolutionary Biology*, **7**, 214.

Drummond, A. J., Ho, S. Y. W., Phillips, M. J. and Rambaut, A. (2006). Relaxed phylogenetics and dating with confidence. *PLoS Biology*, **4**, e88.

Eagles, G. (2007). New angles on South Atlantic opening. *Geophysical Journal International*, **168**, 353–61.

Eagles, G. (2010). The age and origin of the central Scotia Sea. *Geophysical Journal International*, **183**, 587–600.

Evans, S. E., Jones, M. E. H. and Krause, D. W. (2008). A giant frog with South American affinities from the Late Cretaceous of Madagascar. *Proceedings of the National Academy of Sciences of the United States of America*, **105**, 2951–6.

Flynn, J. J., Parrish, J. M., Rakotosamimanana, B., Simpson, W. F. and Wyss, A. R. (1999). A new Middle Jurassic mammal from Madagascar. *Nature*, **401**, 57–60.

Foote, M., Hunter, J. P., Janis, C. M. and Sepkoski, J. J., Jr. (1999). Evolutionary and preservational constraints on origins of biologic groups: divergence times of eutherian mammals. *Science*, **283**, 1310–14.

Gayet, M. (2001). A review of some problems associated with the occurrences of fossil vertebrates in South America. *Journal of South American Earth Sciences*, **14**, 131–45.

Geisler, J. H. and Uhen, M. D. (2003). Morphological support for a close relationship between hippos and whales. *Journal of Vertebrate Paleontology*, **23**, 991–6.

Gelfo, J. N., Goin, F. J., Woodburne, M. O. and de Muizon, C. (2009). Biochronological relationships of the earliest South American Paleogene mammalian faunas. *Palaeontology*, **52**, 251–69.

Gheerbrant, E. (2009). Paleocene emergence of elephant relatives and the rapid radiation of African ungulates. *Proceedings of the National Academy of Sciences of the United States of America*, **106**, 10 717–21.

Giallombardo, A. and AToL Mammal Morphology Team (2010). Postcanine homologies in Mammalia. *Journal of Vertebrate Paleontology*, **30**, 96A.

Gingerich, P. D., Haq, M., Zalmout, I. S., Khan, I. H. and Malkani, M. S. (2001). Origin of whales from early artiodactyls: hands and feet of Eocene Protocetidae from Pakistan. *Science*, **293**, 2239–42.

Godthelp, H., Archer, M., Cifelli, R. L., Hand, S. J. and Gilkeson, C. F. (1992). Earliest known Australian Tertiary mammal fauna. *Nature*, **356**, 514–16.

Godthelp, H., Wroe, S. and Archer, M. (1999). A new marsupial from the Early Eocene Tingamarra local fauna of Murgon, southeastern Queensland: a prototypical Australian marsupial? *Journal of Mammalian Evolution*, **6**, 289–313.

Goswami, A., Prasad, G. V. R., Upchurch, P., *et al.* (2011). A radiation of arboreal basal eutherian mammals beginning in the Late Cretaceous of India. *Proceedings of the National Academy of Sciences of the United States of America*, **108**(39), 16 333–8.

Graur, D. and Higgins, D. G. (1994). Molecular evidence for the inclusion of cetaceans within the order Artiodactyla. *Molecular Biology and Evolution*, **11**, 357–64.

Graur, D., Duret, L. and Gouy, M. (1996). Phylogenetic position of the order Lagomorpha (rabbits, hares, and allies). *Nature*, **379**, 333–5.

Hacking, I. (2001). *An Introduction to Probability and Inductive Logic*. Cambridge, UK: Cambridge University Press.

Hennig, W. (1966). *Phylogenetic Systematics*. Urbana, IL: University of Illinois Press.

Hershkovitz, P. (1968). The recent mammals of the neotropical regions: a zoogeographic and ecological reivew. In *Evolutions, Mammals, and Southern Continents*, ed. A. Keast, F. C. Erk and B. Glass. Albany, NY: State University of New York Press.

Hillis, D. M. (1987). Molecular versus morphological approaches to systematics. *Annual Review of Ecology and Systematics*, **18**, 23–42.

Ho, S. Y. W. (2007). Calibrating molecular estimates of substitution rates and divergence times in birds. *Journal of Avian Biology*, **38**, 409–14.

Ho, S. Y. W. and Phillips, M. J. (2009). Accounting for calibration uncertainty in phylogenetic estimation of evolutionary divergence times. *Systematic Biology*, **58**, 367–80.

Holm, S. (1979). A simple sequentially rejective multiple test procedure. *Scandinavian Journal of Statistics*, **6**, 65–70.

Hooker, J. J. (2001). Tarsals of the extinct insectivoran family Nyctitheriidae (Mammalia): evidence for archontan relationships. *Zoological Journal of the Linnean Society*, **132**, 501–29.

Hug, L. A. and Roger, A. J. (2007). The impact of fossils and taxon sampling on ancient molecular dating analyses. *Molecular Biology and Evolution*, **24**, 1889–97.

Inoue, J., Donoghue, P. C. J. and Yang, Z. (2010). The impact of the representation of fossil calibrations on Bayesian estimation of species divergence times. *Systematic Biology*, **59**, 74–89.

Iturralde-Vinent, M. A. (2006). Meso-Cenozoic Caribbean paleogeography: implications for the historical biogeography of the region. *International Geology Review*, **48**, 791–827.

Ji, Q., Luo, Z.-X., Yuan, C.-X., *et al.* (2002). The earliest known eutherian mammal. *Nature*, **416**, 816–822.

Kay, R. F., Thewissen, J. G. M. and Yoder, A. D. (2001). Cranial anatomy of *Ignacius graybullianus* and the affinities of the Plesiadapiformes. *American Journal of Physical Anthropology*, **89**, 477–98.

Kielan-Jaworowska, Z., Cifelli, R. L. and Luo, Z.-X. (2004). *Mammals from the Age of Dinosaurs: Origins, Evolution, and Structure.* New York: Columbia University Press.

Kimura, M. (1983). *The Neutral Theory of Molecular Evolution.* Cambridge, UK: Cambridge University Press.

Kitazoe, Y., Kishino, H., Waddell, P. J., *et al.* (2007). Robust time estimation reconciles views of the antiquity of placental mammals. *PLoS One*, **2**, e384.

Krause, D. W. (1986). Competitive exclusion and taxonomic displacement in the fossil record: the case of rodents and multituberculates in North America. In *Vertebrates, Phylogeny and Philosophy: A Tribute to George Gaylord Simpson*, ed. K. M. Flanagan and J. A Lillegraven. Laramie, WY: University of Wyoming.

Krause, D. W. and Maas, M. C. (1990). The biogeographic origins of late Paleocene– early Eocene mammalian immigrants to the Western Interior of North America. In *Dawn of the Age of Mammals in the Northern Part of the Rocky Mountain Interior, North America*, ed. T. M. Brown and K. D. Rose. Boulder, CO: Geological Society of America.

Kumar, S. and Hedges, S. B. (1998). A molecular timescale for vertebrate evolution. *Nature*, **392**, 917–20.

Lanfear, R., Welch, J. J. and Bromham, L. (2010). Watching the clock: studying variation in rates of molecular evolution between species. *Trends in Ecology and Evolution*, **25**, 495–503.

Lee, M. S. and Camens, A. B. (2009). Strong morphological support for the molecular evolutionary tree of placental mammals. *Journal of Evolutionary Biology*, **22**, 2243–57.

Luckett, W. P. and Hong, N. (1998). Phylogenetic relationships between the orders Artiodactyla and Cetacea: a combined assessment of morphological and molecular evidence. *Journal of Mammalian Evolution*, **5**, 127–82.

Luo, Z. X. and Wible, J. R. (2005). A Late Jurassic digging mammal and early mammalian diversification. *Science*, **308**, 103–7.

Luo, Z.-X., Cifelli, R. L. and Kielan-Jaworowska, Z. (2001a). Dual origin of tribosphenic mammals. *Nature*, **409**, 53–7.

Luo, Z.-X., Crompton, A. W. and Sun, A.-L. (2001b). A new mammaliaform from the early Jurassic and evolution of mammalian characteristics. *Science*, **292**, 1535–40.

Luo, Z.-X., Kielan-Jaworowska, Z. and Cifelli, R. L. (2002). In quest for a phylogeny of Mesozoic mammals. *Acta Palaeontologica Polonica*, **47**, 1–78.

Luo, Z.-X., Ji, Q., Wible, J. R. and Yuan, C.-X. (2003). An Early Cretaceous tribosphenic mammal and metatherian evolution. *Science*, **302**, 1934–40.

Luo, Z.-X., Kielan-Jaworowska, Z. and Cifelli, R. L. (2004). Evolution of dental replacement in mammals. *Bulletin of the Carnegie Museum of Natural History*, **36**, 159–75.

Luo, Z.-X., Chen, P., Li, G. and Chen, M. (2007). A new eutriconodont mammal and evolutionary development in early mammals. *Nature*, **446**, 288–93.

Luo, Z.-X., Ruf, I., Schultz, J. A. and Martin, T. (2011a). Fossil evidence on evolution of inner ear cochlea in Jurassic mammals. *Proceedings of the Royal Society B*, **278**, 28–34.

Luo, Z.-X., Yuan, C.-X., Meng, Q.-J. and Ji, Q. (2011b). A Jurassic eutherian mammal and divergence of marsupials and placentals. *Nature*, **476**, 442–5.

MacPhee, R. D. E. (1981). Auditory regions of primates and eutherian insectivores: morphology, ontogeny, and character analysis. *Contributions to Primatology*, **18**, 1–282.

MacPhee, R. D. E. (1994). Morphology, adaptations, and relationships of *Plesiorycteropus*, and a diagnosis of a new order of eutherian mammals. *Bulletin of the American Museum of Natural History*, **220**, 1–214.

MacPhee, R. D. E. and Reguero, M. A. (2010). Reinterpretation of a Middle Eocene record of Tardigrada (Pilosa, Xenarthra, Mammalia) from La Meseta Formation, Seymour Island, West Antarctica. *American Museum Novitates*, **3689**, 1–21.

Marjanović, D. and Laurin, M. (2008). Assessing confidence intervals for stratigraphic ranges of higher taxa: the case of Lissamphibia. *Acta Palaeontologica Polonica*, **53**, 413–32.

Marshall, L. G. and de Muizon, C. (1988). The dawn of the age of mammals in South America. *National Geographic Research*, **4**, 23–55.

Martin, A. P. and Palumbi, S. R. (1993). Body size, metabolic rate, generation time, and the molecular clock. *Proceedings of the National Academy of Sciences of the United States of America*, **90**, 4087–91.

Martin, R. D., Soligo, C. and Tavaré, S. (2007). Primate origins: implications of a Cretaceous ancestry. *Folia Primatologica*, **78**, 277–96.

Meng, J. and Wyss, A. R. (2005). Glires (Lagomorpha, Rodentia). In *The Rise of Placental Mammals: Origins and Relationships of the Major Extant Clades*, ed. K. D. Rose and J. D. Archibald. Baltimore, MD: Johns Hopkins University Press.

Misawa, K. and Janke, A. (2003). Revisiting the Glires concept – phylogenetic analysis of nuclear sequences. *Molecular Phylogenetics and Evolution*, **28**, 320–7.

Murphy, W. J., Eizirik, E., Johnson, W. E., *et al.* (2001). Molecular phylogenetics and the origins of placental mammals. *Nature*, **409**, 614–18.

Near, T. J. and Sanderson, M. J. (2004). Assessing the quality of molecular divergence time estimates by fossil calibrations and fossil-based model selection. *Philosophical Transactions of the Royal Society of London B–Biological Sciences*, **359**, 1477–83.

Nei, M. (1987). *Molecular Evolutionary Genetics.* New York: Columbia University Press.

Novacek, M. J. (1992). Mammalian phylogeny: shaking the tree. *Nature*, **356**, 121–5.

Novacek, M. J. and AToL Mammal Morphology Team (2008). A team-based approach yields a new matrix of 4,500 morphological characters for mammalian phylogeny. *Journal of Vertebrate Paleontology*, **28**, 121A.

Novacek, M. J., Rougier, G. W., Wible, J. R., *et al.* (1997). Epipubic bones in eutherian mammals from the Late Cretaceous of Mongolia. *Nature*, **389**, 483–6.

O'Leary, M. A. (2011). MorphoBank: collecting and storing phenomic data for phylogenetic research in the 'cloud'. *American Journal of Physical Anthropology*, **144**, 228–9.

Ohta, T. (1995). Synonymous and nonsynonymous substitutions in mammalian genes and the nearly neutral theory. *Journal of Molecular Evolution*, **40**, 56–63.

Pascual, R. and Ortiz-Jaureguizar, E. (1992). Evolutionary pattern of land mamal faunas during the Late Cretaceous and Paleocene in South America: a comparison with the North American pattern. *Annales Zoologici Fennici*, **28**, 245–52.

Pascual, R. and Ortiz-Jaureguizar, E. (2007). The Gondwanan and South American episodes: two major and unrelated moments in the history of the South American mammals. *Journal of Mammalian Evolution*, **14**, 75–137.

Patterson, C. (ed.) (1987). *Molecules and Morphology in Evolution: Conflict or Compromise?.* Cambridge, UK: Cambridge University Press.

Patterson, C., Williams, D. M. and Humphries, C. J. (1993). Congruence between molecular and morphological phylogenies. *Annual Review of Ecology and Systematics*, **24**, 153–88.

Penny, D., Hasegawa, M., Waddell, P. J. and Hendy, M. D. (1999). Mammalian evolution: timing and implications from using the LogDeterminant transform for proteins of differing amino acid composition. *Systematic Biology*, **48**, 76–93.

Phillips, M. J. (2009). Branch-length estimation bias misleads molecular dating for a vertebrate mitochondrial phylogeny. *Gene*, **441**, 132–40.

Phillips, M. J. and Penny, D. (2003). The root of the mammalian tree inferred from whole mitochondrial genomes. *Molecular Phylogenetics and Evolution*, **28**, 171–85.

Phillips, M. J., Delsuc, F. and Penny, D. (2004). Genome-scale phylogeny and the detection of systematic biases. *Molecular Biology and Evolution*, **21**, 1455–8.

Phillips, M. J., Bennett, T. H. and Lee, M. S. Y. (2009). Molecules, morphology, and ecology indicate a recent, amphibious ancestry for echidnas. *Proceedings of the National Academy of Sciences of the United States of America*, **106**, 17 089–94.

Polly, P. D. (2001). On morphological clocks and paleophylogeography: towards a timescale for *Sorex* hybrid zones. *Genetica*, **112–113**, 339–57.

Posada, D. and Crandall, K. A. (1998). Modeltest: testing the model of DNA substitution. *Bioinformatics*, **14**, 817–18.

Poux, C., Madsen, O., Glos, J., De Jong, W. W. and Vences, M. (2008). Molecular phylogeny and divergence times of Malagasy tenrecs: influence of data partitioning and taxon sampling on dating analyses. *BMC Evolutionary Biology*, **8**, 102.

Prasad, G. V. R., Verma, O., Gheerbrant, E., *et al.* (2010). First mammal evidence from the Late Cretaceous of India for biotic dispersal between India and Africa at the KT transition. *Comptes Rendus Palevol*, **9**, 63–71.

Purvis, A. (1995). A composite estimate of primate phylogeny. *Philosophical Transactions of the Royal Society of London B*, **348**, 405–21.

Pyron, R. A. (2010). A likelihood method for assessing molecular divergence time estimates and the placement of fossil calibrations. *Systematic Biology*, **59**, 185–94.

Rannala, B. and Yang, Z. (2007). Inferring speciation times under an episodic molecular clock. *Systematic Biology*, **56**, 453–66.

Rauhut, O. W. M., Martin, T., Ortiz-Jaureguizar, E. and Puerta, P. (2002). A Jurassic mammal from South America. *Nature*, **416**, 165–8.

Reisz, R. R. and Müller, J. (2004). Molecular timescales and the fossil record: a paleontological perspective. *Trends in Genetics*, **20**, 237–41.

Rich, T. H., Vickers-Rich, P., Constantine, A., *et al.* (1997). A tribosphenic mammal from the Mesozoic of Australia. *Science*, **278**, 1438–42.

Rose, K. D. (1996). On the origin of the order Artiodactyla. *Proceedings of the National Academy of Sciences of the United States of America*, **93**, 1705–9.

Rose, K. D., Deleon, V. B., Missiaen, P., *et al.* (2008). Early Eocene lagomorph (Mammalia) from Western India and the early diversification of Lagomorpha. *Proceedings of the Royal Society B*, **275**, 1203–8.

Rose, K. D., Rana, R. S., Sahni, A., *et al.* (2009). Early Eocene primates from Gujarat, India. *Journal of Human Evolution*, **56**, 366–404.

Rougier, G. W., Wible, J. R. and Novacek, M. J. (1998). Implications of *Deltatheridium* specimens for early marsupial history. *Nature*, **396**, 459–63.

Rougier, G. W., Martinelli, A. G., Forasiepi, A. M. and Novacek, M. J. (2007). New Jurassic mammals from Patagonia, Argentina: a reappraisal of australosphenidan morphology and interrelationships. *American Museum Novitates*, **3566**, 1–54.

Rowe, T., Rich, T. H., Vickers-Rich, P., Springer, M. S. and Woodburne, M. O. (2008). The oldest platypus and its bearing on divergence timing of the platypus and echidna clades. *Proceedings of the National Academy of Sciences of the United States of America*, **105**, 1238–42.

Sánchez-Villagra, M. R., Narita, Y. and Kuratani, S. (2007). Thoracolumbar vertebral number: the first skeletal synapomorphy for afrotherian mammals. *Systematics and Biodiversity*, **5**, 1–7.

Sanderson, M. J. (1997). A nonparametric approach to estimating divergence times in the absence of rate constancy. *Molecular Biology and Evolution*, **14**, 1218–31.

Sanderson, M. J. (2002). Estimating absolute rates of molecular evolution and divergence times: a penalized likelihood approach. *Molecular Biology and Evolution*, **19**, 101–9.

Sansom, R. S., Gabbott, S. E. and Purnell, M. A. (2010). Non-random decay of chordate characters causes bias in fossil interpretation. *Nature*, **463**, 797–800.

Sarich, V. M. and Wilson, A. C. (1973). Generation time and genomic evolution in primates. *Science*, **179**, 1144–7.

Schaeffer, B. (1947). Notes on the origin and function of the artiodactyl tarsus. *American Museum Novitates*, **1356**, 1–24.

Seiffert, E. R. (2007a). Early evolution and biogeography of lorisiform strepsirrhines. *American Journal of Primatology*, **69**, 27–35.

Seiffert, E. R. (2007b). A new estimate of afrotherian phylogeny based on simultaneous analysis of genomic, morphological, and fossil evidence. *BMC Evolutionary Biology*, **7**, 224.

Seiffert, E. R. (2010). The oldest and youngest records of Afrosoricida (Placentalia, Afrotheria) from the Fayum Depression of northern Egypt. *Acta Palaeontologica Polonica*, **55**, 599–616.

Seiffert, E. R., Simons, E. L. and Attia, Y. (2003). Fossil evidence for an ancient divergence of lorises and galagos. *Nature*, **422**, 421–4.

Seiffert, E. R., Simons, E. L., Ryan, T. M., Bown, T. M. and Attia, Y. (2007). New remains of Eocene and Oligocene Afrosoricida (Afrotheria) from Egypt, with implications for the origin(s) of afrosoricid zalambdodonty. *Journal of Vertebrate Paleontology*, **27**, 963–72.

Seligmann, H. (2010). Positive correlations between molecular and morphological rates of evolution. *Journal of Theoretical Biology*, **264**, 799–807.

Shoshani, J., Walter, R. C., Abraha, M., *et al.* (2006). A proboscidean from the late Oligocene of Eritrea, a 'missing link' between early Elephantiformes and Elephantimorpha, and biogeographic implications. *Proceedings of the National Academy of Sciences of the United States of America*, **103**, 17 296–301.

Smith, T., De Bast, E. and Sigé, B. (2010). Euarchontan affinity of Paleocene Afro-European adapisoriculid mammals and their origin in the late Cretaceous Deccan Traps of India. *Naturwissenschaften*, **97**, 417–22.

Spaulding, M., O'Leary, M. A. and Gatesy, J. (2009). Relationships of Cetacea (Artiodactyla) among mammals: increased taxon sampling alters interpretations of key fossils and character evolution. *PLoS One*, **4**, e7602.

Springer, M. S. (1997). Molecular clocks and the timing of the placental and marsupial radiations in relation to the Cretaceous-Tertiary boundary. *Journal of Mammalian Evolution*, **4**, 285–302.

Springer, M. S., Burk-Herrick, A., Meredith, R., *et al.* (2007). The adequacy of morphology for reconstructing the early history of placental mammals. *Systematic Biology*, **56**, 673–84.

Springer, M. S., Murphy, W. J., Eizirik, E. and O'Brien, S. J. (2003). Placental mammal diversification and the Cretaceous-Tertiary boundary. *Proceedings of the National Academy of Sciences of the United States of America*, **100**, 1056–61.

Springer, M. S., Stanhope, M. J., Madsen, O. and de Jong, W. W. (2004). Molecules consolidate the placental mammal tree. *Trends in Ecology and Evolution*, **19**, 430–8.

Stanhope, M. J., Waddell, V. G., Madsen, O., *et al.* (1998). Molecular evidence for multiple origins of Insectivora and for a new order of endemic African insectivore mammals. *Proceedings of the National Academy of Sciences of the United States of America*, **95**, 9967–72.

Storch, G. (2008). Skeletal remains of a diminutive primate from the Paleocene of Germany. *Naturwissenschaften*, **95**, 927–30.

Swofford, D. L. (2002). *PAUP*. Phylogenetic analysis using parsimony (*and other methods). Version 4*. Sunderland, MA: Sinauer Associates.

Tajima, F. (1993). Simple methods for testing the molecular evolutionary clock hypothesis. *Genetics*, **135**, 599–607.

Tavaré, S., Marshall, C. R., Will, O., Soligo, C. and Martin, R. D. (2002). Using the fossil record to estimate the age of the last common ancestor of extant primates. *Nature*, **416**, 726–9.

Teeling, E. C., Springer, M. S., Madsen, O., *et al.* (2005). A molecular phylogeny for bats illuminates biogeography and the fossil record. *Science*, **307**, 580–4.

Thewissen, J. G., Williams, E. M., Roe, L. J. and Hussain, S. T. (2001). Skeletons of terrestrial cetaceans and the relationship of whales to artiodactyls. *Nature*, **413**, 277–81.

Thorne, J. L. and Kishino, H. (2002). Divergence time and evolutionary rate estimation with multilocus data. *Systematic Biology*, **51**, 689–702.

Ting, S.-Y., Meng, J., Qian, L., *et al.* (2007). *Ganungulatum xincunliense*, an artiodactyl-like mammal (Ungulata, Mammalia) from the Paleocene, Chijiang Basin, Jiangxi, China. *Vertebrata PalAsiatica*, **45**, 278–86.

Vickers, J. (2010). The problem of induction. In *The Stanford Encyclopedia of Philosophy*, ed. E. N. Zalta. Stanford, CA: Stanford University.

Vogt, L., Bartolomaeus, T. and Giribet, G. G. (2010). The linguistic problem of morphology: structure versus homology and the standardization of morphological data. *Cladistics*, **26**, 301–25.

Waddell, P. J., Okada, N. and Hasegawa, M. (1999). Towards resolving the interordinal relationships of placental mammals. *Systematic Biology*, **48**, 1–5.

Welch, J. J., Fontanillas, E. and Bromham, L. (2005). Molecular dates for the 'Cambrian Explosion': the influence of prior assumptions. *Systematic Biology*, **54**, 672–8.

Whidden, H. P. (2002). Extrinsic snout musculature in Afrotheria and Lipotyphla. *Journal of Mammalian Evolution*, **9**, 161–84.

Whittaker, J. M., Muller, R. D., Leitchenkov, G., *et al.* (2007). Major Australian-Antarctic plate reorganization at Hawaiian-Emperor bend time. *Science*, **318**, 83–6.

Wible, J. R., Rougier, G. W. and Novacek, M. J. (2005). Anatomical evidence for superordinal/ordinal eutherian taxa in the Cretaceous. In *The Rise of Placental Mammals: Origins and Relationships of the Major Extant Clades*, ed. K. D. Rose and J. D. Archibald. Baltimore, MD: Johns Hopkins University Press.

Wible, J. R., Rougier, G. W., Novacek, M. J. and Asher, R. J. (2007). Cretaceous eutherians and Laurasian origin for placental mammals near the K/T boundary. *Nature*, **447**, 1003–6.

Woodburne, M. O. and Case, J. A. (1996). Dispersal, vicariance, and the late Cretaceous to early Tertiary land mammal biogeography from South America to Australia. *Journal of Mammalian Evolution*, **3**, 121–61.

Woodburne, M. O., Rich, T. H. and Springer, M. S. (2003). The evolution of tribospheny and the antiquity of mammalian clades. *Molecular Phylogenetics and Evolution*, **28**, 360–85.

Yang, Z. and Rannala, B. (2006). Bayesian estimation of species divergence times under a molecular clock using multiple fossil calibrations with soft bounds. *Molecular Biology and Evolution*, **23**, 212–26.

Zuckerkandl, E. and Pauling, L. (1962). Molecular disease, evolution, and genetic heterogeneity. In *Horizons in Biochemistry*, ed. M. Kasha and B. Pullman. New York: Academic Press.

Zuckerkandl, E. and Pauling, L. (1965). Evolutionary divergence and convergence in proteins. In *Evolving Genes and Proteins*, ed. V. Bryson and H. J. Vogel. New York: Academic Press.

4

Morphological largess: can morphology offer more and be modelled as a stochastic evolutionary process?

HANS C. E. LARSSON, T. ALEXANDER
DECECCHI AND LUKE B. HARRISON

Introduction

If 'nothing in biology makes sense except in light of evolution' (Dobzhansky 1973), then evolutionary rates must be the currency exchange rate of biology and species are the currency. Rates of evolutionary change influence ultimate biological factors, such as rates of adaptation, speciation and extinction. These, in turn, determine even higher biological functions, such as ecosystem size and complexity, whose interactions cause truly large-scale biogeographic patterns, such as latitudinal diversity gradients. For example, the higher diversities in the tropics have been hypothesized to be driven by higher rates of mutation and speciation in those climates (Rohde 1992; Allen *et al.* 2002), faster genetic drift in smaller populations in these regions (Fedorov 1966), and more intense biotic interactions driving higher rates of evolution (Dobzhansky 1950).

Despite their great reach, rates of evolutionary change are determined by proximate drivers – the most basic of which are genetic mutations and developmental variation. Genetic mutations are generally considered random, with their rates driven primarily by stochastic perturbations to the processes of genetic replication, such as copy error, point mutations and unequal recombination. Developmental variation is exposed through the phenotypic plasticity present in all morphological traits (West-Eberhard 2003). Phenotypic plasticity involves different drivers which impart local fitness peaks in different populations (Crispo 2007) or are the result of spurious environmental influences on developmental processes. The latter are epigenetic influences on development that may drive the creation of a continuous or discontinuous set of phenotypic traits (Waddington 1942). Both these processes, mutation and

From Clone to Bone: The Synergy of Morphological and Molecular Tools in Palaeobiology,
ed. Robert J. Asher and Johannes Müller. Published by Cambridge University Press.
© Cambridge University Press 2012.

epigenetic developmental variation, provide the ingredients for natural selection. The rate at which the phenotype evolves from these variations is what we are interested in expanding in this chapter. To further complicate matters, the tempo and mode of natural selection upon these proximate heritable variations is determined by a suite of factors, such as population size, body size, generation time and metabolic rate.

It is clear that rates of genetic and phenotypic change are influenced by a range of factors spanning from the genetic level to biome-sized, abiotic influences. With this bewildering set of potential drivers of evolutionary change, it seems a difficult task to untangle any one single mechanism of evolutionary change. One step forward may be to attempt to isolate and estimate the influence each driver has on evolutionary rate. Evolutionary rate becomes central to the question, and may also boil down to what matters most to a lineage in terms of fitness. Optimal fitness can never be achieved simply due to the fact that genetic mutations and environmental changes happen and both are considered to be relatively stochastic. But rate and direction of evolutionary change determine where a population resides on any given adaptive landscape. Determining the velocities of lineage progression across adaptive landscapes is paramount to begin understanding the underlying dynamics of evolutionary change. This method has served microevolutionary studies very well and we suggest it may be extended further to macroevolutionary studies.

Perhaps the most useful way of approaching macroevolutionary walks across adaptive landscapes, and potentially for understanding the raw mechanics of evolution in general, is the genotype–phenotype map. In theory, an organism can be boiled down to a basic genotype–phenotype interaction map overlain by a range of epigenetic effects caused by all possible environments built into it. If known, this purely mechanistic model would allow for the calculation of many fundamental aspects of an organism's phenotype. Although currently unreachable, any steps towards this goal will have value for biologists. One of the key steps in this direction is to calculate a timeframe and rates over which evolutionary change happens.

At the genetic level, methods to estimate rates of molecular evolution have matured in recent decades (see below). But at the phenotypic level, rates of morphological evolution remain more elusive. A great deal of effort has gone into assessing phenotypic evolution using continuous variables (e.g. Hendry and Kinnison 1999; recently reviewed by Gingerich 2009). However, these methods require good sampling from coeval populations to assess rates of evolutionary change on a generational timescale.

We use this chapter to outline a general discussion of assessing rates of genetic and phenotypic change but with application to discrete phenotypic character evolution over large timescales that transcend entire clades, not smaller-scale individual populations. We argue for a renewed focus on discrete phenotypic evolution, using large, less-biased data sets and stochastic models of evolution. By stochastic, we mean models that lie between the extremes of deterministic and complete random models. Clade-level changes are still expected to lie within some bounds determined by several forms of constraints. After a brief review of current methods of measuring molecular and morphological evolutionary rates, we describe why it is appropriate to treat morphology as a stochastic process and propose applications and future directions for study. We present preliminary results of analyses demonstrating that large discrete morphological character data sets do indicate a pattern of stochastic evolutionary change. Coupled with the exciting progress in modelling molecular evolution, a renewed focus on morphological evolution may bring us closer to the elusive goal of understanding the evolution of the genotype–phenotype map over time.

Measuring molecular evolutionary rates and estimating divergence times

Recent research has focused a comparatively large effort on the measurement and causes of variation in the rate of molecular evolution. These methodologies have expanded into the use of these rates to estimate ancient divergence times. Early studies of molecular evolution found that many genetic sequences appear to evolve at a constant rate across many taxa, leading to the proposal of the molecular clock (Zuckerkandl and Pauling 1965; Kimura 1983; see also review in Bromham and Penny 2003). This constant rate permitted the estimation of divergences times by converting genetic distances inferred from molecular sequence data into evolutionary timescales using time calibration points based on fossils or biogeography. Major issues surround the accuracy of fossil calibration points and the errors associated with those dates are often not included in molecular clock divergence estimates (Graur and Martin 2004), although much progress has been made to alleviate some of these errors by incorporating multiple fossil calibrations (Marshall 2008). The use of a clock-like model allows significant deviations from the null expectation to be identified, such as genes potentially under strong directional selection. Recent studies have demonstrated that in most cases, the molecular clock hypothesis is incorrect: the rate of molecular evolution varies considerably, although with a predictable component based on life history (see Bromham 2009; Lanfear et al. 2010). New methods for divergence time estimation can accommodate rate variation, but remain somewhat controversial (reviewed in Welch et al. 2005 and Rutschmann

2006). Here we briefly review how the rate of molecular evolution is measured and how this information is used to estimate divergence times.

The rate of molecular evolution is the amount of evolutionary change over a given period of time in a given lineage. Neither the quantity of evolution in a given lineage nor the length of a lineage is trivially or directly measured. Instead, it is necessary to first use the aligned molecular sequences and phylogenetic methods to measure branch lengths (the quantity of evolution). Branch lengths are the product of substitution rate and time. To disentangle rate and time and obtain absolute substitution rates and divergence time estimates, methods that include time-calibration and explicit modelling of evolutionary rates are needed.

Measuring the quantity of change (branch lengths)

Unlike measuring quantities of morphological evolution (see below), the quantity of molecular evolution is almost always estimated using a stochastic substitution model-based phylogenetic approach (Lanfear *et al.* 2010). Parsimony-based methods are not generally used because they do not account for saturation, or multiple changes in one site on a single branch, nor do they consider constant sequence positions (Yang 2006). Branch lengths are estimated using either a Maximum-Likelihood (ML) or a Bayesian phylogenetic framework and use stochastic markov-chain-based substitution models (for a thorough description and review, see Yang 2006). These models calculate the probability of observing a given sequence position change over time period given a set of parameters that are usually estimated from the data while accounting for the possibility of multiple substitutions.

In a ML framework, a substitution model is used in conjunction with the likelihood function to estimate the set of parameters (branch lengths, topology and substitution model parameters) that maximizes the probability of observing the extant, aligned sequences (see Felsenstein 2003; for common methods see, for example, Guindon and Gascuel 2003; Stamatakis 2006). In studies of molecular rate variation, it is sometimes helpful to fix the topology a-priori in order to reduce the number of parameters to be estimated and potentially derive more accurate estimates of branch length. However, because topology can have a strong effect on other parameter estimates, this is unwise in cases of topological uncertainty, especially for comparative studies (Huelsenbeck *et al.* 2001; see also Bayesian methods, below). Substitution models can be simple or complex; likelihood-ratio tests can be used to compare between different models and to choose the model that best fits the data (reviewed in Sullivan and Joyce 2005).

Bayesian methods are similar to maximum likelihood methods and use the same substitution models. Bayesian methods use prior probability distributions of the parameters and update these using the likelihood calculated from the aligned sequences to produce posterior probability distributions of those parameters (Huelsenbeck *et al.* 2001; Holder and Lewis 2003). There are several advantages of a Bayesian approach, particularly with respect to measuring evolutionary rates.

First, the Bayesian approach natively incorporates prior information into the phylogenetic analysis via the use of priors, or prior probability distributions for the sampled parameters. This allows the incorporation of, for example, fossil-derived speciation and extinction rates, fossil calibrations (see also discussion on dating, below), and alternate (and weighted) phylogenetic topologies. Although this can be advantageous when meaningful information is available (see discussion on dating below), the specification of 'meaningless', or flat, priors is more difficult and can bias the outcome, although this can be mitigated by running analyses only on the priors, without the data. Currently available Bayesian phylogenetic software packages only rarely allow specification of many of these parameters, although there are increasingly more flexible options to specify detailed priors on certain parameters in some programs (e.g. BEAST; Drummond and Rambaut 2007).

Second, because Bayesian methods return the posterior probability distributions of the parameters of interest, they intuitively provide confidence intervals and are able to estimate the marginal probability distribution of one parameter, integrating over all others. The characteristics of Bayesian analyses allow for robust estimations of parameters, even in cases of phylogenetic uncertainty or multiple alternative phylogenetic topologies (Huelsenbeck *et al.* 2000a).

Third, Bayesian methods can test alternative hypotheses concerning substitution models and topologies more generally than a likelihood ratio test by using Bayes factors (Suchard *et al.* 2001). Bayesian methods, coupled with reversible jump markov-chain Monte Carlo methods for sampling the posterior can also be used to treat the substitution model as a sampled parameter and calculate posterior probability distributions across models (Huelsenbeck *et al.* 2004).

Rates of molecular evolution and divergence time estimation

As branch lengths are a measure of the quantity of evolution from a starting point, they are the product of evolutionary rate and time, and therefore cannot be easily decomposed. Early studies that relied on the assumption of rate constancy, or strict molecular clock, were able to use a single calibration point – a known divergence time for a pair of lineages – derived from fossil evidence or biogeography. Using this fixed starting point branch lengths could be scaled revealing the absolute rate of evolution and divergence dates for all the lineages in the analysis. Divergence time estimation is now routinely used to date lineages and to derive the absolute substitution rate (Bromham and Penny 2003). These methods have seen wide application in many areas of biological science, including the testing of hypotheses of biogeography and gene duplications, and even for testing hypotheses concerning the origin of pathogenic viruses (Korber 2000). However, many results based on simple dating techniques tended to conflict with accepted dates from the fossil record and other sources of information (Wray *et al.* 1996; Smith and Peterson 2002). New methods and data have since shown that divergence estimates can be biased in several ways: because of poorly supported phylogenetic topology, the inappropriate use of fossil calibration points, or, most importantly, rate heterogeneity. Likewise, the realization that the rate of molecular evolution was variable has led to new methods to investigate sources and drivers of rate variation. Although absolute rates can be inferred using multiple calibration points (see also below), investigating drivers and correlations of rate variation requires the use of phylogenetic comparative methods (reviewed in Lanfear *et al.* 2010).

Mitigating biases in divergence time estimation

Poorly resolved topology may bias divergence time estimates (Benton *et al.* 2009). This can be somewhat mitigated in a Bayesian framework by integrating over the posterior probability distribution of topologies. The choice of fossil calibration points and how they are treated has also proved controversial (Benton *et al.* 2009). Guidelines and well-characterized published divergences as well as methods that treat the dates as distributions with error rather than point estimates are now available (e.g. Müller and Reisz 2005; Weir and Schluter 2008).

Rate heterogeneity, specifically unequal rates of molecular evolution across lineages in a phylogeny, can, if left unaccounted for, severely bias divergence time estimates. There are several strategies to deal with departures from the strict molecular-clock assumption of equal rates across all lineages (reviewed in Welch and Bromham 2005). One strategy is to remove sequences or lineages that violate the clock by using molecular clock tests (Tajima 1993; Takezaki *et al.* 1995). However, these tests have been shown to be insensitive and frequently do not detect even strong heterogeneity (Bromham *et al.* 2000; Robinson *et al.* 1998). A further option is to model the evolution of the rate of molecular evolution and estimate divergence times using multiple calibrations and rates. Known as *relaxed clock* dating methods, many approaches have been proposed, varying by the method of phylogenetic inference and how they model changes in evolutionary rate (reviewed in Welch and Bromham 2005). First and most simply, local clock methods divide the phylogenetic tree into several lineages within which a strict clock is enforced; these rates are then estimated along with the other parameters (e.g. Yoder and Yang 2000). This is appropriate where rates are very different between but consistent within clades. However, there can be problems with either choosing a few rate groups beforehand, or being unable to choose between equally likely combinations of rate groups (see Rannala 2002). Generalizing local clock methods, fully relaxed clock approaches allow rate changes to be more frequent, or potentially occur at each node or anywhere in the phylogeny. Because there many potential combinations of rate, most methods rely on assumptions about the way evolutionary rates evolve and require several calibration points. Different methods employ different assumptions: some rely on a penalty function that minimizes rate changes across phylogeny (Sanderson 2002), whereas some Bayesian methods emphasize the incorporation of prior knowledge about rate shifts into a prior. Rates are frequently modelled as autocorrelated, or dependent on the parent branch, although this is not necessarily a justifiable assumption (Ho 2009). These relaxed clock methods are generally more suitable for situations of mild heterotachy, where rate is variable, but not to a large degree, or changes gradually.

Finally, some methods conceptually combine both a local clock and a fully relaxed clock, modelling both large between-lineage change and smaller within-lineage changes (e.g. Huelsenbeck *et al.* 2000b). This final class of methods emphasizes the flexibility of the Bayesian approach, which does not necessarily constrain the analysis to a single rigidly defined model. These models do not enforce a gradualist/punctuated dichotomy and can model the occurrence of either in the phylogeny. Indeed, there is potential to extend existing Bayesian methods using informative priors based on independent assessments of speciation and extinction rates, biogeography, etc.

Measuring rates of morphological evolution

Methods to assess rates of morphological evolution have lagged behind molecular research in spite of the long history of research on the rates of morphological evolution. A fundamental confounding factor of measuring morphological rates is the quality of data. It's a well-worn cliché that the rock record is 'imperfect'. Long recognized by palaeontologists, the issue of the inadequacies of the fossil record and the paucity of 'infinitely numerous transitional links' (Darwin 1859, p. 310) was so troubling that Darwin devoted a whole chapter in *On the Origin of Species* to address it (Darwin 1859). Although it is broadly true that a direct reading of the rock record may not truly reflect the diversity of life at any one interval, the fossil record is the only source of data for the vast majority of the history of life. Recently, molecular dating methods (addressed above) have challenged traditional assumptions about relationships between and timing of the origin of modern clades. Yet molecular methods cannot replace fossil evidence because they tell us nothing directly about the context or basic biology (in terms of morphology, ecology, behaviour) of ancestral taxa. Lineages composed of only fossils make up a large proportion of the history of life (see Alroy *et al.* 2008 for a recent global diversity curve through time) but are beyond the purview of molecular dating. Although molecular techniques have an important role to play, the only way to understand the wheres, whys and hows of evolution, and to some extent the whens, is through fossil data.

The palaeontologist reading this may now be wondering why we have so easily abandoned the 'when' question to the realm of the neontologists (researchers whose focus is on extant organisms and processes)? The truth is we have not, nor should we give up so easily on the pursuit of evolutionary rates in deep time, even in lineages without modern representatives. But if we are to take up the challenge of attempting to examine the tempo of evolution (and from that try to determine the mode) we must understand and accept the advantages and limitations of our data source. Geological biases (such as different preservation potential and rock exposure volume) strongly influence signals for taxonomic diversity in the rock record (Alroy *et al.* 2008; Barrett *et al.* 2009; Wall *et al.* 2009). Add to this the fact that in relation to total duration time of a taxon, the origination interval (or indeed its extinction) is brief. Thus the probabilities of catching the precise beginning and endpoints of a lineage are infinitesimally small (Eldredge and Gould 1972; Marshall 1994). This presents a major impediment to the study of evolutionary rates, one which can be overcome using indirect methodologies to estimate these boundaries (Benton *et al.* 2009). Whether these are based on only the rock record (stratigraphic confidence intervals; Marshall 1997), ghost ranges (Nesbitt *et al.* 2009), molecular dates (Eizirik *et al.* 2001) or a combination of methods (Foote *et al.* 1999; Marjanović and Laurin 2008), all methods require assumptions to be made whose processes are currently still poorly understood. Some work has introduced methods to relax or accommodate potential sampling biases in the fossil record (Wagner 2000; Foote 2003). However, their application to low sample numbers (i.e. most terrestrial vertebrates) is uncertain.

A brief history of morphological evolutionary rates

The study of evolutionary rates began with Simpson's *Tempo and Mode in Evolution* (Simpson 1944; Haldane 1949; Gould 1980). Simpson demonstrated that palaeontology could provide more than just the evidence that evolution occurred, but it could also determine what form it took and, more importantly, the speed (Gould 1980). Using morphological change as a yardstick to measure evolutionary rates and the demonstration that those rates vary between groups was ground breaking and influential. Simpson's work to integrate palaeontology into the modern synthesis led to it being accepted as an equal (Olson 1966; Gould 1980; Smocovitis 1992) and his explicit recognition of the power of microevolutionary processes, attempting to integrate them into history of life, modernized the field (Olson 1966). Before this time, many palaeontologists accepted the idea of biological evolution per se, but doubted Darwin's selectionist mechanism for it (Olson 1966). H. F. Osborn, the leading American palaeontologist in the 1930s, was a driving force behind this rejectionist philosophy, as typified by the statement 'We are as remote from adequate explanation of the nature and cause of mechanical evolution of the hard parts of animals as we were when Aristotle first speculated on the subject ...' (Osborn 1922; see also Osborn 1933 for his opinion of genetics and microevolution and Gould 1980 for a more detailed review of palaeontology pre-synthesis). Although *Tempo and Mode* was not universally appreciated (see Wright 1945 for a particularly critical review challenging Simpson to go even further in his acceptance of microevolutionary principles), it was a turning point for both palaeontology and the study of evolutionary rates.

Yet even true giants cannot see all, and in these pre-cladistic days, palaeontologists (Simpson included) tended to view evolution as lineage driven, discounting the importance of cladogenesis in speciation (Cracraft 1987). The importance placed on the lineage and the relatively anagenic nature of evolution influenced Simpson's decision to use either the relative longevity of a lineage (phylogenetic rate) or the number of speciation events within it (taxonomic frequency rate) as the measure of how quickly a group evolves. Thus, according to Simpson,

> The amount of total structural change from *Eohippus* to the recent horse, for instance, can be estimated by the number of genera into which the competent student divides the direct lineage. (Simpson 1949, pp. 98–9)

Using linage duration values, Simpson showed that evolutionary rates varied dramatically between animal groups, most famously the nearly 10× faster (tachytelic) rates of mammals compared with bivalves (bradytelic) (Simpson 1953; Stanley 1985). Following Simpson, palaeontology became not simply an exercise in discovering and cataloguing new finds, but included the use of statistical methods and rate distributions to examine how evolution occurred (also in a temporal sense) (Gould 1980). Palaeontology became intertwined with the greater biological synthesis as researchers brought in concepts from population genetics and ecology (Niklas 1978; Stanley 1985) to help explain these rate differences.

Yet were the conclusions of Simpson and later pupils reflecting reality? Are rabbits really evolving at 10 times the rate of oysters? Did *Didelphis* really stop evolving in the Cretaceous (Simpson 1953)? Of course, as the fossil record is further revealed, some of these questions become moot. For instance, there is a morphological, stratigraphic and taxonomic gap between Cretaceous Herpetotheriidae and opossums (Sánchez-Villagra *et al.* 2007). More importantly are the metrics used actually measuring the rate these groups are evolving? Since the 1970s it has been noted that these large differences in 'rates' mimic the differences in character terminology between clades (Schopf *et al.* 1975). Are these measures simply recording increases in relative complexity (here meaning the number of terms used to describe morphology) of one taxonomic group relative to another with little bearing on the underlying rate of genetic or even phenotypic evolution in each? If we have five times the number of characters for horses as we do for clams, is it surprising we 'find' more horse morphospecies in the fossil record?

Even accounting for researcher biases, does a species origination rate or its mean duration accurately reflect the evolutionary rate within a lineage? Liow *et al.* (2008) demonstrated that large fossil mammals possess higher origination and extinction rates than smaller taxa, a finding that is directly at odds with molecular studies which suggest higher evolutionary rates in smaller taxa (Bromham and Penny 2003). The link between species diversification rates and morphological evolution is also uncertain, as radiations in diversity can occur with little to no morphological disparity (Barraclough and Savolainen 2001; Adams *et al.* 2009). In addition, many of the 'immortal' lineages of Simpson (what we call 'living fossils') have not had 'their evolution slowed down or practically stopped' (Simpson 1953, p. 102). Quite the contrary. Some taxa, like the lepidosaur *Sphenodon punctatus* (or tuatara), display high rates of molecular evolution, higher even than many mammalian groups (Hay *et al.* 2008; Subramanian *et al.* 2009) despite evolving little to no morphological changes since the Miocene (Jones *et al.* 2009). Indeed there are even more extreme cases like that of the freshwater butterfly fish (*Pantodon buchholzi*). An osteoglossomorph with two living populations that differ from each other by over 15% of their sequence, comparable to different genera within the clade, diverged from each other minimally over 50 Ma, yet remain morphologically indistinguishable (Lavoué *et al.* 2011).

Rates in morphology

Building on Simpson's two metrics, speciation rate and lineage duration, researchers sought more practical and less subjective measures of evolutionary rate. Most emphasis has been placed on continuous variables, usually linear dimensions. Their level of success is debatable, but they remain the basic metrics used by evolutionary morphologists for the past 60 years. In the following, the measurements are discussed as two different species (x_1 and x_2); however, they could just as easily be two different times for the same population.

The darwin

First proposed by Haldane (1949), the darwin is the difference between the natural logs of the means of a morphological measurement divided by the time span in millions of years; $|\ln(x_2) - \ln(x_1)|/\Delta t$. The phenotypic measurements must be on a ratio scale: i.e. have an unambiguous zero point and constant interval between adjacent units (Hendry and Kinnison 1999). This measure can be used to compare evolutionary rates between organisms, as long as they are over a similar timescale (Gingerich 1993), or to compare different traits within the same organism. The darwin attempted to represent the proportional rate of change for a trait on an absolute timescale, thus giving a universal metric applicable to both fossil and living organisms. This measure has serious philosophical and practical limitations (Gould 1984; Gingerich 1993; Hendry and Kinnison 1999), mostly concerned with the arbitrary nature of both the natural log transformation and the use of an absolute timescale. For instance, Haldane himself understood this, and believed that if selection was the primary driver of evolution, then a relative timescale (generations), not an absolute timescale (years), would be a more appropriate denominator (Haldane 1949).

The haldane

Popularized by Gingerich (1993) and Hendry and Kinnison (1999), the haldane is the difference in the means of a trait between two populations (often as natural logs), standardized by the pooled standard deviation within those populations, per generation; $|x_2/S(x)_{pooled} - x_1/S(x)_{pooled}|/(\Delta t/\text{generation length})$. The numerator is thus unitless and the denominator is more realistic to biological timescales of evolution (Gingerich 1993). Using the standard deviation accounts for the phenotypic variance seen in natural populations, whereas the use of a generation replaces an absolute delineation (the year) with a relative one (generations). Yet these modifications, by addressing some of the limitations of the darwin, impose new ones. Unlike the year, which is a relatively stable concept (though the number of solar days it represents has shifted; Zhao et al. 2006), determining the generation time in extinct organisms is difficult. Generations within extant mammals can range over two orders of magnitude and vary widely in similar sized and even closely related taxa (Western 1979; Millar and Zammuto 1983). Practically, this means workers can only use taxa closely related to extant groups to estimate generational turnover with any precision. Recent work by Lee and Werning (2008) has attempted to circumvent this handicap by producing generation data for dinosaur taxa based on histology. Finally, although the standard deviation can reliably measure phenotypic disparity (Haldane 1949; Gingrich 1993), it requires a large sample size, something lacking in many taxa, extinct and extant, and clear boundaries for what constitutes a single population, something that is also generally unresolved in the fossil record.

The felsen

Ackerly (2009) introduced a new measure of continuous trait evolutionary divergence rate called the felsen, in honour of Joe Felsenstein. A felsen is based on a phylogenetic topology, rather than a difference in time, like the above two measures. A felsen is an increase of one unit of variance among sister taxa of natural log transformed trait measures per million years. This is represented as $(\ln(x_2) - \ln(x_1))^2/\Delta t$; Δt is here twice the Δt normally calculated because it is the sum of both branches to x_1 and x_2 from their most recent common ancestor, recognizing that Δt should be the sum of evolution along the two independent branches. The time is absolute and suggested to be represented in millions of years. The darwin for the same sister-pair of taxa would be $|\ln(x_2) - \ln(x_1)|/\Delta t$. The numerator is unitless and measures the evolution of variance among sister taxa. In maintaining an annual denominator, the felsen is applicable to macro-evolutionary studies, where generation times are usually not known. This metric has thus far only been used to estimate evolutionary rates among extant sister-pairs. However, modifications could easily be made for use of this method with non-coeval sister-pairs and Brownian motion models.

Patristic distance analysis

Derived from the numerical taxonomic approach championed by Sneath and Sokal (1972), patristic distance analysis (which is the same as cladistic distance *sensu* Jackson and Cheetham 1994) is the sum total of all character changes (including reversals) that occur along each phylogenetic pathway between any two nodes on a tree. It is a pairwise methodology, allowing for comparisons of relative rates of evolution between branches (Smith 1994). A modification of the general methodology was proposed by Wagner (1997) to account for the potential for unequal character space between taxa, called patristic dissimilarity. Patristic dissimilarity is the total branch length divided by the number of scorable characters for each node. This accounts for increases in potential characters and character states that accrue along a phylogeny, as many derived characters are either elaborations of basal traits or only available after the acquisition of one or more evolutionary novelties (nested states).

Both patristic methodologies are intimately linked to an a-priori phylogenetic hypothesis and any modification of this hypothesis will alter the calculated rate scores (Wagner 1997; Dececchi and Larsson 2009). This reliance on a particular phylogenetic reconstruction limits the use of this methodology to well-resolved clades. Additionally, patristic methods rely on the premise that all character changes are equally probable and equally weighted. This uniform approach assumes that any state change in character 1 and character 100, or the change from states 0 to 1 and 1 to 2 within character 100 are equivalent and independent of time. The inherent problem in this assumption can be particularly acute when comparing the rates of change between anatomical modules (i.e. head, axial skeleton, forelimbs or hind limbs) or between basal and highly derived members of a clade where nested states can represent major shifts in morphology. In reality there may be significantly different amounts of information parcelled between each state within the same character and states across different characters. This presents us with a similar philosophical problem to that addressed by Schopf and colleagues over 30 years ago (1975) – the lack of equivalence, in terms of information, between units. This becomes especially critical when uniform, ecologically significant or clade-defining characters are compared with highly variable ones. Is the presence of asymmetrically veined forelimb feathers really phylogenetically, ecologically, and temporally equivalent to the number of teeth in the premaxilla of a theropod? And within a single module, can we really assume that all missing data are equal or should regional phenotypic plasticity be weighed when counting characters? Some attempts have been made to begin to answer questions of the limits of character spaces within lineages (Wagner et al. 2006), but, in general, little attention is paid to these fundamental problems.

Because patristic scores have no temporal component, absolute rates can only be achieved by using outside information to date nodes, typically either time slice binning or the taxa themselves (Sidor and Hopson 1998; Ruta et al. 2006; Brusatte et al. 2008). The inclusion of real dates is both helpful and problematic. For better or for worse, parsimony does not take unequal stratigraphic fits into account. The reconstruction of ancestral states, whether or not one of the sister clades is markedly younger (the so-called 'temporal paradox' that plagues origin dates of many clades), allows us to capture changes not preserved in the rock record. The cost of this is the loss of any potential data generated by the stratigraphic sequence of taxa occurrence (Wagner 1997) as well as the introduction of large ghost lineages (Norell 1993), which haunt many modern phylogenies. Costs of missing stratigraphic data have been used to estimate 'stratigraphic debt', a metric that summarizes the amount of missing stratigraphic time for a given phylogenetic topology (Fox *et al*. 1999; Bodenbender and Fisher 2001) and this debt could be incorporated into evolutionary rates if we allow for mixed rate models.

In clades with high degrees of stratigraphic congruency, indicating a close relationship between stratigraphic occurrence data and tree topology, the use of nodal dating constraints is relatively simple. In these cases one can examine not just the tempo of morphological evolution, but how that tempo itself changes over time (Ruta *et al*. 2006; Brusatte *et al*. 2008), bearing in mind the reservation mentioned above. When this is not the case, such as the diversification of maniraptoran theropods (Dececchi and Larsson 2009), methods have been devised to subdivide range extensions based on branch length itself (see Ruta *et al*. 2006). These methods usually involve either a uniform division of branch time between all clades, using an older out-group, or involve using branch length itself as a proxy for time accrued.

Disparity analysis

Disparity analysis is the categorization of the range of morphologies displayed by a clade and the changes in that range along time or phylogeny. Disparity data use the differences in morphospace occupation, typically using a Euclidean distance matrix to generate the loading values, to determine the relative distance between all taxa (Foote 1991, 1994; Wagner 1997). Although these data can be derived from direct measurements (Harmon *et al*. 2003), in palaeontology it is generally derived from a character matrix. Patristic-based methods are reliant on a specific tree topology, but disparity analysis is relatively free of this constraint. Disparity analysis records both the absolute range of variation and the average dissimilarity between members of a clade as proxies for the amounts of evolution occurring (see Brusatte *et al*. 2008 for a more detailed discussion). These data can then be binned by age and clade to generate reconstructions of diversification within a group.

Although disparity analysis is robust against phylogenetic perturbations and sampling effects (Wagner 1997), it does have its limitations. As stated previously, the level of morphological diversification is not always correlated to evolutionary rates of change as many specious and rapidly evolving clades show little evidence of occupying large morphospace (Barraclough and Savolainen 2001; Adams *et al*. 2009). Although not directly tied to a phylogeny, disparity analysis is still reliant on character choice and information content, similar to patristic-based methodologies. The difficulty in capturing fine-scale morphological differences, which may or may not have great ecological or behavioural importance in an expansive and wide reaching data set, means that all morphological diversity estimates are underestimation. In addition, key characters which become canalized in a lineage, and may be the basis for later ecological and taxonomic success (such as seen in birds), are undervalued. For example, comparing birds with mammals would indicate that the latter show higher rates of evolution, as physical requirements for flight place constraints on the avian bauplan that most mammalian clades lack.

Integrating genotypic and phenotypic evolutionary rates

Evolutionary rates of genetic and phenotypic change have been estimated for decades (see above). In a classic selectionist view, there is no reason to suspect that the rate of evolution of the phenotype and the genotype should necessarily be correlated. For specific morphologies and genes, correlations may be more likely: studies have in some cases shown that the rate of molecular evolution in phenotype-associated genes and their related morphologies can covary (Dorus *et al.* 2004; Herlyn and Zischler 2007) although the robustness of correlations is unclear (Hurle *et al.* 2007). At an aggregate level, whether genotype and phenotype should be correlated is more controversial. The neutral theory implies a correlation is less likely, as most mutations should have no effect on fitness (and thus phenotype) (Kimura 1983). Certain living fossils appear to prove the point. As noted above, tuataras (*Sphenodon* sp.) have very high rates of molecular evolution despite relative morphological stasis (Hay *et al.* 2008). However, rates could correlate if either an appreciable percentage of substitutions had phenotypic effect or if both rates varied based on some biotic variable, for example body size (Bromham *et al.* 2002; see also our argument for neutral evolution of phenotype, below). Empirically, workers have found evidence for cases of both highly correlated genotypic and phenotypic evolution (Smith *et al.* 1992; Omland 1997) and others without such correlations (Davies and Savolainen 2006). These differing results are dependent on methods, sampled clades, and data quality, and all are based on relatively small morphological data sets (see our argument for very large data sets, below).

Morphological largess

Morphology is complex. Today's discrete morphological phylogenetic databases are beginning to capture some of this complexity and are a far cry from those even a decade ago. Phylogenetic databases during the 1990s were typically of the order of 50 taxa and 100 characters, or approximately 5000 data cells. Today's matrices have grown with many in the realm of 100 000 data cells and some, such as that of Livezey and Zusi (2006, 2007), approach totals of around 500 000. The massive growth of morphological data, by nearly two orders of magnitude, may come with some statistical advantages that have a historical precedent.

This large growth in morphological data is reminiscent of the explosive growth of molecular data. Molecular data sets grew exponentially during the late 1990s and now continue to grow at a very steep linear rate (www.ncbi.nlm. nih.gov/genbank/genbankstats.html). Current rates of morphological data growth are hopefully at the early exponential phase. We anticipate growth in discrete data for some clades to levels achieved by Livezey and Zusi (2006, 2007), of approximately half a million data cells. We anticipate an even greater growth in multivariate data, such as 2D and 3D geometric morphometric data.

Mendelian genetics arose from observations on a limited number of traits and assuming a limited number of genes. Its predictions are well founded and suggest a discontinuous range of phenotypic traits. Yet, as any breeder knows, selected traits are often continuous in nature. It was not until Fisher's work over 60 years ago that geneticists were able to demonstrate that a continuous distribution of traits could be derived from enough genes. From this developed the stochastic process models used to the present day. And what of morphological data? Is there evidence that it too is more continuous in nature than traditionally believed? Can we now begin treating morphological evolution within the context of a stochastic process model? Darwin (1859) hinted at this intriguing idea when he suggested that '[v]ariations neither useful nor injurious would not be affected by natural selection, and would be left either a fluctuating element, as perhaps we see in certain polymorphic species, or would ultimately become fixed . . .'. This statement is a concise and prescient statement of a neutral theory of evolution: that neutral mutations ('variations neither useful nor injurious') would result in polymorphisms ('fluctuating elements') that could become fixed in the population.

An argument for stochastic evolution of discrete morphologies

Biologists may initially regard the use of discrete morphological char-acter data as incompatible with a stochastic model of evolution. Morphology is assumed to be under selection at all times. Countless researchers have documented directional and stabilizing selection on discrete and continuous morphological data in short timescales, ranging from years (e.g. Grant 1999) to thousands of years (e.g. Gingerich 1993; Hunt et al. 2008). Are not patterns of phenotypic evolution on even longer timescales of millions of years generally assumed to be adaptive? If a trait is fixed and present for millions of years, must it not have to contribute to fitness? Well, caution to this line of thinking was famously raised by Gould and Lewontin (1979) in the sense that not all phenotypic traits are easily explained by adaptive evolution. We now understand a great deal about mechanisms that constrain biological evolution that range from genetic, developmental, biomechanical, ecological, to complex systems constraints.

Empirical work on selection in the vast majority of extant populations is not directional. These populations are well adapted for their context, presumed to reside near the peaks of their fitness landscapes, and thus rest on low-sloping fitness plateaus. Nearly all data suggest extant populations are under very weak stabilizing selection (Kingsolver *et al.* 2001; Estes and Arnold 2007; Siepielski *et al.* 2009). The next questions that follow are how long, on average, are lineages under weak stabilizing selection, and are there periods that strong directional selection may be dominant? These patterns are not yet known, but probably depend largely on scale, as the modern biota only represent one tiny time slice.

If selection on extant populations is weakly stabilizing, and thus 'nearly neutral', are all modules within it also so? If we consider the entire organism, not specific traits, will we get different results? And even if a specific morphological trait's evolution were purely directional, this in no way dictates that other traits within an organism or a lineage must be. Trade-offs are the norm in laboratory-based evolutionary experiments. Refinement in one trait is generally tied with a reduction, or loss, in another (Bell 2008). Could trade-offs within the entire organism produce a neutral mode of evolutionary change if the entire, or a sufficient portion, of the organism were studied? Some phenotypes may follow this 'nearly neutral' model. Palaeobiologists may not yet realize the role of stochastic processes in phenotypic evolution.

What's the null?

An important issue that has not been addressed in most macroevolutionary studies is the creation of a proper null model. Parsimony has been the mainstay for the past several decades, but this does not provide a null model in the sense of testing for deviations from a normally distributed range of expected results. This model of testing implies parametric data, but large sets of discrete data may also behave parametrically, similar to how multiple alleles may contribute to continuous phenotypic variation when some single alleles yield discrete phenotypic traits (see above). With this in mind, perhaps parsimony should not be the null model of large morphological data sets, be they discrete or continuous. We propose that large morphological data sets should instead be compared against a stochastic, neutral model of evolutionary change. This would assume uniform rates of character change and uniform rates of novel character state (or continuous limits) acquisition within characters. Testing of such hypotheses then becomes easily manageable, such as for rate variations within and among characters over time and between clades. This treatment would not only establish a more appropriate statistical testing framework, but also provide a simpler model of evolutionary change. If we are concerned about measuring rates of morphological change, shouldn't we be measuring against a stochastic, null model with a presumed normal distribution of change? More complex models of change could be examined after this initial, simplest stochastic model. These include directional change and stasis models (e.g. Hunt 2007).

A test of first principles

If morphological evolution can be modelled as a stochastic process, it must fulfil a basic, underlying principle. Its aggregate evolutionary change must be relatively linear with respect to time. This is because the neutral component of phenotypic evolution, like the molecular clock, should 'tick' with time. Molecular-clock analyses implicitly use nearly neutral rates of change. We present a few examples of data sets dominated by fossil taxa. These data sets are comprehensive analyses of Mesozoic mammals (Luo *et al.* 2007), crocodylomorphs (Larsson 2000), non-avian theropods (Holtz *et al.* 2004) and early birds (Clarke *et al.* 2006). Common to all these data sets are large taxonomic breadth (up to more than 100 taxa), large morphological sampling (greater than 200 discrete characters), and large temporal breadth (ranging from 100 Ma to greater than 200 Ma). We use the original published phylogenetic topologies for each of these matrices. The stratigraphic age of the most basal taxon is used as the initial estimate of the basal divergence of the entire phylogeny. Although this assumption is certainly flawed, the basal taxon for each of these data sets is most probably within the ballpark range of the base node divergence. In any event, this potential flaw does not seem to have adverse effects on our results. The stratigraphic age for all other taxa is used to estimate the time of divergence of each taxon from the base node. The amount of morphological evolution is calculated using parsimony methods. The total number of unambiguous and delayed transformations were counted from the base node to each of the terminal taxa. These values were then plotted to assess, in this conservative method, the correlation of discrete character change to time. Regressions were derived without forcing them through the plot's origin. This preliminary method is simplistic and open to a number of biases, including oversampling of character changes about the base node, omission of unique autapomorphies, and missing data. However, as a first principles approach, we consider it methodologically and conceptually conservative. The absence of unique autapomorphies should affect each terminal taxon but is not expected to create a significant bias in the results if we assume a relatively equal distribution of unique autapomorphies across all terminal taxa – something that may or may not be valid, but is not examinable until data sets include this kind of character data.

The results are compelling (Table 4.1, Figure 4.1). In all cases, a linear regression model has the best fit. We present the full data without trimming taxa with large amounts of missing data. When these taxa are removed from the analysis, the regression r-squared values are improved but the fits remain linear and relatively unchanged (data not shown). The r-squared values range widely from 0.38 to 0.88, implying a significant, linear relationship between morphological evolution and time. The lowest r-squared values are for Holtz et al. (2004), a data set which included taxa with large amounts of missing data. When taxa with greater than 50% missing data are excluded, the linear regression r-squared values are similar to the others presented here (data not shown). The linear regression implies a relatively consistent relationship between time elapsed and quantity of evolutionary change, which is in line with a stochastic process model of change. The best-fit data are the large Mesozoic mammal and crocodylomorph data sets. We attribute these better fits to their larger taxonomic, stratigraphic and phenotypic breadth. An overwhelming majority of these data sets are composed of cranial characters. When partitioned, cranial characters yield similar or better linear regression fits. These character matrices sample the skull more thoroughly than any other part of the skeleton. This high degree of sampling may better saturate the total possible morphological variation in this region of the skeleton and reveal a clearer stochastic signal. The smallest data set, from Clarke et al. (2006), has only 20 fossil taxa and spans a relatively smaller stratigraphic range. In spite of this, the results of this data also yield a fairly good linear fit. Additional tests may include gamma distributed site rates in a Bayesian Inference analysis with no branch length constraints.

These significant linear correlations support a nearly neutral model of evolution. There are cases in each regression of outliers and these may indicate isolated instances of relatively rapid and slowed evolutionary rates; however, on average over these long timescales, evolutionary rates appear nearly neutral. Moreover, this nearly neutral model suggests that the adaptive landscape in

Table 4.1 Reduced major axis linear regression statistics for the evolutionary rates of discrete morphological characters from the four fossil-dominated phylogenetic data sets discussed in the text. Only unambiguous and delayed transformed optimizations of character state changes are shown. Results from the total character sets and only cranial characters are given for each original data set. Regression data when excluding extant taxa in Luo *et al.* (2007) are also included.

Data		No. taxa	No. chars	Unambiguous regression	r^2	Delayed transformed regression	r^2
Clarke *et al.* (2006)	total	24	205	$y = 0.432x - 21.61$	0.69	$y = 1.319x - 28.72$	0.77
	skull			$y = 0.081x - 0.77$	0.85	$y = 0.326x - 0.56$	0.82
Holtz *et al.* (2004)	total	73	638	$y = 1.427x - 37.24$	0.38	$y = 2.788x - 49.49$	0.48
	skull			$y = 0.475x - 14.69$	0.31	$y = 0.985x - 27.14$	0.50
Luo *et al.* (2007)	total	102	436	$y = 1.171x + 11.963$	0.79	$y = 2.261x + 6.45$	0.76
	skull			$y = 0.887x - 16.44$	0.82	$y = 1.691x - 18.32$	0.79
Luo *et al.* (2007) excluding extant taxa	total	76	436	$y = 1.388x - 5.62$	0.84	$y = 2.800x - 39.45$	0.79
	skull			$y = 1.052x - 249$	0.86	$y = 2.073x - 14.75$	0.81
Larsson (2000)	total	115	316	$y = 0.785x + 34.29$	0.86	$y = 1.484x + 50.53$	0.87
	skull			$y = 0.590x + 14.42$	0.88	$y = 1.078x + 22.59$	0.88

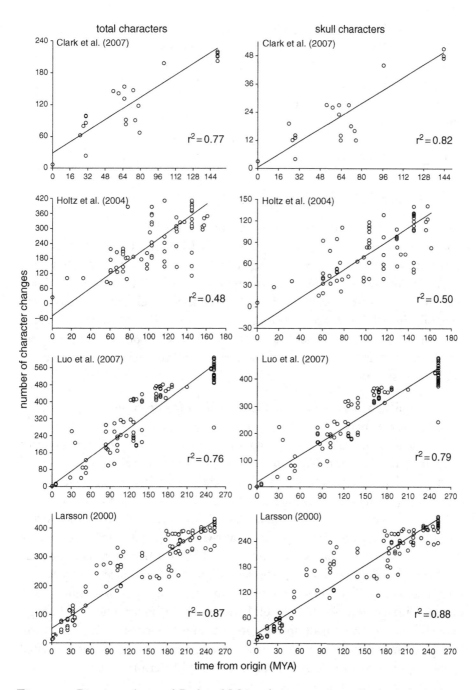

Figure 4.1 Bivariate plots and Reduced Major Axis regressions of morphological character evolution. Examples of four discrete morphological phylogenetic analyses are shown for their complete data and a partition of only cranial characters. All data are derived from delayed transformed optimizations on the original, published phylogenies. The time from origin dates are calculated as the difference of the stratigraphic ages of the outgroup from that of the oldest representative of all other terminal taxa in each phylogeny. All points are graphed as overlapping open circles.

these lineages has, on average, also been relatively flat throughout these long timescales, similar to the measured adaptive landscapes of today. This sounds provocative, given the dramatic environmental and biotic changes over these timescales, and we stress these results are only preliminary, open to future testing and scrutiny. Our focus in the near future will be to examine these patterns in additional data sets, across particularly dramatic environmental shifts, and selected morphological changes to see if and how further patterns can be gleaned from large-scale patterns of evolutionary rates.

Morphological divergence times?

We believe that morphology may represent an untapped and valuable resource for the estimation of divergence times. Certainly, there are theoretical and practical arguments against using morphology in this manner. However, keeping in mind our arguments above, morphology, if measured and analysed carefully in a statistical framework, can provide a novel and in many ways potentially important source of divergence time estimates. The empirical results shown above hint that there is at least a neutral component to phenotypic evolution in big data sets, further arguing that extracting temporal information should be possible.

Limitations of molecular sequence data for divergence time estimates

At first glance, molecular sequence data may seem ideal for the estimation of divergence times. Molecular sequences can be sampled relatively easily and objectively, although the subjectivity of gene choice remains. Complex stochastic models have been developed that incorporate biological realities of sequence evolution, as have statistical frameworks to test such hypotheses. However, molecular sequence data suffer from one overwhelming limitation: they can only be measured in extant (or only recently extinct) taxa. Apart from multigenerational viral and single cell lineage studies, all measured sequences are treated as contemporaneous. Therefore, because all sequences are modern, dating deeply divergent lineages will also be difficult because of long-branch effects. Although some efforts have been made to mitigate these problems (e.g. Magallón 2010), these problems are reflected in the widely divergent divergence times estimated for the origin of Metazoa, for example (Smith and Peterson, 2002).

What morphology can contribute

Most obviously, morphology is measurable in extinct taxa. This has several important advantages. First, long branches can be broken. Second, because fossils can be temporally situated using the rock record, morphology allows for the use of tip dates, as well as fossil calibrations for known divergences. A similar methodology has been successfully used to study evolutionary rates in heterochronous micro-organism sequences For example, Rambaut (2000) implemented algorithms to estimate evolutionary rates of molecular sequence evolution among non-contemporaneous taxa and demonstrated its use with a series of dengue virus gene sequences from populations isolated over the past half century. The use of morphology would also allow the dating of divergences between lineages with no extant relatives. Finally, new methods to estimate the leaf age in extant heterochronous sequences (Shapiro *et al.* 2011) might produce alternate estimates of age if a fossil cannot be stratigraphically situated.

Morphology has also proven useful resolving difficult phylogenetic problems when it is used together with molecular sequence data in a partitioned phylogenetic analysis (Nylander *et al.* 2004). The use of an appropriate morphological data set in partitioned divergence time analysis may improve the accuracy of divergence time estimation.

A roadmap to using morphology to date divergences

Divergence times are generally estimated from either extant taxa using molecular-clock methods, based on well-sampled (e.g. many marine hard invertebrates) to poorly (e.g. most terrestrial invertebrates and vertebrates) sampled fossil records, or a combination of both. These yield broad ranges of estimated divergence times that will influence any study of evolutionary rate. If we can treat phenotypic evolution in a stochastic process model, fossil morphologies may provide a new tool to estimate ancient divergence times. In light of the strong body of research on divergence time estimation using molecular sequence data and of the arguments presented above, we propose a road map to using discrete morphological character data for divergence time estimation.

1. *Morphological data sets used for divergence time estimation need to be large, objectively chosen, with large taxonomic breadth.* To reasonably capture as much phenotypic variation as possible, data sets must evenly sample as many independent morphologies as possible, including constant and autapomorphic characters. These very large data sets of morphology mitigate lineage-specific selection biases and reduce biases of excluded unique autapomorphies because these traits are more probably convergently evolved in other taxa.

2. *The tempo and mode of evolution in these large data sets must be studied empirically.* Both older parsimony-based patristic and time-calibrated methods (see our analysis, above) should be used along with stochastic model-based methodology used for the study of molecular evolutionary

rates (see above and Lanfear *et al.* 2010). This should inform the choice of model for the evolution of the rate of evolution (e.g. autocorrelated, or not) as well as prior specification in Bayesian analyses. These large data sets could also serve to test alternate stochastic models for morphological evolution, moving beyond the Mk1 model (Lewis 2001) and attempt to incorporate biological realities, including, for example, pleiotropic constraints on character developmental evolution. Currently, phenotypic character state changes are treated as equal to one another, although this is very probably not true. Some characters are probably more 'burdened' and pleiotropic (Riedl 1978) and functional constraints limit variations within these characters. An example of this kind of character would be the notochord. Individual states within a character are also not likely the same either, for similar reasons. Empirical testing of these assumptions is needed to refine models of phenotypic evolution. Empirical study should also shed light on the mode and tempo of phenotypic evolution in general and would allow for statistical testing of rate correlates, as in molecular sequence studies.

3. *These morphological data sets should be analysed in a Bayesian or Maximum Likelihood framework that models stochastic phenotypic evolution, potential rate heterogeneity and branch lengths via time calibrations and tip dates.* Like current divergence time software for molecular sequence data (e.g. Drummond and Rambaut 2007), dating methods must treat these large data sets as potentially non-clock-like, with model selection and prior specification based on empirical research (see above). The inclusion of tip dates should mirror previous work in molecular dating of micro-organism lineages and ancient sequences (Rambaut 2000).

Other applications

Other uses to this method of modelling phenotypic evolutionary rates are to derive short and long-term patterns of evolutionary rates. These have been done to varying degrees with continuous data (e.g. Adams and Collyer 2009) and discrete data (e.g. Dececchi and Larsson 2009). However, most continuous phenotypic data sets are generally small (limited to a handful of measurements) and subject to allometric influences. Discrete morphological data are surely also influenced by allometry, but are expected to be less so, given the coarse nature of character state definitions. However, with the framework presented here, we can begin to test these potentially confounding factors and how they contribute to large-scale patterns of phenotypic evolution.

Evolutionary rate of phenotypic change is expected to yield insights into developmental and evolutionary processes. If rates of phenotypic evolution are rather constant through time, what does this say about their underlying processes? Are there limits to how fast phenotypes can evolve over long periods of time? Are relatively complex phenotypes at some limit of evolutionary rate (Orr 2000)? Do large-scale patterns of phenotypic evolutionary rates correlate with equally large-scale patterns of phylogenetic diversification? Refining our study of these patterns will lead, hopefully, to a better understanding of what levels of constraints may be acting on particular classes of morphologies over long timescales. General patterns of constraints to phenotypic variation will be critical to deriving the macroevolutionary patterns of the 'phenotypic landscape'. Only then can more fundamental progress be made to integrating genotypic and phenotypic evolution to derive evolutionary topographies of genotype–phenotype maps.

Conclusions

The application of stochastic process models to morphological evolution has many advantages, many applications, and may reflect many potential underlying biological phenomena. Although morphological evolution over large timescales is generally considered to be adaptive, we show that rates of morphological evolution are essentially linear, given data sets that capture large-enough temporal, taxonomic and phenotypic ranges. The relatively large timescales and large data matrix sizes may be a key factor underlying these predictable rates of change, with clock-like patterns resulting from a 'Law of Large Numbers' phenomenon. We also touch on a number of other reasons why discrete characters may evolve in a stochastic pattern, including trade-offs within the total phenotype of a lineage and the predominance of weak stabilizing selection in studies of extant populations. Continuous phenotypic data would be, intuitively, more amenable to a stochastic process model, because these data have the potential to capture minor degrees of variation. Discrete morphological data would seem to be less so because the discrete nature of the data masks all but the most obvious variations in morphology. Discrete data have great value though because these are the traits that morphologists have been defining and coding for several decades. Many fossils are taphonomically distorted, to some degree, so discrete data may prove to offer the clearest signal of phenotypic evolution.

We outline a roadmap to developing a more sophisticated treatment of phenotypic data and elevating its application to methodologies developed for molecular data. These include applying different gamma distributions to rate variations, rate heterogeneity methods to maximum likelihood, and Bayesian methods of phylogenetic estimations.

The long-term goal of this research trajectory will be to refine discrete, and ultimately continuous, phenotypic character definitions and coding, and to accurately estimate the rates of phenotypic evolution. Ultimately, we hope to bridge the gap between genotypic and phenotypic evolution. These associations may, in turn, tell us more about the evolution of intermediate realms between the genome and anatomy. These intermediate scales include gene regulatory networks that govern gene expressions, and morphogenesis of the phenotype. Only then will the hypothetical links within the genotype–phenotype map be testable.

Summary

Evolutionary rates are paramount in biology. Periods of high rates suggest intense periods of directional selection and offer unpredicted insights into how evolution shapes biodiversity. Tracing the genotype–phenotype map across evolutionary trajectories helps us to explain and describe existing and extinct biodiversity. One step toward tracing the genotype–phenotype map across time is to correlate genetic evolutionary rates with morphological evolutionary rates. We briefly review current methods to assess genetic and morphological evolutionary rates and present a novel method to estimate evolutionary rates of discrete morphological characters using stratigraphically calibrated fossils. The examples we present best fit a linear correlation and therefore imply a relatively constant, clock-like evolutionary rate for discrete morphological characters. This pattern may be due, in part, to the relatively large size of the data matrices we used, but opens doors to the possibility of assessing discrete morphological character evolution as a stochastic process. We discuss potential applications to this, including using morphological clocks to estimate ancient divergence times and also biologically relevant reasons as to why morphological evolution may be relatively constant over large timescales. The importance of this approach is that peaks in evolutionary rates of genes and morphologies may now be correlated to better estimate and understand the evolution of genotype–phenotype maps.

Acknowledgements

We thank Johannes Müller and Robert Asher for inviting us to contribute to this volume and giving us the opportunity to formally introduce and expand our ideas. We thank Pete Wagner (especially for his insightful discussion of determined versus stochastic versus random processes) and an anonymous reviewer for excellent comments that improved this manuscript. We also thank

R. Carroll, N. Frobisch, A. Hendry, F. Lando, E. Maxwell and R. O'Keefe for valuable discussions that helped develop this topic. Funding for this research was supported by a NSERC Discovery Grant and Canada Research Chair to HCEL, a FQRNT graduate fellowship to TAD, and a NSERC graduate fellowship to LBH.

REFERENCES

Ackerly, D. (2009). Conservatism and diversification of plant functional traits: evolutionary rates versus phylogenetic signal. *Proceedings of the National Academy of Sciences of the United States of America*, **106** Suppl 2, 19 699–706.

Adams, D. C. and Collyer, M. L. (2009). A general framework for the analysis of phenotypic trajectories in evolutionary studies. *Evolution*, **63**, 1–12.

Adams, D. C., Berns, C. M., Kozak, K. H. and Wiens, J. J. (2009). Are rates of species diversification correlated with rates of morphological evolution? *Proceedings of the Royal Society B–Biological Sciences*, **276**, 2729–38.

Allen, A. P., Brown, J. H. and Gillooly, J. F. (2002). Global biodiversity, biochemical kinetics, and the energetic-equivalence rule. *Science*, **297**, 1545–8.

Alroy, J., Aberhan, M., Bottjer, D. J., *et al.* (2008). Phanerozoic trends in the global diversity of marine invertebrates. *Science*, **321**, 97–100.

Barraclough, T. G. and Savolainen, V. (2001). Evolutionary rates and species diversity in flowering plants. *Evolution*, **55**, 677–83.

Barrett, P. M., McGowan, A. J. and Page, V. (2009). Dinosaur diversity and the rock record. *Proceedings of the Royal Society B–Biological Sciences*, **276**, 2667–74.

Bell, G. (2008). *Selection: The Mechanism of Evolution*. Oxford, UK: Oxford University Press.

Benton, M. J., Donoghue, P. C. J. and Asher, R. J. (2009). Calibrating and constraining molecular clocks. In *The Timetree of Life*, ed. S. B. Hedges and S. Kumar, pp. 35–86. Oxford, UK: Oxford University Press.

Bodenbender, B. E. and Fisher, D. C. (2001). Stratocladistic analysis of blastoid phylogeny. *Journal of Paleontology*, **75**, 351–69.

Bromham, L. (2009). Why do species vary in their rate of molecular evolution? *Biology Letters*, **5**, 401–4.

Bromham, L. and Penny, D. (2003). The modern molecular clock. *Nature Reviews Genetics*, **4**, 216–24.

Bromham, L., Penny, D., Rambaut, A. and Hendy, M. D. (2000). The power of relative rates tests depends on the data. *Journal of Molecular Evolution*, **50**, 296–301.

Bromham, L., Woolfit, M., Lee, M. S. Y. and Rambaut, A. (2002). Testing the relationship between morphological and molecular rates of change along phylogenies. *Evolution*, **56**, 1921–30.

Brusatte, S. L., Benton, M. J., Ruta, M. and Lloyd, G. T. (2008). Superiority, competition, and opportunism in the evolutionary radiation of dinosaurs. *Science*, **276**, 1485–8.

Clarke, J. A., Zhou, Z. and Zhang, F. (2006). Insight into the evolution of avian flight from a new clade of Early Cretaceous ornithurines from China and the morphology of *Yixianornis grabaui. Journal of Anatomy*, 287–308.

Cracraft, J. (1987). Species concepts and the ontology of evolution. *Biology and Philosophy*, **2**, 329–46.

Crispo, E. (2007). The Baldwin effect and genetic assimilation: revisiting two mechanisms of evolutionary change mediated by phenotypic plasticity. *Evolution*, **61**, 2469–79.

Darwin, C. (1859). *On the Origin of Species by Means of Natural Selection, or the Preservation of Favoured Races in the Struggle for Life*. London: John Murray.

Davies, T. J. and Savolainen, V. (2006). Neutral theory, phylogenies, and the relationship between phenotypic change and evolutionary rates. *Evolution*, **60**, 476–83.

Dececchi, T. A. and Larsson, H. C. E. (2009). Patristic evolutionary rates suggest a punctuated pattern in forelimb evolution before and after the origin of birds. *Paleobiology*, **35**, 1–12.

Dobzhansky, T. (1950). Evolution in the tropics. *American Scientist*, **38**, 209–21.

Dobzhansky, T. (1973). Nothing in biology makes sense except in the light of evolution. *The American Biology Teacher*, **35**, 125–9.

Dorus, S., Evans, P. D., Wyckoff, G. J., Choi, S. S. and Lahn, B. T. (2004). Rate of molecular evolution of the seminal protein gene SEMG2 correlates with levels of female promiscuity. *Nature Genetics*, **36**, 1326–9.

Drummond, A. J. and Rambaut, A. (2007). BEAST: Bayesian evolutionary analysis by sampling trees. *BMC Evolutionary Biology*, **7**, 214.

Eizirik, E., Murphy, W. J. and O'Brien, S. J. (2001). Molecular dating and biogeography of the early placental mammal radiation. *Journal of Heredity*, **92**, 212–9.

Eldredge, N. and Gould, S. J. (1972). Punctuated equilibria: an alternative to phyletic gradualism. In *Models in Paleobiology*, ed. T. J. M. Schopf, pp. 82–115. San Francisco, CA: Freeman, Cooper.

Estes, S. and Arnold, S. J. (2007). Resolving the paradox of stasis: models with stabilizing selection explain evolutionary divergence on all timescales. *American Naturalist*, **169**, 227–44.

Fedorov, A. A. (1966). The structure of the tropical rain forest and speciation in the humid tropics. *Journal of Ecology*, **54**, 1–11.

Felsenstein, J. (2003). *Inferring Phylogenies*. Sunderland, MA: Sinauer Associates.

Foote, M. (1991). Morphologic patterns of diversification: examples from trilobites. *Palaeontology*, **34**, 461–85.

Foote, M. (1994). Morphological disparity in Ordovician-Devonian crinoids and the early saturation of morphological space. *Paleobiology*, **20**, 320–44.

Foote, M. (2003). Origination and extinction through the Phanerozoic: a new approach. *Journal of Geology*, **111**, 125–48.

Foote, M., Hunter, J. P., Janis, C. M. and Sepkoski, J. J. (1999). Evolutionary and preservational constraints on origins of biologic groups: divergence times of eutherian mammals. *Science*, **283**, 1310–14.

Fox, D. L., Fisher, D. C., Leighton, L. R. (1999). Reconstructing phylogeny with and without temporal data. *Science*, **284**, 1816–19.

Gingerich, P. D. (1993). Quantification and comparison of evolutionary rates. *American Journal of Science*, **293**, 453–78.

Gingerich, P. D. (2009). Rates of evolution. *Annual Review of Ecology, Evolution, and Systematics*, **40**, 657–75.

Gould, S. J. (1980). Is a new and general theory of evolution emerging? *Paleobiology*, **6**, 119–30.

Gould, S. J. (1984). Smooth curve of evolutionary rate: a psychological and mathematical artefact. *Science*, **226**, 994–5.

Gould, S. J. and Lewontin, R. C. (1979). The spandrels of San Marco and the panglossian paradigm: a critique of the adaptationist programme. *Proceedings of the Royal Society of London B–Biological Sciences*, **205**, 581–98.

Grant, P. R. (1999). *Ecology and Evolution of Darwin's Finches*. Princeton, NJ: Princeton University Press.

Graur, D., Martin, W. (2004). Reading the entrails of chickens: molecular timescales of evolution and the illusion of precision. *Trends in Genetics*, **20**, 80–6.

Guindon, S. and Gascuel, O. (2003). A simple, fast, and accurate algorithm to estimate large phylogenies by maximum likelihood. *Systematic Biology*, **52**, 696–704.

Haldane, J. B. S. (1949). Suggestions as to the quantitative measurement of rates of evolution. *Evolution*, **3**, 51–6.

Harmon, L. J., Schulte, J. A., Larson, A. and Losos, J. B. (2003). Tempo and mode of evolutionary radiation in iguanian lizards. *Science*, **301**, 961–4.

Hay, J. M., Subramanian, S., Millar, C. D., Mohandesan, E. and Lambert, D. M. (2008). Rapid molecular evolution in a living fossil. *Trends in Genetics*, **24**, 106–9.

Hendry, A. P. and Kinnison, M. T. (1999). The pace of modern life: measuring rates of contemporary microevolution. *Evolution*, **53**, 1637–53.

Herlyn, H. and Zischler, H. (2007). Sequence evolution of the sperm ligand zonadhesin correlates negatively with body weight dimorphism in primates. *Evolution*, **61**, 289–98.

Ho, S. Y. W. (2009). An examination of phylogenetic models of substitution rate variation among lineages. *Biology Letters*, **5**, 421–4.

Holder, M. and Lewis, P. O. (2003). Phylogeny estimation: traditional and Bayesian approaches. *Nature Reviews Genetics*, **4**, 275–84.

Holtz, T. R., Molnar, R. E. and Currie, P. J. (2004). Basal Tetanurae. In *The Dinosauria*, 2nd edn., ed. D. B. Weishampel, P. Dodson and H. Osmólska, pp. 71–110. Berkeley, CA: University of California Press.

Huelsenbeck, J. P., Larget, B. and Swofford, D. (2000b). A compound Poisson process for relaxing the molecular clock. *Genetics*, **154**, 1879–92.

Huelsenbeck, J. P., Rannala, B. and Masly, J. P. (2000a). Accommodating phylogenetic uncertainty in evolutionary studies. *Science*, **288**, 2349–50.

Huelsenbeck, J. P., Ronquist, F., Nielsen, R. and Bollback, J. P. (2001). Bayesian inference of phylogeny and its impact on evolutionary biology. *Science*, **294**, 2310–14.

Huelsenbeck, J. P., Larget, B. and Alfaro, M. E. (2004). Bayesian phylogenetic model selection using reversible jump Markov chain Monte Carlo. *Molecular Biology and Evolution*, **21**, 1123–33.

Hunt, G. (2007). The relative importance of directional change, random walks, and stasis in the evolution of fossil lineages. *Proceedings of the National Academy of Sciences*, **104**, 18 404–8.

Hunt, G., Bell, M. A. and Travis, M. P. (2008). Evolution toward a new adaptive optimum: phenotypic evolution in a fossil stickleback lineage. *Evolution*, **62**, 700–10.

Hurle, B., Swanson, W. and Green, E. D. (2007). Comparative sequence analyses reveal rapid and divergent evolutionary changes of the WFDC locus in the primate lineage. *Genome Research*, **17**, 276–86.

Jackson, J. B. C. and Cheetham, A. H. (1994). Phylogeny reconstruction and the tempo of speciation in cheilostome Bryozoa. *Paleobiology*, **20**, 407–23.

Jones, M. E. H., Tennyson, A. J. D., Worthy, J. P., Evans, S. E. and Worthy, T. H. (2009). A sphenodontine (Rhynchocephalia) from the Miocene of New Zealand and the paleobiology of the Tuatara (Sphenodon). *Proceedings of the Royal Society B–Biological Sciences*, **276**, 1385–90.

Kimura, M. (1983). *The Neutral Theory of Molecular Evolution*. Cambridge, UK: Cambridge University Press.

Kingsolver, J. G., Hoekstra, H. E., Hoekstra, J. M., *et al.* (2001). The strength of phenotypic selection in natural populations. *American Naturalist*, **157**, 245–61.

Korber, B. (2000). Timing the ancestor of the HIV-1 pandemic strains. *Science*, **288**, 1789–96.

Lanfear, R., Welch, J. J. and Bromham, L. (2010). Watching the clock: studying variation in rates of molecular evolution between species. *Trends in Ecology and Evolution*, **25**, 495–503.

Larsson, H. C. E. (2000). Ontogeny and phylogeny of the archosauriform skeleton. Ph.D. Thesis, University of Chicago.

Lavoué, S., Miya, M., Arnegard, M. E., *et al.* (2011). Remarkable morphological stasis in an extant vertebrate despite tens of millions of years of divergence. *Proceedings of the Royal Society B–Biological Sciences*, **278**, 1003–8.

Lee, A. H. and Werning, S. (2008). Sexual maturity in growing dinosaurs does not fit reptilian growth models. *Proceedings of the National Academy of Sciences of the United States of America*, **105**, 582–7.

Lewis, P. O. (2001). A likelihood approach to estimating phylogeny from discrete morphological character data. *Systematic Biology*, **50**, 913–25.

Liow, L. H., Fortelius, M., Bingham, E., *et al.* (2008). Higher origination and extinction rates in larger mammals. *Proceedings of the National Academy of Sciences of the United States of America*, **105**, 6097–102.

Livezey, B. C. and Zusi, R. (2006). Higher-order phylogeny of modern birds (Theropoda, Aves: Neornithes) based on comparative anatomy.I. Methods and characters. *Bulletin of Carnegie Museum of Natural History*, **37**, 1–544.

Livezey, B. C. and Zusi, R. L. (2007). Higher-order phylogeny of modern birds (Theropoda, Aves: Neornithes) based on comparative anatomy. II. Analysis and discussion. *Zoological Journal of the Linnean Society*, **149**, 1–95.

Luo, Z.-X., Chen, P., Li, G. and Chen, M. (2007). A new eutriconodont mammal and evolutionary development in early mammals. *Nature*, **446**, 288–93.

Magallón, S. (2010). Using fossils to break long branches in molecular dating: a comparison of relaxed clocks applied to the origin of angiosperms. *Systematic Biology*, **59**, 384–99.

Marjanović, D. and Laurin, M. (2008). Assessing confidence intervals for stratigraphic ranges of higher taxa: the case of Lissamphibia. *Acta Palaeontologica Polonica*, **53**, 413–32.

Marshall, C. R. (1994). Confidence intervals on stratigraphic ranges: partial relaxation of the assumption of randomly distributed fossil horizons. *Paleobiology*, **20**, 459–69.

Marshall, C. R. (1997). Confidence intervals on stratigraphic ranges with nonrandom distributions of fossil horizons. *Paleobiology*, **23**, 165–173.

Marshall, C. R. (2008). A simple method for bracketing absolute divergence times on molecular phylogenies using multiple fossil calibration points. *American Naturalist*, **171**, 726–42.

Millar, J. S. and Zammuto, R. M. (1983). Life histories of mammals: an analysis of life tables. *Ecology*, **64**, 631–5.

Müller, J. and Reisz, R. R. (2005). Four well-constrained calibration points from the vertebrate fossil record for molecular clock estimates. *BioEssays*, **27**, 1069–75.

Nesbitt, S. J., Smith, N. D., Irmis, R. B., *et al.* (2009). A complete skeleton of a Late Triassic saurischian and the early evolution of dinosaurs. *Science*, **326**, 1530–3.

Niklas, K. J. (1978). Coupled evolutionary rates and the fossil record. *Brittonia*, **30**, 373–94.

Norell, M. A. (1993). Tree-based approaches to understanding history: comments on ranks, rules, and the quality of the fossil record. *American Journal of Science*, **293**, 407–17.

Nylander, J. A. A., Ronquist, F., Huelsenbeck, J. P. and Nieves-Aldrey, J. L. (2004). Bayesian phylogenetic analysis of combined data. *Systematic Biology*, **53**, 47–67.

Olson, E. C. (1966). Community evolution and the origin of mammals. *Ecology*, **47**, 291–302.

Omland, K. E. (1997). Correlated rates of molecular and morphological evolution. *Evolution*, **51**, 1381–93.

Orr, H. A. (2000). Adaptation and the cost of complexity. *Evolution*, **54**, 13–20.

Osborn, H. F. (1922). Orthogenesis as observed from paleontological evidence beginning in the year 1889. *American Naturalist*, **56**, 134–43.

Osborn, H. F. (1933). Biological inductions from the evolution of the Proboscidea. *Proceedings of the National Academy of Sciences of the United States of America*, **19**, 159–63.

Rambaut, A. (2000). Estimating the rate of molecular evolution: incorporating non-contemporaneous sequences into maximum likelihood phylogenies. *Bioinformatics*, **16**, 395–9.

Rannala, B. (2002). Identifiability of parameters in MCMC Bayesian inference of phylogeny. *Systematic Biology*, **51**, 754–60.

Riedl, R. (1978). *Order in Living Organisms: A Systems Analysis of Evolution*. New York: Wiley.

Robinson, M., Gouy, M., Gautier, C. and Mouchiroud, D. (1998). Sensitivity of the relative-rate test to taxonomic sampling. *Molecular Biology and Evolution*, **15**, 1091–8.

Rohde, K. (1992). Latitudinal gradients in species diversity: the search for the primary cause. *Oikos*, **65**, 514–27.

Ruta, M., Wagner, P. J. and Coates, M. I. (2006). Evolutionary patterns in early tetrapods. I. Rapid initial diversification followed by decrease in rates of character change. *Proceedings of the Royal Society B–Biological Sciences*, **273**, 2107–11.

Rutschmann, F. (2006). Molecular dating of phylogenetic trees: a brief review of current methods that estimate divergence times. *Diversity Distributions*, **12**, 35–48.

Sánchez-Villagra, M., Ladevèze, S., Horovitz, I., *et al.* (2007). Exceptionally preserved North American Paleogene metatherians: adaptations and discovery of a major gap in the opossum fossil record. *Biology Letters*, **3**, 318–22.

Sanderson, M. J. (2002). Estimating absolute rates of molecular evolution and divergence times: a penalized likelihood approach. *Molecular Biology and Evolution*, **19**, 101–9.

Schopf, T. J. M., Raup, D. M., Gould, S. J. and Simberloff, D. S. (1975). Genomic versus morphologic rates of evolution: influence of morphologic complexity. *Paleobiology*, **1**, 63–70.

Shapiro, B., Ho, S. Y. W., Drummond, A. J., *et al.* (2011) A Bayesian phylogenetic method to estimate unknown sequence ages. *Molecular Biology and Evolution*, **28**, 879–87.

Sidor, C. A. and Hopson, J. A. (1998). Ghost lineages and 'mammalness': assessing the temporal pattern of character acquisition in the Synapsida. *Paleobiology*, **24**, 254–73.

Siepielski, A. M., DiBattista, J. D. and Carlson, S. M. (2009). It's about time: the temporal dynamics of phenotypic selection in the wild. *Ecology Letters*, **12**, 1261–76.

Simpson, G. G. (1944). *Tempo and Mode in Evolution*. New York: Columbia University Press.

Simpson, G. G. (1949). *Meaning of Evolution*. New Haven, CT: Yale University Press.

Simpson, G. G. (1953). *The Major Features of Evolution*. New York: Columbia University Press.

Smith, A. B. (1994). Rooting molecular trees: problems and strategies. *Biological Journal of the Linnean Society*, **51**, 279–92.

Smith, A. B. and Peterson, K. J. (2002). Dating the time of origin of major clades: molecular clocks and the fossil record. *Annual Review of Earth and Planetary Sciences*, **30**, 65–88.

Smith A. B., Lafay B. and Christen, R. (1992). Comparative variation of morphological and molecular evolution through geologic time: 28S ribosomal RNA versus morphology in echinoids. *Philosophical Transactions of the Royal Society of London B–Biological Sciences*, **338**, 365–82.

Smocovitis, V. B. (1992). Unifying biology: the evolutionary synthesis and evolutionary biology. *Journal of the History of Biology*, **25**, 1–65.

Sneath, P. H. and Sokal, R. R. (1972). *Numerical Taxonomy: The Principles and Practice of Numerical Classification*. San Francisco, CA: Freeman.

Stamatakis, A. (2006). RAxML-VI-HPC: maximum likelihood-based phylogenetic analyses with thousands of taxa and mixed models. *Bioinformatics* (Oxford, UK), **22**, 2688–90.

Stanley, S. M. (1985). Rates of evolution. *Paleobiology*, **11**, 13–26.

Subramanian, S., Hay, J. M., Mohandesan, E., Millar, C. D. and Lambert, D. M. (2009). Molecular and morphological evolution in the tuatara are decoupled. *Trends in Genetics*, **25**, 16–18.

Suchard, M. A, Weiss, R. E. and Sinsheimer, J. S. (2001). Bayesian selection of continuous-time Markov chain evolutionary models. *Molecular Biology and Evolution*, **18**, 1001–13.

Sullivan, J. and Joyce, P. (2005). Model selection in phylogenetics. *Annual Review of Ecology, Evolution, and Systematics*, **36**, 445–66.

Tajima, F. (1993). Simple methods for testing the molecular evolutionary clock hypothesis. *Genetics*, **135**, 599–607.

Takezaki, N., Rzhetsky, A and Nei, M. (1995). Phylogenetic test of the molecular clock and linearized trees. *Molecular Biology and Evolution*, **12**, 823–33.

Waddington, C. H. (1942). The canalization of development and the inheritance of acquired characters. *Nature*, **150**, 563.

Wagner, P. J. (1997). Patterns of morphologic diversification among the Rostroconchia. *Paleobiology*, **23**, 115–50.

Wagner, P. J. (2000). Likelihood tests of hypothesized durations: determining and accommodating biasing factors. *Paleobiology*, **26**, 431–49.

Wagner, P. J., Ruta, M. and Coates, M. I. (2006). Evolutionary patterns in early tetrapods. II. Differing constraints on available character space among clades. *Proceedings of the Royal Society B–Biological Sciences*, **273**, 2113–18.

Wall, P. D., Ivany, L. C. and Wilkinson, B. H. (2009). Revisiting Raup: exploring the influence of outcrop area on diversity in light of modern sample-standardization techniques. *Paleobiology*, **35**, 146–67.

Weir, J. T. and Schluter, D. (2008). Calibrating the avian molecular clock. *Molecular Ecology*, **17**, 2321–8.

Welch, J. J. and Bromham, L. (2005). Molecular dating when rates vary. *Trends in Ecology and Evolution*, **20**, 320–7.

West-Eberhard, M. J. (2003). *Developmental Plasticity and Evolution*. New York: Oxford University Press.

Western, D. (1979). Size, life history and ecology in mammals. *African Journal of Ecology*, **17**, 185–204.

Wray, G. A., Levinton, J. S. and Shapiro, L. H. (1996). Molecular evidence for deep Precambrian divergences among metazoan phyla. *Science*, **274**, 568–73.

Wright, S. (1945). Tempo and mode in evolution: a critical review. *Ecology*, **26**, 415–19.

Yang, Z. (2006). *Computational Molecular Evolution*. Oxford, UK: Oxford University Press.

Yoder, A. D. and Yang, Z. (2000). Estimation of primate speciation dates using local molecular clocks. *Molecular Biology and Evolution*, **17**, 1081–90.

Zhao, Z., Zhou, Y. and Ji, G. (2006). The periodic growth increments of biological shells and the orbital parameters of Earth-Moon system. *Environmental Geology*, **51**, 1271–7.

Zuckerkandl, E. and Pauling, L. (1965). Evolutionary divergence and convergence, in proteins. In *Evolving Genes and Proteins*, ed. V. Bryson and H. Vogel, pp. 97–166. New York: Academic Press.

5

Species selection in the molecular age

CARL SIMPSON AND JOHANNES MÜLLER

Introduction

Everything biological varies. Without variation, evolution would not be possible. This is a truism in macroevolution as much as it is within and between organisms. Species vary in their phenotypic and macroecological traits (Brown 1995) and variation also exists in the taxonomic rates of speciation and extinction over time (Alroy 2008), among taxa (Van Valen 1973; Sepkoski 1981; Raup and Boyajian 1988), and within taxa (Van Valen 1973, 1975; Liow *et al.* 2008; McPeek 2008; Simpson and Harnik 2009; Simpson 2010). Variation in diversification rates produces the major patterns of diversification we observe in the fossil record. Understanding the patterns and causes of variation in diversification rates has been the focus of palaeobiology for decades (Simpson 1944, 1953; Van Valen 1973; Raup 1978; Gould and Calloway 1980; Sepkoski 1981; Raup 1991a, 1991b).

Palaeobiologists, however, are not the only ones interested in understanding the patterns and causes of diversification. Diversification is also interesting to ecologists for at least two reasons. Major spatial patterns of diversity such as the latitudinal diversity gradient are likely to be underpinned by historical patterns of speciation and extinction (Jablonski and Hunt 2006; Krug *et al.* 2007, 2008; Kiessling *et al.* 2010). Also, many distributions of ecologically important traits, for example body size, may be, in part, a product of the historical patterns of differential diversification (Stanley 1975; Van Valen 1975). The second interest is the issue of diversity limitation. Data from the fossil record and molecular phylogenetics of extant organisms have been brought in to study this issue and evidence is accumulating that diversity is, in fact, constrained (Cracraft 1982; Nee *et al.* 1992b; Paradis 1997; Pybus and Harvey 2000; Nee 2001; Ricklefs 2007; Alroy 2008, 2009, 2010; Alroy *et al.* 2008; McPeek 2008; Phillimore and Price 2008; Rabosky and Lovette 2008a, 2009b; Phillimore and Price 2009; Quental and Marshall 2009, 2010; Rabosky 2009b).

From Clone to Bone: The Synergy of Morphological and Molecular Tools in Palaeobiology, ed. Robert J. Asher and Johannes Müller. Published by Cambridge University Press. © Cambridge University Press 2012.

Given the many ways that rates vary, we know little about the causes of diversification rates. While proponents of the concept of 'constrained diversification' argue for ecological control (Alroy 2008), variation in rates within and among taxa could also be caused by morphological traits. The fact is that the causes of diversification rates will be many, variable, and possibly nonlinear, and this multivariate complexity makes synthesis difficult. But there is a way to cut through the complexity. Variation in diversification rates creates the potential for evolution to occur at or above the species level due to species selection.

Species selection occurs when there is a causal relationship between a trait and fitness, which in species selection is diversification rate. In multilevel selection, and species selection in particular, there is more than one level of fitness (Arnold and Fristrup 1982; Jablonski 2008; Simpson 2010). Fitnesses that occur at hierarchical levels above the familiar organismal level, such as the colonial, are understood in general detail (Simpson 2011, 2012). But in species selection there are levels of demography – speciation and extinction – that constitute higher-level fitness. In species selection, the traits themselves need not even be expressed at the organismal level. So in contrast to a phenotypic trait relevant to the fitness of an individual, higher-level fitness traits can also be emergent, expressed only at higher hierarchical levels and not reducible to organismal traits. Much of the discussion of species selection has focused on these definitions (see Grantham 1995 for a review). However, as we will see below, species selection can simply be understood as the causal covariance between traits (at any level) and diversification rates (which is a higher-level fitness).

If selection is to occur, a simple mathematical relationship between fitness and causal traits must be obeyed. This theorem, known as the Price's theorem, is an exact description of evolutionary change over time, and is particularly useful if multiple levels of selection co-occur (Hamilton 1975; Arnold and Fristrup 1982; Rice 2004; Okasha 2006; Rice 2008; Simpson 2010). The change in the mean of a trait over time can be decomposed into the change attributable to the process of natural selection and the change attributable to the process of reproduction (which includes selection at lower levels and a multitude of other processes). The magnitude and direction of selection (S) is a function of the covariance between phenotype (ϕ) and fitness (W): $S = \frac{1}{\bar{W}} \text{cov}(W, \phi)$, or equivalently as a function of the linear regression of fitness on phenotype and the variance in the phenotype $S = \frac{1}{\bar{W}} \beta_{\phi,W} \text{var}(\phi)$ (Rice 2004). The fact that the important relationship between fitness and phenotype is simply their linear regression makes the empirical detection of species selection straightforward because any nonlinearities in the relationship between traits and diversification rates can safely be ignored (Simpson 2010). The covariance approach to species selection also helps clarify how multiple traits may evolve by species selection given the covariation among traits and their heritability (Rice 2004; P. G. Harnik *et al.* unpublished data).

Species selection has a long conceptual history (Gould 2002; Jablonski 2008) going back to Lyell (Lyell 1832; Van Valen 1975). Within palaeobiology, species selection was largely discussed and accepted as a hypothetical evolutionary process with the ability to be an effective force contingent on the pattern of punctuated equilibrium (Gould 2002). Gould and others used the pattern of punctuated equilibrium as a way to derive species selection through the process of elimination. If there is a trend in a clade, and species within the clade are static over time, then one possible way to reconcile these patterns is if diversification rates varied in such a way as to produce the trend. Gould (2002) bundled together species selection and punctuated equilibrium largely to deal with Williams' (1966) criticism of group selection. Williams argued that group selection could only be invoked if other selective forces are opposing or have zero strength. Although it is easier to think about species selection when punctuated equilibrium occurs, there is no need for restricting the operation of species selection to those scenarios (McShea 2004; Jablonski 2008; Simpson 2010). This is because speciation and extinction can co-occur with the birth and death of organisms. As a consequence, the species-level selective vector can also co-occur with the organismal-level selective vector and they can have any angle between them (Slatkin 1981; Arnold and Fristrup 1982; Rice 1995; Simpson 2010). Species selection is still largely an open empirical issue – we do not know its relative frequency, how strong it can be, or what sorts of traits may evolve due to its action.

Many discussions of species selection focus on the types of traits that cause differential diversification rates. Two basic types of traits are discussed: organismal-level and emergent traits. Often species selection (in the strictest sense) has been restricted to those times where differential diversification rates are caused only by traits that are emergent at the same level (Grantham 1995; Jablonski 2008). Emergent traits cannot be expressed in the phenotype of a single organism, as they are properties of populations of organisms. The majority of these traits are macroecological: geographic range, abundance, and population structure are good examples. Other traits, like body size or dispersal ability are somewhat unclear if they are emergent since they can influence diversification rates and organismal-level fitness (Van Valen 1971, 1975). Traits like body size illustrate the importance of not restricting species selection to being caused by only emergent traits a-priori – diversification rates may be caused by traits at any hierarchical level. The only requirement for the operation of species selection is that differential diversification rates are caused by some (biotic or abiotic) factor.

Many palaeontologists have suggested that large-scale morphological trends have been caused by species selection so that traits change in frequency over time due to their covariance with diversification rates (Arnold and Fristrup 1982; Vrba 1984; Vrba and Gould 1986; Gould and Eldredge 1988; Lloyd and Gould 1993; Grantham 1995; Gould and Lloyd 1998; Gould 2002; Okasha 2003; McShea 2004; Jablonski 2008). Adaptive, organismal-level explanations for these trends are not possible because the trends occur over such vast amounts of time, and organismal-level selection must surely fluctuate. However, some recent empirical work on the subject has found that the effect of species selection is not uniform either, at least during a large-scale trend in crinoid morphology (Simpson 2010). Not surprisingly, the pattern of selection varies in both magnitude and direction over the history of crinoids in a way very similar to the fluctuations observed in selection in recent organisms (Grant and Grant 2002). Fluctuating selection makes sense both biologically and in light of large-scale trends; trends are not common across taxa nor are they generally persistent when they do occur. The strategy of identifying species selection only in trends will tend to miss many examples of real species selection.

Time-calibrated molecular phylogenies are a potentially rich source of data on the species selection that has not been explored to its full potential (Rabosky and McCune 2009). But species selection has always been of interest to those working on diversification in molecular trees (Nee et al. 1992b). The major strategy most of this work uses is to develop models of the evolutionary process that can then be fit to the patterns of species diversification, usually in the form of various types of branching models (Nee et al. 1992a, 1994; Nee 2001, 2004, 2006; Maddison et al. 2007; Ricklefs 2007; McPeek 2008; Phillimore and Price 2008; Alfaro et al. 2009, 2010; FitzJohn et al. 2009; Quental and Marshall 2009, 2010; Rabosky 2009a, 2009b; FitzJohn 2010). An example is the Yule model, a time-homogeneous pure-birth branching process. In this model, speciation rate is constant and there is no extinction. Additional complexity is incorporated in the birth–death processes, which factors extinction into the Yule model. Variations in diversification rates are identified either temporally (Simpson et al. 2011) or by looking for variations in species richness or character state that indicate rate shifts (Maddison et al. 2007; Alfaro et al. 2009, 2010; FitzJohn et al. 2009, 2010).

Our goal in this chapter is to reboot the discussion of inferring diversification from molecular phylogenies with the goal of establishing not only the potential for identifying temporal variation in rates in a molecular phylogeny, but also its causes. The full potential of studying species selection is more than conceptual; the simple pattern of selection provides an easy entry into the complex interaction between traits and rates. We focus our attention away from how to use various branching models towards the statistical description of rate variation. Additionally, we use Price's theorem to interpret variations in diversification rates biologically. We will illustrate a macroevolutionary-minded way to construct temporal patterns of diversification. An additional issue, about the pattern of covariance between taxonomic rates and phenotypic or macro-ecological traits, opens the door to answering an important macroevolutionary question – what factors control the rates of diversification?

Measuring speciation and extinction rates

It is standard practice in palaeobiology to estimate time series of taxo-nomic rates from only those species or genera that are observed to cross a temporal boundary between stages or other temporal bins. Estimating boundary crosser rates (Foote 2000; Alroy 2008; Alroy et al. 2008) involves tabulating the numbers of four fundamental types of taxa: (1) the number of taxa that both enter in and cross out of an interval, N_{bt}; (2) those that enter in and go extinct in the interval, N_{bL}; (3) those that originate in the interval and cross out of it, N_{Ft}; and (4) the taxa that are restricted to the interval, N_{FL}. Of these four types of taxa, only three are used to estimate rates. Maximum likelihood origination rates ($\hat{p}$) are a function of the number of taxa that cross through an interval and the number originating there: $\hat{p} = -\ln(N_{bt}/N_t)$, whereas extinction rates ($\hat{q}$) are a function of the number of taxa crossing though an interval and entering into it: $\hat{q} = -\ln(N_{bt}/N_b)$ (Foote 2003; Kiessling and Aberhan 2007). These rates are derived from an exponential model, where the probability of a lineage leaving an interval that was already extant at the start is: $N_{bt} = N_b e^{-q_i} = N_t e^{-p_i}$ (Foote 2003).

Although these 'boundary crosser' rates are thought to be equivalent to those that are calculated from molecular data (Nee et al. 1992b; Alroy 2009), there is a tendency for boundary crosser rates to be much larger than rates estimated only from molecular data alone. This is due to the fact that counts of one (N_{bL}) of the three fundamental types of taxa used to measure rates cannot be made in molecular phylogenies. This makes direct estimation of extinction rates impos-sible and also biases estimates of speciation rates. If taxa that contribute to N_{bL} in one interval are long lived, they will not be able to contribute to N_{bt} in prior intervals in which they range though either.

Simpson et al. (2011) proposed a new method for estimating diversification rates from molecular phylogenies and results in a time series of rates. Given a phylogeny, diversification rates within an interval of time can be estimated from the number of branching events (k_s) and the sum of branch lengths in the interval – including branches that range through without speciating. The number of lineages ranging through, but not branching, is equal to the total number of lineages in the interval, n, minus the number of branching events, k_s, or $n - k_s$. Note that branching event i occurs at time t_i. The time span represented by each stage is denoted Δt_s and the youngest age of the stage is denoted t_s. The maximum likelihood estimator of diversification rate is equal to

$$\hat{\delta} = k_s \bigg/ \left[(n - k_s)\Delta t_s + \sum_{t=1}^{k_s}(t_i - t_s) \right].$$

Repeating this equation over time intervals, such as geological stages, provides a time series of rates directly comparable to rates estimated from the fossil record. Using this approach, Simpson *et al.* (2011) found a significant positive correlation between changes in diversification rates estimated in fossils and molecular phylogenies in reef corals (see also Larsson *et al.*, this volume, for a comparison of morphological rates of evolution against those derived from molecular studies).

Despite this high correlation in pattern, the magnitudes of the molecular diversification rate estimates are biased by the lack of species that contribute to N_{bt}. This bias means that speciation and extinction rates cannot be meaningfully estimated directly from molecularly derived phylogenies of only extant species (Simpson *et al.* 2011). But because the trajectory of those rates is robust, any differential rates of diversification can also be inferred. This ability to measure differential rates allows species selection to be studied using molecular phylogenies.

Detecting species selection

All species selection is a covariance between diversification rates and various traits that a species or members of a species possess (Simpson 2010). There are two major possibilities for how this covariance can occur. The first is that variation in rates can be largely random and uncaused. This would result in drift at the species level if some minor phenotypic change accumulated (Gould 2002). Species-level drift (a higher level version of drift in microevolution) is particularly important if the number of species is small. This is because chance events, such as the survival of one species with a particular set of traits, will over time have a large effect on the frequency distribution of those traits in the species pool. On this basis, Gould (2002) argued that species drift may be common in the history of life because the numbers of species involved are low. But species drift may actually be relatively rare if trends are uncommon and heritability of macroecological traits is high (Jablonski 1987; Hunt *et al.* 2005), because some selective vector would be needed to oppose any morphological change due to correlations with macroecological traits (P. G. Harnik *et al.* unpublished data).

The covariance between diversification rate and traits may also be causal. Many macroecological traits correlate with extinction and speciation rates (Jablonski 1987; Payne and Finnegan 2007; Simpson and Harnik 2009; Simpson 2010) and many workers consider these traits to be causally linked even if the actual causal pathways remain unknown. Macroecological traits often covary with each other (Brown 1995) and with extinction (Purvis *et al.* 2000) in complex ways, so trends in macroecological traits are not expected to produce trends consistently. This complexity of causes provides us with an opportunity to use species selection and the predictive ability of Price's theorem as entry point to discovering and untangling the causes of evolutionary rates.

It has often been argued that tree asymmetry (where some basal branches are species poor and others are species rich) is diagnostic of species selection (Lieberman *et al.* 1993; Rabosky and McCune 2009). But this is not true, as it has been shown that changes in character states can influence diversification even if these traits are sprinkled thoughout a phylogeny (Maddison *et al.* 2007; FitzJohn *et al.* 2009; FitzJohn 2010; Simpson 2010). Species selection is a process that is independent of tree topology, but many possible traits may still have characteristic qualitative patterns detectable from the distribution of traits on trees alone. From this point of view, there are three possible ways that species selection can occur (Figure 5.1). (1) Species selection can occur among higher taxa if divergences among their traits cause differences in their diversification rates. (2) Species selection can occur within higher taxa when a monophyletic subclade differs in rates from other subclades. (3) But species selection can also occur among unrelated members of a higher taxon that share the same trait values. In analogy with organismal-level selection, what matters for fitness is the trait value of individuals, not their pedigree. If large bears have more offspring, even large bears from small parents will have more offspring. Now at the species level, if large geographic range prevents extinction, it does not matter if a species is descended from a small-ranging parent.

The only thing that distinguishes between these three forms of species selection is how the causal traits vary across the phylogeny. In the first type, traits are invariant within a taxon but vary among them. In the second type of species selection, a monophyletic clade within a higher taxon possesses a trait which then influences diversification. An extreme example of this would be when a key innovation causes an adaptive radiation, where high rates are associated with the evolution of a single trait. Macroecological traits, however, are much more variable and so may commonly underlay the third type of species selection. Not only do all taxa share traits (for example, every species has a geographic range), but often species with a particular macroecological trait,

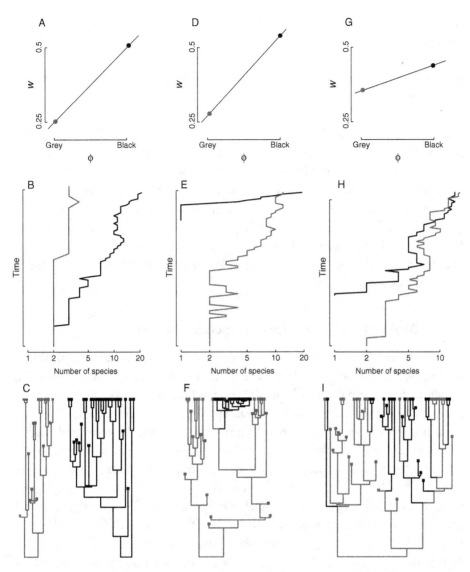

Figure 5.1 Three types of species selection distinguished by the distribution of traits across a phylogeny. In all panels, the black group has higher diversification rates than the grey group. In the first column (panels A, B and C), traits are invariant within clades but vary among them. The middle column (D, E and F) shows species selection where a single clade has a higher rate of diversification than the rest of the clade. And the right column (G, H and I) shows species selection where the high diversification trait is distributed across the tree. The diversity over time for the high and low rate groups are plotted for each type of species selection (B, E and H). The selection gradient for each type of species selection is shown in the top panels (A, D and G). Note that from the point of view of selection, each type of species selection is equivalent.

such as large geographic range, are not directly related. If macroecological traits were used for taxonomic purposes, then membership would be polyphyletic. As a consequence, groups with members of mixed ancestry are not a problem for species selection because the only thing that matters is the covariance between traits and rates.

Unfortunately there is no ideal solution to the problem of how to detect rate variation within a clade. A recent inverse modelling approach has been developed to identify the effects of binary or quantitative characters on taxonomic rates (Maddison *et al.* 2007; FitzJohn *et al.* 2009; FitzJohn 2010), but this approach, as of yet, does not allow for temporal variation in rates. Alternatively, a model selection approach has been used successfully recently that identifies time intervals where differential rates occur (Simpson and Harnik 2009; Simpson 2010; Simpson and Kiessling 2010; Kiessling and Simpson 2011). This second, model selection approach is easily integrated into Price's theorem (Simpson 2010) and is what we will use in the example below.

Molecular phylogenies and species selection

A major hurdle to detecting species selection has been the difficulty in untangling alternative mechanisms that could influence species-level evolution, for which stasis (or punctuated equilibrium) has been seen to control (Lieberman *et al.* 1993; Gould 2002). Price's theorem approach to species selection helps to unravel it from other processes that lead to a phylogenetically correlated change (like organismal-level selection) by partitioning the change in mean phenotype attributable to each process (Arnold and Fristrup 1982; Rice 2004; Simpson 2010). When Price's theorem has been used in the fossil record, no phylogeny was available, so changes not attributable to species selection had to be inferred from the frequency distribution of phenotypes (Simpson 2010). Time-calibrated molecular phylogenies provide the best opportunity to directly measure the contribution of species selection and other processes because they contain a record of both rate variation and phylogenetic relationships.

Let us now focus on one specific example: the evolution of pharyngeal jaws in labrid fishes (Alfaro *et al.* 2009). Modifications to the pharyngeal jaw apparatus are thought to be a key innovation (see Smith and Johanson, this volume) that drives high diversification within the labrids (Liem and Greenwood 1981; Kaufman and Liem 1982; Stiassny and Jensen 1987). Alfaro *et al.* (2009) found that a subsequent diversification of parrotfishes is associated more strongly with diversification driven by sexual selection in the *Scarus–Chlorurus* clade than with the evolution of the parrotfish pharyngeal mill. We can demonstrate the utility of Price's theorem approach to species selection by reanalysing the labrid data.

Although we largely replicate the results presented in Alfaro *et al.* (2009), we can take the analysis one step further by identifying if it is sexual selection or jaw structure that influences diversification within the parrotfish lineage to a greater extent.

The simple Price's theorem formulation of selection described above is for only a single trait, but it can easily be extended to a multivariate situation. Recall that the selection differential (S) is defined, $S = \frac{1}{\bar{W}}\text{cov}(W,\phi) = \frac{1}{\bar{W}}\beta_{\phi,W}\text{var}(\phi)$. Incorporating multiple traits involves measuring the linear regression of fitness on each trait independently with its partial regression, $\beta_{\phi i|,W}$, where $\phi_{i|}$ indicates variation in trait i that is independent of other traits. This can be done using multiple regression or path analysis depending on what underlying model of covariation among traits we are interested in. This results in one partial regression on fitness per trait (in the vector, $\overrightarrow{\beta_{\phi|,W}}$) and therefore a vector $(\overrightarrow{S})$ represents selection in all traits. If the variance–covariance matrix of traits is denoted P, then the multivariate selection vector is equal to: $\overrightarrow{S} = \frac{1}{\bar{W}}P\cdot\beta_{\phi|,W}$ (Rice, 2004). In labrid fishes, the parrotfish pharyngeal jaw mill does not cause sexual dichromatism (which is a colour difference between the sexes) because sexual dichromatism is not limited to the parrotfishes. If there is a tight association, then a causally explicit path model could be used to specify the partial regressions in $\overrightarrow{\beta_{\phi|,W}}$. Instead, we use a simple multiple regression approach to capture the covariation between our traits of interest.

A temporal pattern of diversification rates is estimated using the method of Simpson *et al.* (2011), which we outlined above. In Figure 5.2 we present a time series of selection gradients for the parrotfish pharyngeal mill and sexual dichromatism. On average, selection is much stronger for sexual dichromatism ($\bar{S}_{SD} = 0.279$) than for jaw structure ($\bar{S}_{PPM} = 0.016$) when controlled for their covariance. From the magnitude of the selection coefficients, we can infer that sexual dichromatism and sexual selection played a considerably larger role in parrotfish diversification than did their innovative jaw mechanism. Relative to other labrid fishes, selection is actually against the parrotfish pharyngeal mill for a period of approximately 10 million years in the Miocene (Figure 5.2). This decline in relative diversification rates occurs near the time when modern coral–algal reefs start to decline in volume from their early Miocene peak (Kiessling 2009).

Controlling for the covariance between traits makes a difference in the inferred patterns of diversification. In Figure 5.2 there is a peak in the diversification rate of species with the parrotfish pharyngeal mill in the early Miocene that is not seen in the selection coefficients. When covariance between jaw structure and dichromatism is taken into account, the diversification rate of the parrotfish is relatively lower than the rates for the clade with sexual

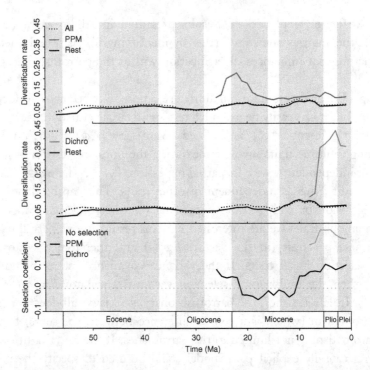

Figure 5.2 Temporal patterns of diversification in labrid fishes. Branching lengths are partitioned into those with and without parrotfish pharyngeal mills (top panel) and those with and without sexual dichromatism (middle panel). Model averaging is used to estimate diversification trajectories from a pool of four models (see text for discussion). A model selection approach is then used to test if a two-rate model (where rates differ between lineages with each character state) is supported over a single-rate model independent of character state. In both plots, the dotted line is the diversification rate for all labrids. Estimates of selection coefficients (bottom panel) for jaw structure and sexual dichromatism in labrid fishes over time. Selection coefficients are estimated from the diversification rates in the top panels and account for any covariance between jaw structure and sexual dichromatism.

dichromatism and so the Miocene peak is tempered. Converting to relative rates and controlling for the covariance among traits allows us to compare the selection acting on individual trait states. The parrotfish jaw is temporarily selected against, which means that relative to the rest of the labrids, parrotfish diversify at a comparatively lower rate.

Sexual dichromatism may not directly increase diversification rates. If not, some other character that is precisely codistributed with it must. This may sound unsatisfying at first, but the fact that we know that the trait causing high rates in parrotfish is codistributed with sexual dichromatism, means that many potential characteristics can be rejected outright once they are mapped onto the phylogeny if they do not vary with sexual dichromatism. In the worst case, the cause of selection is just a correlation away.

How common and strong is species selection?

Now that we know that species selection is the variation in diversification rates within a clade, we can ask an important question: What is the relative frequency and average strength of species selection in nature (Jablonski 2008)? Using molecular phylogenies, we can provide an approximate answer by looking for an indicator of species selection. In the simplest case, if species selection operates within a clade, then the frequency distribution of branch lengths of that clade will be the result of a composite of many diversification rates. If rates vary among character states, then there will be one rate per state. One problem with measuring the relative frequency of species selection is that the phenotypic groupings are unknown a-priori. Binary, multistate or quantitative characters may cause species selection so the number of groups in a clade (and thus the number of different rates within it) is unknown prior to analysis. What we can measure a-priori – and without knowing the causal traits – is the variance in rates within a clade. A simple way to measure the variance among all the potential groupings is to estimate the inverse of the shape parameter ($1/\alpha$) of the gamma distribution (Holman 1983; Venditti $et\ al.$ 2010; C. Simpson, unpublished data) fit to the distribution of branch lengths in a clade. The benefit of this approach is that the variance in the rates can be estimated without defining the number or identity of groups. If there is no variation in rates within a clade, the gamma distribution becomes equivalent to an exponential, so there is no tendency to find rate variation if there is none present. Multiple rates are inferred to be present if $1/\alpha > 0$ and their variance is given by $1/\alpha$. This is an indirect test so may be confounded by other processes such as the temporal variation of rates of the group as a whole without species selection. A non-zero variance is only necessary for species selection, not sufficient. If we find zero or small variances, there is little chance for species selection to operate. Keep in mind that with this approach we can reject the operation of species selection in cases where variances in rates are small, but not prove its operation in those cases where the variance is high. A non-zero variance indicates that the raw material for species selection, variation in rates, is present.

In order to empirically investigate this issue we used the set of 245 time-calibrated trees of chordates, arthropods, molluscs and plants compiled by

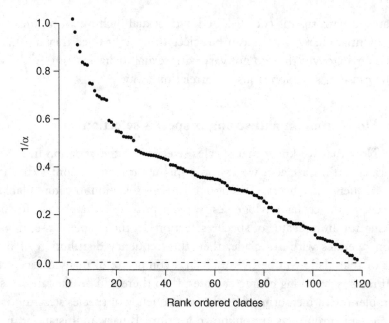

Figure 5.3 The inferred necessary conditions for species selection in time-calibrated molecular phylogenies of extant organisms. If a clade diversifies with more than one rate, the shape parameter ($1/\alpha$) of the gamma distribution will estimate the variance of the constituent rates, which measures the strength of species selection. The greater the variance in rates, the stronger the selective vector will be. Plotted here in rank order is the variance in diversification rate of chordates, arthropods, molluscs and plants derived from 120 time-calibrated molecular phylogenies (data from McPeek 2008 and McPeek and Brown 2007). The raw material for species selection is inferred to operate within clades when $1/\alpha$ is larger than zero. If there is only a single rate, and consequently no species selection, $1/\alpha$ will be equal to zero and equivalent to an exponential distribution.

McPeek (McPeek 2008; McPeek and Brown 2007) as an example. By applying non-parametric rate smoothing (Sanderson 1997) we convert the trees into ultrametric chronograms and then scale the branch lengths to units of time following the calibration protocol of McPeek and Brown (2007).

Out of the 245 trees in the data set, we were able to estimate a shape parameter ($1/\alpha$) for 118 clades. We found that 100% of the clades have a non-zero variance, suggesting that the raw material for species selection is ubiquitous in these clades (Figure 5.3). Our result is not surprising given that rate variation among clades has been long observed, with each higher taxon having its own rate (Van Valen 1973; Sepkoski 1981; Raup and Boyajian 1988; Alroy 2004, 2008). However, until we have an estimate of the relative importance of macroecological or morphological traits, we will be unable to know what the most common cause of species selection is. Even so, we can conclude that species selection in some form is ubiquitous in nature, both within and among clades.

Conclusions

Species selection is not just of theoretical interest to palaeobiologists. It can be used to cut through complex interactions between traits and rates to identify each trait's relative contribution to diversification. As we were able to show, Price's theorem is a powerful way to organize measurements and to help ask the right questions about the causes of evolutionary rates. Also, it is the temporal pattern of selectivity that is the key to identifying the traits that cause species selection. In our example, the labrid fishes, it is sexual selection that inflates diversification rates consistently and strongly. In comparison, the parrotfish pharyngeal jaw apparatus can be rejected as a direct cause of high diversification because its association with diversification rate is variable in magnitude and direction. Although it is clearly an important trait, diversification is not directly caused by its presence.

When measured temporally, the magnitude and direction of species selection on organismal-level traits varies widely (Figure 5.2 and Simpson 2010). Conversely, many emergent traits, such as sexual selection (as it is the resultant of interactions among organisms) and geographic range, have been observed to be associated with species selection that is consistent in direction but variable in magnitude (Figure 5.2 and Payne and Finnegan 2007). It is too early to tell if this is a general result, but the inference we can make from these patterns is that emergent traits may be more important for diversification than organismal-level traits.

The question about the relative frequency and strength of species selection can, however, be made without knowledge of the number or kind of traits involved. The variance in rates, estimated by the shape parameter of a gamma distribution fit to the distribution of branch lengths, is positive in a broad sample of molecular phylogenies. This observation, while not proof positive that species selection is operating, does show that the raw material for species selection is present.

Species selection is no mere palaeontological oddity. It in fact seems to be both common and strong, although we do not know much about it yet. The fossil and molecular phylogenetic records can be used to measure the strength and direction of species selection. With that knowledge we can understand the types of traits and particular circumstances that promote or diminish diversification.

Summary

Diversification rates are not uniform across time, species or clades. When they vary systematically with one or more traits, it is known as species selection and may influence the change in frequencies of taxa or traits over time. Biologists working with either the fossil record or molecular systematics (using comparative methods) are interested in how diversification and traits covary to produce the biological patterns we observe today. Traits can cause diversification which can in turn influence the frequency of traits. Multiple traits can interact or influence diversification in complex ways. Price's theorem is a simple statement of how the change in the mean trait values over time is caused by selection along with any other evolutionary process and can be used to untangle selective differences in many traits and over time. The independent contribution of multiple traits to rate variation over time can be measured in both the fossil and molecular phylogenetic record. We demonstrate this approach using a time-calibrated phylogeny of labrid fish and show that sexual selection is a consistent cause of high diversification but morphological innovation is not. We also provide evidence that species selection is common in nature.

Acknowledgements

We wish to thank Mike Alfaro and Francesco Santini (UCLA) for providing us with the data set on labrid fishes. This study was funded by the Deutsche Forschungsgemeinschaft (grant number PAK 602).

REFERENCES

Alfaro, M. E., Brock, C. D., Banbury, B. L. and Wainwright, P. C. (2009). Does evolutionary innovation in pharyngeal jaws lead to rapid lineage diversification in labrid fishes? *BMC Evolutionary Biology*, **9**, 255.

Alfaro, M. E., Santini, F., Brock, C. D., *et al.* (2010). Eleven exceptional radiations plus high turnover explain species diversity in jawed vertebrates. *PNAS*, **106**, 13 410–14.

Alroy, J. (2004). Are Sepkoski's evolutionary faunas dynamically coherent? *Evolutionary Ecology Research*, **6**, 1–32.

Alroy, J. (2008). Dynamics of origination and extinction in the marine fossil record. *Proceedings of the National Academy of Sciences of the United States of America*, **105**, 11 536.

Alroy, J. (2009). Speciation and extinction in the fossil record of North American mammals. In *Speciation and Patterns of Diversity*, ed. J. B. R. Butlin and D. Schluter, pp. 301–23. Cambridge, UK: Cambridge University Press.

Alroy, J. (2010). Geographic, environmental, and intrinsic biotic controls on phanerozoic marine diversification. *Palaeontology*, **53**, 1211–35.

Alroy, J., Aberhan, M., Bottjer, D. J., *et al.* (2008). Phanerozoic trends in the global diversity of marine invertebrates. *Science*, **321**, 97.

Arnold, A. and Fristrup, K. (1982). The theory of evolution by natural selection: a hierarchical expansion. *Paleobiology*, **8**, 113–29.

Brown, J. H. (1995). *Macroecology.* Chicago, IL: University of Chicago Press.

Cracraft, J. (1982). A non-equilibrium theory for the rate-control of speciation and extinction and the origin of macroevolutionary patterns. *Systematic Zoology*, **31**, 348–65.

FitzJohn, R. G. (2010). Quantitative traits and diversification. *Systematic Biology*, **59**, 619–33.

FitzJohn, R. G., Maddison, W. P. and Otto, S. P. (2009). Estimating trait-dependent speciation and extinction rates from incompletely resolved phylogenies. *Systematic Biology*, **58**, 595–611.

Foote, M. (2000). Origination and extinction components of taxonomic diversity: general problems. *Paleobiology*, **26**, 74–102.

Foote, M. (2003). Origination and extinction through the Phanerozoic: a new approach. *Journal of Geology*, **111**, 125–48.

Gould, S. J. (2002). *The Structure of Evolutionary Theory.* Cambridge, MA: Beknap Press of Harvard University Press.

Gould, S. J. and Calloway, C. B. (1980). Clams and brachiopods: ships that pass in the night. *Paleobiology*, **6**, 383–96.

Gould, S. J. and Eldredge, N. (1988). Species selection: its range and power. *Nature*, **334**, 19.

Gould, S. J. and Lloyd, E. (1998). Individuality and adaptation across levels of selection: how shall we name and generalize the unit of Darwinism? *PNAS*, **96**, 11 904–9.

Grant, P. R. and Grant, B. R. (2002). Unpredictable evolution in a 30-year study of Darwin's finches. *Science*, **296**, 707–11.

Grantham, T. A. (1995). Hierarchical approaches to macroevolution: recent work on species selection and the 'effect hypothesis'. *Annual Review of Ecology, Evolution, and Systematics*, **26**, 301–21.

Hamilton, W. D. (1975). Innate social aptitudes of man: an approach from evolutionary genetics. In *Biosocial Anthropology*, ed. R. Fox, pp. 133–155. New York: Wiley.

Holman, E. (1983). Time scales and taxonomic survivorship. *Paleobiology*, **9**, 20–5.

Hunt, G., Roy, K. and Jablonski, D. (2005). Species-level heritability reaffirmed: a comment on 'On the heritability of geographic range sizes'. *American Naturalist*, **166**, 129–35.

Jablonski, D. (1987). Heritability at the species level: analysis of geographic ranges of cretaceous mollusks. *Science*, **238**, 360–3.

Jablonski, D. (2008). Species selection: theory and data. *Annual Review of Ecology, Evolution, and Systematics*, **39**, 501–24.

Jablonski, D. and Hunt, G. (2006). Larval ecology, geographic range, and species survivorship in Cretaceous mollusks: organismic versus species-level explanations. *American Naturalist*, **168**, 556–64.

Kaufman, L. S. and Liem, K. F. (1982). Fishes of the suborder Labroidei (Pisces: Perciformes): phylogeny, ecology and evolutionary significance. *Breviora*, **472**, 1–19.

Kiessling, W. (2009). Geologic and biologic controls on the evolution of reefs. *Annual Review of Ecology, Evolution, and Systematics*, **40**, 173–92.

Kiessling, W. and Aberhan, M. (2007). Geographical distribution and extinction risk: lessons from Triassic–Jurassic marine benthic organisms. *Journal of Biogeography*, **34**, 1473–89.

Kiessling, W. and Simpson, C. (2011). On the potential for ocean acidification to be a general cause of ancient reef crises. *Global Change Biology*, **17**, 56–67.

Kiessling, W., Simpson, C. and Foote, M. (2010). Reefs as cradles of evolution and sources of biodiversity in the Phanerozoic. *Science*, **327**, 196–8.

Krug, A. Z., Jablonski, D. and Valentine, J. W. (2007). Contrarian clade confirms the ubiquity of spatial origination patterns in the production of latitudinal diversity gradients. *Proceedings of the National Academy of Sciences of the United States of America*, **104**, 18 129–34.

Krug, A. Z., Jablonski, D. and Valentine, J. W. (2008). Species-genus ratios reflect a global history of diversification and range expansion in marine bivalves. *Proceedings of the Royal Society of London B*, **275**, 1117–23.

Lieberman, B., Allmon, W. and Eldredge, N. (1993). Levels of selection and macroevolutionary patterns in the turritellid gastropods. *Paleobiology*, **19**, 205–15.

Liem, K. F. and Greenwood, P. H. (1981). A functional approach to the phylogeny of pharyngognath teleosts. *American Zoology*, **21**, 83–101.

Liow, L. H., Fortelius, M., Bingham, E., *et al.* (2008). Higher origination and extinction rates in larger mammals. *PNAS*, **105**, 6097.

Lloyd, E. and Gould, S. J. (1993). Species selection on variability. *PNAS*, **90**, 595–9.

Lyell, C. (1832). *Principles of Geology*, Vol. 2. London: John Murray.

Maddison, W. P., Midford, P. E. and Otto, S. P. (2007). Estimating a binary character's effect on speciation and extinction. *Systematic Biology*, **56**, 701–10.

McPeek, M. A. (2008). The ecological dynamics of clade diversification and community assembly. *American Naturalist*, **172**, E270–84.

McPeek, M. A. and Brown, J. M. (2007). Clade age and not diversification rate explains species richness among animal taxa. *American Naturalist*, **169**, 97–106.

McShea, D. W. (2004). A revised Darwinism. *Biology and Philosophy*, **19**, 45–53.

Nee, S. (2001). Inferring speciation rates from phylogenies. *Evolution*, 661–8.

Nee, S. (2004). Extinct meets extant: simple models in paleontology and molecular phylogenetics. *Paleobiology*, **30**, 172–8.

Nee, S. (2006). Birth-death models in macroevolution. *Annual Review of Ecology, Evolution, and Systematics*, **37**, 1–17.

Nee, S., Holmes, E. C., May, R. M. and Harvey, P. H. (1992a). Estimating extinction from molecular phylogenies. In *Extinction Rates*, ed. J. L. Lawton and R. M. May. Oxford, UK: Oxford University Press.

Nee, S., Mooers, A. O. and Harvey, P. H. (1992b). Tempo and mode of evolution revealed from molecular phylogenies. *Proceedings of the National Academy of Sciences of the United States of America*, **89**, 8322–6.

Nee, S., Holmes, E., May, R. and Harvey, P. (1994). Extinction rates can be estimated from molecular phylogenies. *Philosophical Transactions B*, **344**, 77–82.

Okasha, S. (2003). *Multi-Level Selection, Price's Equation and Causality*. Causality: Metaphysics and Methods, Technical Report 12/03, 33 pp. London: London School of Economics, Centre for Philosophy of Natural and Social Science.

Okasha, S. (2006). *Evolution and the Levels of Selection*. Oxford, UK: Oxford University Press.

Paradis, E. (1997). Assessing temporal variations in diversification rates from phylogenies: estimation and hypothesis testing. *Proceedings of the Royal Society of London B*, **264**, 1141.

Payne, J. L. and Finnegan, S. (2007). The effect of geographic range on extinction risk during background and mass extinction. *Proceedings of the National Academy of Sciences of the United States of America*, **104**, 10 506–11.

Phillimore, A. B. and Price, T. D. (2008). Density-dependent cladogenesis in birds. *PLoS Biology*, **6**, e71.

Phillimore, A. B. and Price, T. (2009). Ecological influences on the temporal pattern of speciation. In *Speciation and Patterns of Diversity*, ed. R. Butlin, J. Bridle and D. Schluter, eds. Speciation and Patterns of Diversity. Cambridge, UK: Cambridge University Press.

Purvis, A., Jones, K. E. and Mace, G. M. (2000). Extinction. *BioEssays*, **22**, 1123–33.

Pybus, O. G. and Harvey, P. H. (2000). Testing macro-evolutionary models using incomplete molecular phylogenies. *Philosophical Transactions of the Royal Society B*, **267**, 2267–72.

Quental, T. B. and Marshall, C. R. (2009). Extinction during evolutionary radiations: reconciling the fossil record with molecular phylogenies. *Evolution*, **63**, 3158–67.

Quental, T. B. and Marshall, C. R. (2010). Diversity dynamics: molecular phylogenies need the fossil record. *Trends in Ecology and Evolution*, **25**, 434–41.

Rabosky, D. L. (2009a). Ecological limits and diversification rate: alternative paradigms to explain the variation in species richness among clades and regions. *Ecology Letters*, **12**, 735–43.

Rabosky, D. L. (2009b). Ecological limits on clade diversification in higher taxa. *American Naturalist*, **173**, 662–74.

Rabosky, D. L. and Lovette, I. (2008a). Density-dependent diversification in North American wood warblers. *Proceedings of the Royal Society of London B*, **275**, 2363–71.

Rabosky, D. L. and Lovette, I. (2008b). Explosive evolutionary radiations: decreasing speciation or increasing extinction through time? *Evolution*, **62**, 1866–75.

Rabosky, D. L. and McCune, A. (2009). Reinventing species selection with molecular phylogenies. *Trends in Ecology and Evolution*, **25**, 68–74.

Raup, D. (1978). Cohort analysis of generic survivorship. *Paleobiology*, **4**, 1–15.

Raup, D. (1991a). A kill curve for Phanerozoic marine species. *Paleobiology*, **17**, 37–48.

Raup, D. M. (1991b). *Bad Genes of Bad Luck?* New York: W.W. Norton.

Raup, D. M. and Boyajian, G. E. (1988). Patterns of generic extinction in the fossil record. *Paleobiology*, **14**, 109–25.

Rice, S. H. (1995). A genetical theory of species selection. *Journal of Theoretical Biology*, **177**, 237–45.

Rice, S. H. (2004). *Evolutionary Theory: Mathematical and Conceptual Foundations*. Sunderland, MA: Sinauer Associates.

Rice, S. H. (2008). A stochastic version of the Price equation reveals the interplay of deterministic and stochastic processes in evolution. *BMC Evolutionary Biology*, **8**, 262.

Ricklefs, R. (2007). Estimating diversification rates from phylogenetic information. *Trends in Ecology and Evolution*, **22**, 601–10.

Sanderson, M. J. (1997). A nonparametric approach to estimating divergence times in the absence of rate constancy. *Molecular Biology and Evolution*, **14**, 1218.

Sepkoski, J. J., Jr. (1981). A factor analytic description of the Phanerozoic marine fossil record. *Paleobiology*, **7**, 36–53.

Simpson, C. (2010). Species selection and driven mechanisms jointly generate a large-scale morphological trend in monobathrid crinoids. *Paleobiology*, **36**, 481–96.

Simpson, C. (2011). How many levels are there? How insights from evolutionary transitions in individuality help measure the hierarchical complexity of life. In *The Major Transitions in Evolution Revisited*, ed. B. Calcott and K. Sterelney, pp. 199–226. Cambridge, MA: MIT Press.

Simpson, C. (2012). The evolutionary history of division of labour. *Proceedings of the Royal Society B: Biological Sciences*. doi: 10.1098/rspb.2011.0766.

Simpson, C. and Harnik, P. G. (2009). Assessing the role of abundance in marine bivalve extinction over the post-Paleozoic. *Paleobiology*, **35**, 631–47.

Simpson, C. and Kiessling, W. (2010). The role of extinction in large-scale diversity-stability relationships. *Proceedings of the Royal Society B–Biological Sciences*, **277**, 1451–6.

Simpson, C., Kiessling, W., Mewis, H., Baron-Szabo, R. C. and Müller, J. (2011). Evolutionary diversification of reef corals: a comparison of the molecular and fossil records. *Evolution*. doi: 10.1111/j.1558–5646.2011.01365.x.

Simpson, G. G. (1944). *Tempo and Mode in Evolution*. New York: Columbia University Press.

Simpson, G. G. (1953). *The Major Features of Evolution*. New York: Columbia University Press.

Slatkin, M. (1981). A diffusion model of species selection. *Paleobiology*, **7**, 421–5.

Stanley, S. M. (1975). A theory of evolution above the species level. *Proceedings of the National Academy of Sciences of the United States of America*, **72**, 646–50.

Stiassny, M. L. J. and Jensen, J. S. (1987). Labroid intrarelationships revisited: morphological complexity, key innovations, and the study of comparative diversity. *Bulletin of the Museum of Comparative Zoology, Harvard University*, **151**, 269–319.

Van Valen, L. (1973). A new evolutionary law. *Evolutionary Theory*, **1**, 1–30.

Van Valen, L. (1975). Group selection, sex, and fossils. *Evolution*, **29**, 87–94.

Van Valen, L. M. (1971). Group selection and the evolution of dispersal. *Evolution*, **25**, 591–8.

Venditti, C., Meade, A. and Pagel, M. (2010). Phylogenies reveal new interpretation of speciation and the Red Queen. *Nature*, **463**, 349–52.

Vrba, E. S. (1984). What is species selection? *Systematic Zoology*, **33**, 318–28.

Vrba, E. S. and Gould, S. J. (1986). The hierarchical expansion of sorting and selection – sorting and selection cannot be equated. *Paleobiology*, **12**, 217–28.

Williams, G. C. (1966). *Adaptation and Natural Selection: A Critique of Some Current Evolutionary Thought*. Princeton, NJ: Princeton University Press.

6

Reconstructing the molecular underpinnings of morphological diversification: a case study of the Triassic fish *Saurichthys*

LEONHARD SCHMID

Introduction

Humanity has always been fascinated by the abundance of beautiful forms in nature. Throughout our history, myths, religions and science have tried to give explanations for the great diversity in the living world. Darwin (1859) still had to admit 'our ignorance of the laws of variation is profound'. In recent decades, research has brought some light into the molecular mechanisms of development that produce that variation and which are ultimately responsible for morphological diversity. Today's biodiversity, however, is only an estimated two to four per cent of the diversity that ever existed on earth (Benton 2009), the rest being extinct, in most cases leaving no traces of DNA or proteins for analyses. Is it possible to get insights into the molecular underpinnings of the diversity of organisms that have been extinct for millions of years?

To deal with this question I will first review how, in general, inferences on the molecular basis of morphological diversification in geological time can be drawn and I will assess the foundations of these inferences. The issues of reconstruction, evidence and predictability of evolution will be discussed. Second, I will present a case study on the Triassic ray-finned (actinopterygian) fish *Saurichthys*. Based on research on extant organisms reported in the literature I will develop inferences on the molecular underpinnings of the most prominent features of this fish, namely body shape and fin position, jaw elongation and scale reduction.

From Clone to Bone: The Synergy of Morphological and Molecular Tools in Palaeobiology, ed. Robert J. Asher and Johannes Müller. Published by Cambridge University Press.

Evidence for the molecular reconstruction of morphological diversification of extinct organisms

Molecular reconstruction

In palaeontology 'reconstruction' is a key principle with a long tradition (Benton 2010). Based on fossil remains the anatomy of organisms is reconstructed, not only its hard parts but its soft parts too. Furthermore, the function of the parts and the behaviour of the organisms are reconstructed, as well as their phylogenetic relationship to other organisms. Benton (2010) recently coined the term 'evidence-based reconstruction' in order to clearly distinguish it from 'scientific hypothesis'. Reconstructions are not directly testable, but they are more than speculation or fanciful contentions. Evidence-based reconstruction is an inference to the best explanation (an argument used to justify scientific realism; Van Fraassen 1980). It rests on the best current knowledge of the matter and, with progress of the latter, it can be refuted in a similar fashion as scientific hypotheses (Popper 1963). In palaeontology, reconstructions rely on the prerequisites of parsimony, uniformitarianism and actualism (Stanley 2005). As an example of contention, Benton (2010) mentioned the speculation that a certain dinosaur exhibited red skin colour. By contrast, as an example of evidence-based reconstruction, he cited the inference on the mass of a dinosaur's muscles from the structure of its bones.

The reconstruction of the molecular underpinnings of morphological diversification of extinct organisms is just an extension of the traditional fields of palaeontology. But whereas the general principle is the same, details of the extension are problematic. We can only partially refer to actualism to postulate that fossils had the same genes as their living relatives. Their genomes certainly must have differed somehow. The possibility that genes and genetic regulatory mechanisms of a given fossil are held in common with a specific living counterpart has to be considered on a case by case basis. Molecular reconstructions are complex and require detailed justifications.

An example is given by the following quotation from a paper dealing with the molecular reconstruction of the skeleton of a Devonian fish (*Palaeospondylus*) (Johanson *et al*. 2010a): 'we are following the axioms of Hall (1975), who proposed that developmental processes of fossil and living bone remain the same . . . From these axioms we believe it is justified to apply the results of gene knockout studies in tetrapods to interpret processes of evolution in the broader vertebrate fossil record.' This justification referring to actualism for the extrapolation of genetic mechanisms from mice and chicken to a fossil basal fish can be complemented by other lines of evidence. Indeed, these authors add later: 'Of course, it is impossible to test for gene expression in fossils, but what is important with respect to the *Palaeospondylus* skeleton is that the endochondral bone pathway can be divided into a series of steps and that particular genes are involved in these steps. Moreover, several known morphologies produced by the misexpression of certain of these genes are similar to *Palaeospondylus*.' The (histological) comparison of the fossil's skeleton with stages of development for which the genetic regulation is known in extant model organisms adds evidence to the inference that the same genes are involved.

Evidence

Where does evidence for molecular reconstruction come from? Evidence is not an 'all or nothing' or 'present or absent' situation. Rather, it exists on a continuum ranging from little to overwhelming.

A molecular mechanism of morphological change elucidated on extant organisms by methods such as gene knockout, gene misexpression, *in situ* hybridization, genetic engineering, implantation of beads soaked with morphogenic substances, genetic analysis of mutants and variations within and between species can be assumed to have been active also in extinct organisms. Without any additional information this simple **induction** conveys little evidence. Intuitively it is better when the organisms compared are phylogenetically close (but see below).

More evidence can be reached by **extrapolation**, the process of constructing new data points outside a set of known data points. The greater the number and the consistency of known data the better the evidence for the extrapolation. Again evidence should be improved by inclusion of closely related organisms (but see below; Arendt and Reznick 2008). An example illustrating this idea is the statement that the genome of the elephant shark could be used to test the suggestion that certain genes involved in humans, mice and chickens were expressed in a placoderm, an extinct stem group gnathostome (Johanson *et al.* 2010b).

Even better evidence results from '**extant phylogenetic bracketing**'. The method has been used for soft tissue reconstruction (Witmer 1995). It can be extended to molecular mechanisms: when a molecular process exists in extant groups which form a clade within which the fossil of interest is phylogenetically resolved, one can infer that the molecular process existed in the common ancestor of the clade and consequently in the fossil taxon of interest. A good example is the inference of late-phase *Hox* gene expression in pectoral fins of extinct sarcopterygians based on observations in extant actinopterygians on the one hand and tetrapods on the other (Shubin *et al.* 2009). In our case study (see below) we used extant phylogenetic bracketing to infer the presence of the ectodysplasin signalling pathway in a basal actinopterygian fish (Schmid and Sánchez-Villagra 2010 and below). It has been shown (Pantalacci *et al.* 2008) that this pathway exists in chondrichthyans (represented by the dogfish shark *Squalus* and the chimaera *Callorhinchus*) as well as in many teleosts (e.g. zebrafish, medaka, salmon, stickleback). From this phylogenetic bracket it can be concluded that the ectodysplasin pathway is common to all gnathostomes (verified in over 50 species), including the fossil of our case study.

Comparative morphology is important for molecular reconstructions particularly in combination with the methods discussed above. This idea was expressed, for example, by Donoghue and Purnell (2005): 'when attempts are made to compare or integrate analyses of genomic and morphological evolution, there is no logical reason for excluding extinct taxa [from the hypothesis linking gen(om)e duplication events in vertebrates with evolutionary jumps]'.

Inferences from morphological comparisons nevertheless pose some problems. The simple rationale that the same genetic processes underlie the same morphological changes in extinct and in extant organisms, especially when the latter form a phylogenetical bracket around the former, is not exempt from pitfalls.

At first sight it seems important that the features compared in extinct and in extant organisms are homologous to permit the inference that similar molecular processes are involved. Homology, however, depends on the hierarchical level on which the structures are compared. Analogous structures at one level (for instance bat wings and bird wings as wings) can be homologous at a deeper level (bat wings and bird wings as forelimbs). Genetic analyses of the last decades have revealed that molecular processes controlling morphological features often are highly conserved which means that they are homologous at a very deep level ('deep homology' Shubin *et al.* 1997). In the words of Shubin *et al.* (2009): 'Deep homology [describes] the sharing of the genetic regulatory apparatus that is used to build *morphologically* and phylogenetically *disparate* animal features.' (emphasis L.S.). In other words: 'deep homology stems from the disassociation between homologous regulatory mechanisms in relation to morphological traits' (Scotland 2010). Concerning molecular reconstruction, this understanding shifts the focus from morphological comparison to the question of how deep is the genetic homology for the (possibly disparate) features compared. Genetic reconstruction of morphological change must be aware that morphological novelties are often not due to new genes but to the co-option not only of pre-existing cis-regulatory elements (Carroll 2008) but of whole regulatory networks (Davidson 2006). An example is the demonstration that oral teeth of jawed vertebrates have co-opted an ancient gene network existing in pharyngeal teeth of agnathans (Fraser *et al.* 2009; Smith and Johanson this volume).

But though it is correct (as outlined just above) that disparate morphologies can be controlled by similar genetic mechanisms, the contrary may be true too: similar structures may be controlled by very different genetic mechanisms if they evolved by convergence in phylogenetically distant lines (see further discussion below).

Another issue complicating molecular reconstruction is gene or genome duplication. Several evolutionary novelties have been achieved by gene duplication and consecutive gene subfunctionalization or neofunctionalization (Donoghue and Purnell 2005; Gao and Lynch 2009). Molecular reconstruction has to take into account this possibility and track down its morphological consequences (Durand and Hoberman 2006).

Thus, the generalization of the conclusion that similar genetic processes would underlie similar morphologies in extant and in extinct organisms is not valid in every case and by itself is not sufficient to justify a specific mechanistic inference. Morphological comparisons have to be used with great caution and inferences based on them can only be drawn when evidence is derived from additional information.

Predictability of evolution

Best evidence (certainty) for reconstructions of the molecular underpinnings of morphological diversification of extinct taxa would be reached if evolution were predictable.

There are two opposing opinions about this issue. One, prominently advocated by Stephen Jay Gould (1990), insists on the contingency of evolution. The successive coincidences of historically unique circumstances and events are embodied in his famous metaphor of the 'tape of life'. Replaying this tape from any given time and place in the past 'would lead evolution down a pathway radically different from the road actually taken ... The divergent route of the replay would be just as interpretable, just as explainable *after the fact*, as the actual road. But the diversity of possible itineraries does demonstrate that eventual results cannot be predicted at the outset.' The opposing position is defended by Simon Conway Morris (2003, 2010). Its essence is expressed in the title of one of his papers: 'Evolution: like any other science it is predictable' (Conway Morris 2010).

In fact, more and more examples in favour of the predictability of evolution accumulate (e.g. Gompel and Prud'homme 2009; Mundy 2009; Stern and

Orgogozo 2009). Expressions like 'toolkit genes' (Gerhart and Kirschner 2007), 'ancestral complexity' (Carroll 2008) and 'deep homology' (Shubin *et al.* 2009; Scotland 2010) point to the now well-known fact mentioned above that a core of developmental genes and regulatory circuits are highly conserved and shared by morphologically and phylogenetically very disparate species. Moreover, recent investigations provide growing evidence that adaptation to similar environments favours not only convergent evolution of phenotypic traits but also genetic convergence, the evolution of similar genetic mechanisms in independent lineages (Conant and Wagner 2003; Wood *et al.* 2005; Weinreich *et al.* 2006; Wray 2007; Arendt and Reznick 2008; Gompel and Prud'homme 2009).

As mentioned above, there has been a considerable shift in the last decades away from the idea that evolutionary innovations were due to new genes, and towards the empirical demonstration that, instead, ancient genes are redeployed and co-opted in new regulatory networks (Davidson 2006; Fraser *et al.* 2009; Woltering and Duboule 2010). Therefore, reconstructions of the molecular underpinnings of diversification cannot rely on single gene mutations, though they are frequently the source of early insights. The consideration of regulatory genetic networks adds a whole dimension of complexity. It raises the question of how close is close enough to consider the roles of individual genes of pathways to be the same or different (Arendt and Reznick 2008). Nevertheless, the modular hierarchical architecture ('topology'; Barabási and Oltvai 2004) of gene regulatory networks, in part empirically demonstrated, permits conclusions to be drawn from single gene alterations. Regulatory genes occupying nodal positions in networks ('hubs'; Barabási and Oltvai 2004) function as morphogenetic switches. They are prime targets for evolutionary changes and are related to repeated evolution (Gompel and Prud'homme 2009). A classical example is trichome patterning in *Drosophila* larvae. Though dependent on very complex interactions of many gene products, a cis-regulatory change in a single, so-called input–output gene (*shaven baby/ovo*) controls the whole phenotypic trait due to its nodal position between patterning and differentiation genes (Stern and Orgogozo 2009).

Though the number of examples in favour of the predictability of evolution is growing, the counter-examples must not be overlooked. A famous one is the convergent adaptive melanism in different populations of lava-dwelling populations of pocket mice which is realized by different genes (Hoekstra and Nachmann 2003), as is the convergent bright pigment pattern in beach mice (Steiner *et al.* 2009). As summarized by Arendt and Reznick (2008), 'closely related organisms often evolve the same phenotype via different mechanisms and distantly related organisms often evolve the same phenotype via the same mechanism'. The authors compiled many examples for this statement and they conclude: 'At best, the association between taxonomic affinity and the similarity of the mechanism that causes the independent evolution of phenotypic similarity might be a probabilistic one – more closely related species might be more likely to evolve phenotypic similarity via the same mechanism than more distantly related species.'

To sum up it can be stated that the reconstruction of the morphological diversity of extinct clades is a complex venture and has to overcome several obstacles. Evolution is predictable in some contexts, but not in others. Therefore, molecular reconstructions obviously face some level of uncertainty. Evidence for an adequate reconstruction can nevertheless be evaluated on a case by case basis. It is dependent on different factors and methods. Among these, bracketing and extrapolation from many consistent data of extant, closely related species are the most important. Comparative morphology, by contrast, requires the recognition that morphologically disparate features can be controlled by similar molecular processes and, conversely, similar traits can be dependent on different genes.

The case study

This study focuses on the fish *Saurichthys* (Figures 6.1, 6.2) because it is well documented and because it shows some unique traits which make it well suited to tracing its morphological diversification. These traits are (1) an elongate body with pelvic, anal and dorsal fins positioned far posteriorly, (2) a

Figure 6.1 *Saurichthys curionii* from Monte San Giorgio, southern Switzerland, Cassina beds of Lower Meride Limestone, Ladinian, Middle Triassic, about 240 million years old. (From Museo Cantonale di Storia Naturale Lugano, Switzerland. Specimen MCSN 8013, preparator Sergio Rampinelli, photograph courtesy of Rudolf Stockar.) Scale bar represents 10 cm. (See also colour plate.)

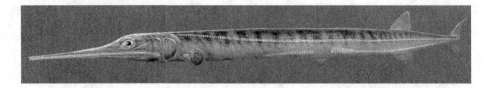

Figure 6.2 *Saurichthys curionii*. (Reconstruction from Furrer 2003; illustration by B. Scheffold. Copyright Palaeontological Institute and Museum University of Zürich/Museo Cantonale di Storia Naturale Lugano with permission H. Furrer and R. Stockar.)

very long snout, and (3) in most species a reduction of scales to longitudinal rows (Rieppel 1985, 1992). After a brief overview of *Saurichthys*, reconstructions of molecular underpinnings of each of these features will be presented.

The functional aspects of the morphological changes of *Saurichthys* will not be examined. Consequently, questions of fitness, adaptation and selection are beyond the scope of this chapter. Also omitted are phylogenetic considerations and the temporal succession of morphological change. Accordingly, the issue of possible evolutionary trends (Rieppel 1985, 1992; Mutter *et al.* 2008) will not be discussed.

Saurichthys and its diversification

Saurichthys is a genus in the clade Saurichthyidae which contains three other genera (Mutter *et al.* 2008; Wu *et al.* 2009). Phylogenetically it belongs to the basal actinopterygians and is related to extant sturgeons (acipenserids), paddlefishes (polyodontids), and bichirs (polypterids) (Gardiner *et al.* 2005; Nelson 2006) (Figure 6.3). Its occurrence is confirmed throughout the Triassic and Early Jurassic. Reports from the Permian are doubtful (Mutter *et al.* 2008). In the Early Triassic it already had a nearly worldwide distribution in marine environments as well as in a few freshwater habitats (Mutter *et al.* 2008).

The genus is characterized by a streamlined, elongate body with relatively small pelvic, anal and dorsal fins in an extremely posterior position. The most prominent trait is the elongation of the jaws which form a beak-like, long, tapered and pointed snout (Figures 6.1, 6.2). Numerous pointed teeth of variable size are present. Viviparity is documented by intracorporeal embryos of different stages (Renesto and Stockar 2009) and the male copulatory organ (Bürgin 1990).

The genus displayed a considerable diversity with over 30 species (Mutter *et al.* 2008). These differ in the number and shape of the dermal bones of the skull, the proportions of the rostrum, the robustness of the dentition, the

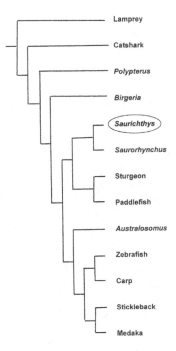

Figure 6.3 Phylogeny showing the position of *Saurichthys* among basal actinopterygians. (After Gardiner *et al.* 2005 and Nelson 2006.)

ossification of the sclerotic ring, and the shape of the operculum (Rieppel 1985, 1992; Mutter *et al.* 2008). Some species such as *S. madagascariensis* are fully scaled (Rieppel 1980). Their midlateral scales may be dorsoventrally elongated as so-called deep or deepened scales (Rieppel 1980; Mutter *et al.* 2008). Most species, however, have scales reduced to isolated longitudinal rows with unscaled regions in between. The number of 6, 4 or 2 rows of scales is diagnostic for some species, as is the extreme dorsoventral elongation of scales of the midlateral row to a rib-like aspect (Rieppel 1985, 1992). Other diagnostic criteria are the presence or absence of fringing fulcra on the unpaired fins as well as the number and the segmentation of the fin rays (Rieppel 1992).

Molecular underpinnings of body shape and fin position of *Saurichthys*

Body elongation

The considerable elongation of the body of *Saurichthys* is realized not by an anterior–posterior enlargement of individual skeletal and muscular segments as, for example, in the fish *Sphyraena barracuda* or in the salamander *Lineatriton* (Ward and Brainerd 2007) but by an increase in their number. Body elongation

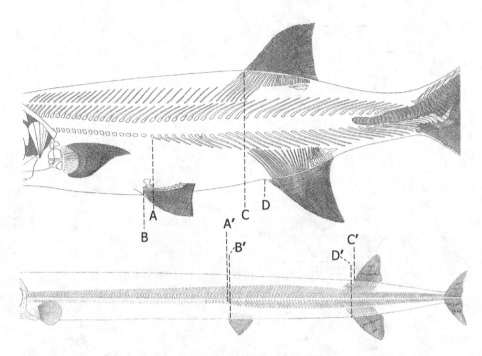

Figure 6.4 Comparison of the vertebral columns and the positions of fins between *Birgeria stensiöi* (top) and *Saurichthys curionii* (bottom). The vertebral columns are aligned to the same lengths to show the relative posteriorization of the precaudal–caudal transition and the even more posteriorized anterior insertions of the fins of *Saurichthys*. Note the unossified centra of the vertebrae and the more than three times higher number of vertebrae in *Saurichthys*. Abbreviations are as follows: A and A′ indicate the boundary between precaudal and caudal vertebral column; B and B′ the anterior insertion of the pelvic fin; C and C′ the anterior insertion of the dorsal fin; and D and D′ the anterior insertion of the anal fin of *Birgeria* and *Saurichthys*, respectively. (*Birgeria* from Schwarz 1970; *Saurichthys* from Rieppel 1985.)

via large vertebral numbers has been accomplished independently numerous times in fish and is well known in caecilian amphibians and squamate reptiles (Ward and Brainerd 2007). That the number of vertebrae of *Saurichthys* is extraordinary is illustrated by a comparison with its coeval, extinct close relative *Birgeria* (Figure 6.4). Fortunately, vertebral counts are readily possible in both genera, in contrast to most fossil fish, due to absent or sparse scalation. Also, a comparison with the closely related extant polypterids (bichirs), famous for their eel-like shape and their high number of vertebrae (Ward and Brainerd 2007, Suzuki *et al.* 2010), is revealing. Whereas *Birgeria stensiöi* has a total number of 43–46 vertebrae (without the vertebrae of the caudal fin) (Schwarz 1970), the corresponding number is approximately 190 for *S. krambergeri*

(Griffith 1962), 170–180 for *S. curionii* and *S. macrocephalus* (Rieppel 1985), about 165 for *S. costasquamosus* (Rieppel 1985), 159–161 for *S. dawaziensis* (Wu *et al.* 2009), an estimated 135 for *S. toxolepis* (Mutter *et al.* 2008) and approximately 127 for *S. paucitrichus* (Rieppel 1992). Species of *Polypterus* reach between 53 and 65 vertebrae, and the record holder of this clade, *Erpetoichthys calabaricus*, has 110 to 113 (Suzuki *et al.* 2010). These comparisons demonstrate that the vertebrae of *Saurichthys* are also relatively small, otherwise it would look like a snake or an eel.

Saurichthys must have diversified from an ancestor with much fewer vertebrae. How could it augment its vertebral number to such an extent?

The somites from which the vertebrae originate are formed early in development by somitogenesis. This rhythmically repeated morphogenic process results in progressive segmentation of the presomitic mesoderm situated on both sides of the notochord (Holley 2007; Alexander *et al.* 2009; Gomez and Pourquié 2009; Gibb *et al.* 2010). A clock and wave front model consisting of an oscillator coupled with a morphogen gradient is widely accepted as driving this process (Holley 2007; Alexander *et al.* 2009; Gomez and Pourquié 2009; Gibb *et al.* 2010). Oscillating gene transcriptions in presomitic cells function as a clock. A somite unit forms when the front of the maturation wave reaches a group of cells in the appropriate phase of oscillation (Alexander *et al.* 2009; Gomez and Pourquié 2009; Gibb *et al.* 2010). Therefore the size of a somite depends on the speed of the wave front, and the rate of somite formation is determined by the frequency of the oscillator.

Thus, the somite and consequently the vertebral number may be enhanced either by a higher oscillator frequency or by prolongation of the overall time for somitogenesis. The first possibility, however, does not change significantly the length of the organism as the higher number is compensated for by a smaller size of the somites. Another possibility is the prolongation of somitogenesis by a longer activity of the tail bud, which results in a posterior growth of the embryo. This possibility obviously predominated in the development of *Saurichthys*.

The maintenance of tail bud activity requires that its cells are kept in an undifferentiated state and prevented from reaching maturation. Maturation of the cells of the presomitic mesenchyme is known to depend on retinoic acid (Alexander *et al.* 2009; Gomez and Pourquié 2009; Aulehla and Pourquié 2010). Enzymes synthesizing retinoic acid (including aldehyde dehydrogenase) are expressed in the anterior presomitic mesenchyme and in the newly pinched off somites (Aulehla and Pourquié 2010). Retinoic acid diffuses over long distances but Cyp26A1, a cytochrome p450 enzyme involved in its degradation, is expressed in the tail bud under the influence of *caudal*-related genes (*cdx*, a gene family known to be required for posterior development in *Drosophila* and mice; Aulehla and Pourquié 2010). Thus, a source–sink mechanism with spatially separated enzymes responsible for producing and degrading products maintains a retinoic acid signalling gradient (Aulehla and Pourquié 2010). Additionally, high levels of Fgf and Wnt signalling are necessary to maintain cells in an undifferentiated state (Aulehla and Pourquié 2010). Their de novo transcription is confined to the tail bud region (Wnt again under control of *caudal*; Aulehla and Pourquié 2010). A posterior–anterior gradient of Fgf and Wnt across the presomitic mesenchyme is created by decay of the corresponding mRNAs as cells leave the progenitor zone of the tail bud (Aulehla and Pourquié 2010). Consequently, a similar gradient of their downstream target gene products can be observed. It is apparent that a shift in the equilibrium of the three signalling pathways (and the expression of their downstream target genes) determines when – after how many somites – the tail bud activity stops. A limiting factor might also be the proliferation capacity of the progenitor zone of the tail bud after exhaustion of the supply of new cells at the end of gastrulation (Aulehla and Pourquié 2010). The ratio between the speed of somitogenesis and posterior axis growth determines how fast the presomitic mesenchyme shrinks. As Fgf and Wnt production are likely to be proportional to the remaining tissue size, shrinking leads rapidly to the end of somitogenesis. Moreover, apoptosis – as observed in chicken tail bud – can lead to a precocious termination of somitogenesis (Gomez and Pourquié 2009). Whatever the reason, the termination of somitogenesis must be under strict genetic control as it leads to species-specific numbers of vertebrae with relatively low intraspecific variance.

In sum, changes in the regulation of retinoic acid, Fgf or Wnt pathways and in the expression of their downstream target genes must be involved in the diversification of vertebral numbers and thus in body shape.

Extrapolating these findings to *Saurichthys*, we can infer a downregulation of retinoic acid synthesis (as by aldehyde dehydrogenase), or an upregulation of its degradation (as by Cyp26A1), or an upregulation in the signalling pathways of Fgf or Wnt in the tail bud. This regulatory network must be combined with a high capacity of cell proliferation in the tail bud's progenitor zone which ultimately has to be fed by sufficient yolk supply.

Regionalization of the body

A superficial examination of *Saurichthys* suggests that it is the anterior part of its body and not its rear that is elongated (Figure 6.2). Thus, posterior growth alone as suggested by prolonged tail bud activity would not explain the particular shape of this fish. The impression of anterior elongation of *Saurichthys*, however, is misleading due to a posterior shift of the fins. This can be verified by precise anatomical indicators. Ichthyological convention classifies the vertebrae of fish as precaudal and caudal (Helfman *et al.* 2009). Caudal vertebrae are defined by the presence of haemal arches, a structure absent in precaudal vertebrae. In fossils, observing this difference might be problematic because of poor preservation of these fragile, tiny structures, which in some cases are not ossified. Fortunately we can rely on another marker in *Saurichthys*. A single row of scales on the median ventral line has to become paired or doubled to create space for the median outlet of the anus. This 'anal loop' of scales can be observed frequently. In contrast to many fish species, and in contrast to what the nomenclature suggests, it is not located near the anal fin but between and slightly behind the pelvic fins. In the anatomical reconstruction of *S. curionii* (Rieppel 1985), the anteriormost haemal arch, anterior pelvic fin insertion and anal loop fall approximately onto the same line, so that for comparison the landmarks may be interchanged.

Again a comparison with *Birgeria stensiöi* is useful (Figure 6.4). Whereas in this fish more than 60% of the vertebrae are caudal (Schwarz 1970), less than 50% are in *S. curionii*, and less than 45% in *S. macrocephalus* (Rieppel 1985). The relative anterior elongation of *Saurichthys* is not as great as expected by external inspection. But it probably evolved from an ancestor with a lower proportion between precaudal and caudal vertebrae. How could it develop more precaudal vertebrae after posterior growth?

Several phylogenetic studies have demonstrated that the vertebral number in vertebrates varies in a regionally specific way. This suggests a modularity in the development of the regions (Johanson *et al.* 2005; Ward and Brainerd 2007; Müller *et al.* 2010; Schröter and Oates 2010; Asher *et al.* 2011; Buchholtz, this volume). For amniotes, a survey including a large number of extant and fossil species confirms that segmentation and homeotic regionalization are decoupled (Müller *et al.* 2010). A similar analysis of extant fish showed that there are groups that increase only caudal vertebral number, groups that increase equally in both regions, and groups that increase only precaudal vertebral number, demonstrating that also in ray-finned fish segmentation and regionalization are decoupled (Ward and Brainerd 2007).

Regional vertebral identity is known to depend on *Hox* genes, which encode homeobox-containing transcription factors (Wellik 2007; Mallo *et al.* 2010). The anterior limit of expression of successive *Hox* genes is relevant. Caudally, expression usually continues until the end of the organism so that in caudal direction a growing number of genes are expressed (Wellik 2007; Mallo *et al.* 2010). Often several *Hox* genes together contribute to the identity of vertebrae at a particular axial level. In mice a homeotic transformation corresponding to an anterior elongation results when the whole paralogous *Hox* group 10 is inactivated: more thoracic (rib bearing) vertebrae are formed posteriorly, with a corresponding reduction in lumbar ones (Mallo *et al.* 2010). While in mouse and chick the anterior boundary of *Hox9* paralogue group coincides with the thoracolumbar transition (Mallo *et al.* 2010), this boundary falls in the middle of the precaudal region in zebrafish (Holley 2007). Here, the anterior limit of *Hoxd12a* corresponds to the beginning of caudal vertebrae but there are some transitional vertebrae complicating an exact correlation (Bird and Mabee 2003; Holley 2007). Thus, the last true precaudal vertebra falls into the anterior limit of *Hoxc10a* (Holley 2007). It is likely a shift in the most anterior expression of some *Hox* gene(s) that causes the homeotic transformation which results in the great differences in the regionalization of the vertebral column of fishes (Ward and Mehta 2010).

Hox genes in turn are regulated by other upstream genes. Progenitors of precaudal and caudal somites are already specified in the blastula before gastrulation (Holley 2007). The subdivision in the somite primordia depends on various *t-box* genes, *nodals*, *Wnts* and *Fgfs* (Holley 2007). Again, retinoic acid, Wnt and Fgf signalling levels also control *Hox* gene expression by intermediary of *caudal-like* homeobox genes (*cdx*) (Aulehla and Pourquié 2010). Although in somitogenesis there exists an antagonistic relationship between them, all three pathways are considered to have a posteriorizing effect on axial development (Aulehla and Pourquié 2010).

Extrapolating these data to *Saurichthys*, we can infer that its anterior elongation must be caused by a homeotic shift in the genes expressed in the vertebral column. The shift might be due to alterations somewhere within the regulatory circuitry comprising retinoic acid, Wnt, and Fgf among others controlling the expression of *Hox* genes. Ultimately, a shift of the anterior expression boundary of certain *Hox* genes can be postulated. For example, by extrapolation from the zebrafish, *Hoxd12a* would be a likely candidate (Bird and Mabee 2003; Holley 2007).

Fin position

A remarkable and diagnostic feature of the genus *Saurichthys* is the posterior position of its dorsal, pelvic and anal fins (Figures 6.2, 6.4). An important question in this context is whether the position of fins is determined by somitic axial patterning, that is, connected to a defined vertebral region like the limbs are in tetrapods, or independent of vertebral column regionalization. A consideration of fish diversity shows a great variability in fin positions. The pelvic fins can be dislocated from their abdominal position to a thoracic one directly below the pectoral fins, and in some cases even anterior to these in a so-called jugular position (Moyle and Cech 2000; Helfman *et al.* 2009). Posteriorly, they can be positioned at the level of the vent, displacing the anal fin from its usual position behind the anus. This is the situation observed in *Saurichthys*. How can it be explained on a molecular level?

Some inferences on paired fin position in *Saurichthys* can be drawn from limb localization in mice and chickens (Mallo *et al.* 2010). Research on zebrafish and stickleback confirm that the molecular processes involved are largely conserved (Mercader 2007) so that an extrapolation to basal actinopterygians is reasonable. Investigations on sharks and lampreys shed light on the development of unpaired fins but contribute also to the understanding of the evolution of paired fins (Freitas *et al.* 2006). Thus, we have here another example of extant phylogenetic bracketing (Figure 6.3).

Generally, primordia are induced at a specific location in the embryo in response to pre-existing combinatorial positional cues (Mercader 2007). For paired appendages in vertebrates the first step is the establishment of a limb (or fin) field which is dependent on interactions of ectoderm and lateral plate mesoderm (Mercader 2007). It was demonstrated that the appendage-inducing signal originates in the paraxial somitic mesoderm and is relayed from there to the lateral plates (Mercader 2007). One of the first genes promoting fin outgrowth expressed in the pelvic fin field is the T-box transcription factor gene *Tbx4* (Mercader 2007). Observations in stickleback show that it is dependent on pitx1 transcription factor (Shapiro *et al.* 2006). More importantly, it is controlled (via the intermediary signal of Wnt) by retinoic acid which, as mentioned above, is synthesized in somites and anterior presomitic mesenchyme (Mercader 2007). Thus, retinoic acid synthesis is at least one factor for the axial position of paired fins.

Axial position of limbs is defined by differential *Hox* gene expression (Wellik 2007). Although the limbs of different tetrapods differ with respect to the somite level at which they arise, their position is constant with respect to the level of *Hox* gene expression along the anterior–posterior axis (Burke *et al.* 1995). *Gdf11* expression in the posterior end of chick and mouse is known to be involved in positioning forelimbs and hind limbs via an axial shift of *Hox* expression (Murata *et al.* 2010). Reduction of the expression of *Gdf11* in zebrafish leads to a caudal shift of *Hoxc10a* expression and concomitantly to a caudal displacement of pelvic fins (Murata *et al.* 2010). The highly variable position of fins in fish, however, cannot be explained completely by axial shifts in *Hox* gene expression (Murata *et al.* 2010). The presumptive pelvic fin cells do reside originally in the lateral plate next to the anterior border of *Hoxc10a* expression where the hind limbs also develop in tetrapods (Murata *et al.* 2010). But as demonstrated on teleosts, they may move considerably to their particular final position (Murata *et al.* 2010). Their specific migration may be caused by differences in allometric growth of the trunk with respect to the lateral plate mesoderm, by variations in the protrusion of the trunk from the yolk, and by other unknown mechanisms (Murata *et al.* 2010).

Positioning of median fins has been investigated in catsharks and lampreys (Freitas *et al.* 2006). The first dorsal fin region is characterized by expression of *Hoxd9* and *Hoxd10* (Freitas *et al.* 2006). Additional *Hoxd12* is expressed in the second dorsal and anal fin regions (Freitas *et al.* 2006). Later in development *Tbx18* is expressed in all median fin folds (Freitas *et al.* 2006). In lamprey, orthologues of the same genes are expressed in the median fins (Freitas *et al.* 2006). The molecular programme of development of unpaired fins in lamprey and sharks turned out to be very similar to the programme known from paired fin development (Freitas *et al.* 2006). From these findings the conclusion was drawn that this programme had originated in conjunction with paraxial mesodermally derived median fins before paired fins evolved from lateral plate mesoderm and was subsequently co-opted by these (Freitas *et al.* 2006).

Thus, for *Saurichthys*, we can propose a posterior shift of the anterior limits of fin-inducing *Hox* gene expression (*Hoxc10a* for pelvic fins, probably *Hoxd12* for anal and dorsal fins), possibly controlled by reduced *Gdf11* expression in the caudal region. If these were the only positional signals involved, fin position would be connected to a certain vertebral number, as vertebral regionalization is also determined by *Hox* genes. But, as mentioned above (Murata *et al.* 2010), the fin primordia can move independently of vertebral column regionalization after their determination by *Hox* genes. Such a migration in caudal direction in addition to a *Hox* gene shift towards the tail must be assumed to explain the even more posterior position of the fins compared with the general poster-iorization of the body.

Elongation of the jaws

The long, tapering beak-like snout, the rostrum, is very characteristic of all species of the genus *Saurichthys* (Figures 6.1, 6.2). Measured from the anterior margin of the orbita it accounts for 73% on average of the head length in *S. curionii* (Rieppel 1985), 70% in *S. costasquamosus* (Rieppel 1985) and *S. toxolepis* (Mutter *et al.* 2008), and 63% in *S. macrocephalus* (Rieppel 1985). Extant fishes that independently developed similarly elongated jaws include the needlefish Belonidae and the gars Lepisosteidae.

The reconstruction of the molecular bases of the development of this peculiar feature of *Saurichthys* is challenging. Considerable research has been performed on the genetics and evolution of jaws in cichlids, which recently have undergone a tremendous radiation in east African and neotropical crater lakes (Salzburger and Meyer 2004; Elmer *et al.* 2010). The diversification of their feeding habits and feeding apparatus is an important field of investigation (Salzburger and Meyer 2004; Elmer *et al.* 2010). Unfortunately, the evolution of cichlid jaws has not led to an elongated or beak-like jaw morphology. Thus, we must draw inferences mainly from true beaks (of birds) and face lengths of carnivorans and assume the conservation of the correspondent genetic mechanisms.

The cells that primarily contribute to the craniofacial skeleton, ectomesenchymal derivatives of neural crest cells, are unique to vertebrates (Donoghue *et al.* 2008). They provide this clade with a great variability as they are migratory and pluripotent. A key innovation in vertebrate evolution was the appearance of jaws consisting of separate dorsal and ventral skeletal elements connected by a joint (Minoux and Rijli 2010). It was a prerequisite for an extensive radiation. The wide-ranging morphological craniofacial diversification was enabled by the multiple interactions between the numerous tissues of various origins that are in contact in this area (Gitton *et al.* 2010; Minoux and Rijli 2010). Cranial neural crest cells receive positioning information from external head epithelium, paraxial head mesoderm, developing brain, ectoderm of the stomodeum and endoderm of the foregut (Minoux and Rijli 2010; Gitton *et al.* 2010). Correspondingly complex are the interactions and consequences of signalling molecules. Despite this complex patterning environment, a few single gene effects can be isolated to explain snout elongation of *Saurichthys*.

Investigations into the legendary diversity of beak shapes of Darwin's finches revealed that species with deeper and broader beaks express *Bmp4* in their beak prominences earlier and at higher levels than do species with narrow and shallow beaks (Abzhanov *et al.* 2004). Conversely, calmodulin, a molecule involved in mediating calcium ion signalling, is expressed at higher levels in long and pointed beaks than in more robust beak types (Abzhanov *et al.* 2006). Thus, two different pathways act to shape the beak along different axes: calmodulin controls the length, whereas Bmp4 controls the width and depth. The effect of Bmp4 can be reproduced in cichlids and in zebrafish: overexpression of *Bmp4* or implantation of beads soaked with this signalling molecule result in an increase in the depth of their jaws (Albertson *et al.* 2005). Calmodulin however, only shows a linkage to the width of the jaws of cichlids (Parsons and Albertson 2009). Concerning jaw length, the authors state that either the effect of calmodulin is so small as not to be detected by their experimental design or that it is involved as a component of a larger signalling pathway.

Face length of Carnivora has been shown to correlate with the transcriptional activity of the runt-related transcription factor 2 (*Runx2*) (Sears *et al.* 2007). This transcription factor is a regulator of bone development and can control both its developmental rate and timing by enhancement of early osteoblast differentiation and inhibition of terminal osteoblast differentiation (Sears *et al.* 2007).

Thus, for the time being, one can infer that a very high expression of calmodulin was involved in the shaping of the jaws of *Saurichthys*, possibly as a component of other, still unknown, signalling pathways. The continuous tapering of the rostrum could be explained by a posterior–anterior gradient of Bmp4. The diversification to different forms of the rostrum would then be explained by variations in the relative proportions of the two morphogens. In addition, differential but generally high activities of the runt-related transcription factor 2 might have been involved.

Scale reduction and deep scales

Scale reduction is a particularly interesting feature in *Saurichthys* because several transitional states can be observed within this genus. In the Jurassic genus *Saurorhynchus* the reduction continues to a completely scaleless species (*S. brevirostris*) which therefore is included in our consideration. However, the transition from fully scaled to scaleless is not gradual. It is a stepwise reduction to species-specific numbers (6, 4, 2, 0) of distinct longitudinal rows of scales (Figure 6.5). Scale reduction is not unique to *Saurichthys*. It appears independently in several distantly related clades (e.g. muraena in Anguilliformes, catfish in Siluriformes, ragfish in Perciformes, pearlfish in Ophidiiformes, pufferfish in Tetraodontiformes, etc.). It must not be confused with so-called deciduous scales, which are lost secondarily during the lifetime of certain fishes. Arrangement of scales in separated rows is found not only in sticklebacks (as lateral plates), domesticated carps and certain minnows, which will be described below, but also in the killifish *Aphanius*, for example. In this genus there exist fully scaled species, others with reduced scales, and nearly naked ones (Wildekamp *et al.* 1999). Within the species *Aphanius anatoliae* scale reduction goes gradually from fully scaled to scales arranged in the midlateral line (Grimm 1979) (Figure 6.6). Originally, different species were distinguished. Later analysis demonstrated that the reduction is caused by an unstable variation in the genes not favoured or suppressed by stabilizing selection, so that mutants with all variations in scaling survive within a single

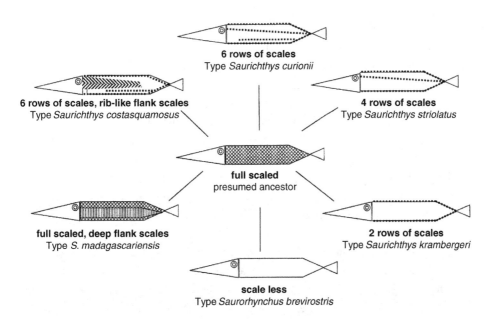

Figure 6.5 Diversification of scale reduction among saurichthyids. Originating from a fully scaled hypothetical ancestor, the species within this clade fall into different types characterized by the degree of scale reduction and by the presence or absence of deep scales or rib-like scales.

Figure 6.6 *Aphanius anatoliae* (modern killifish, Cyprinodontidae). Reduction of scales from fully scaled (left) to a midlateral row of remaining scales (right) in different phenotypes of the same species (Grimm 1979). The variation of the scale pattern in this species parallels a feature of the morphological diversification of *Saurichthys*.

species (Wildekamp *et al.* 1999). Unfortunately we do not know which genes exactly are involved in this case. Another example is provided by the evolution of acipenserids. A species from the Upper Cretaceous is completely covered by scales (Grande and Hilton 2006) whereas extant sturgeons have five rows of scutes (modified scales). The arrangement of scales in rows is reflected in the ontogeny of fishes. Scales usually develop first along the lateral line in posterior–anterior direction, then in rows dorsal and ventral to it (Donoghue 2002; Helfman *et al.* 2009; Cloutier 2010). In acanthodians from the Carboniferous this developmental pattern is more accentuated than in extant actinopterygians such as zebrafish (Cloutier 2010). The direction of the development of the first row is reversed in some Carboniferous and in some extant actinopterygians (Sire and Akimenko 2004; Cloutier 2010). In any case scales in the ventral and dorsal median lines are the last to develop. It was suggested that tension transmitted to the skin, an epigenetic factor, would delimit the areas in which scales develop (Sire and Akimenko 2004). This is contradicted by the scale pattern in the genus *Saurichthys* where species exist that have scales exclusively in the median lines where tension is least. This scale pattern also contradicts the rule that the last regions to develop scales in ontogeny are the first to lose them in phylogeny (Helfman *et al.* 2009).

Deep, dorsoventrally elongated scales are found in *S. madagascariensis* (Rieppel 1980) and in an extreme version designated as 'rib-like scales' in *S. costasquamosus* and *S. paucitrichus* (Rieppel 1985, 1992). This feature also has evolved several times independently. Deep scales occur in *Australosomus*, a Triassic fish closely related to *Saurichthys* (Figure 6.3) (Nielsen 1949; Mutter *et al.* 2008), and in several taxa of fusiform actinopterygians of the Carboniferous, Triassic and Jurassic (Bürgin 1999; Mutter and Herzog 2004). Deep scales might be homologous to lateral plates or scutes we find in such disparate extant clades as sturgeons (Acipenseridae), sticklebacks (Gasterosteidae), pipefishes (Syngnathidae), armoured searobins (Peristediidae), armoured catfishes (Loricariidae, Callichthyidae), and seahorses (Syngnathidae) (Nelson 2006; Helfman *et al.* 2009).

What might be the molecular underpinnings of the evolutionarily recurrent features of deep scales and scale reduction with arrangement in longitudinal rows?

Surprisingly, similar patterns of scale reduction and dorsoventral enlargement were observed in several mutants of the model organisms zebrafish (Harris *et al.* 2008; Rohner *et al.* 2009) and medaka (Kondo *et al.* 2001). They had either mutations in the ectodysplasin signalling pathway (Figure 6.7) or in the fibroblast growth factor signalling pathway (Figure 6.8). Mutations in such signalling pathways are bound to have pleiotropic effects as the same pathways are used in the development of many organs and body parts. Usually pleiotropic effects are lethal or have severe negative effects on fitness. Therefore they are generally not considered to contribute to morphological diversification. However, pleiotropy may be overcome by gene duplication and subfunctionalization. This is the case for the mutations of the fibroblast growth factor pathway in zebrafish (Rohner *et al.* 2009). Similar mutations in medaka, which has no duplication of the genes of this pathway, are lethal (Rohner *et al.* 2009). That mutations in paralogues of the fibroblast growth factor pathway components are not just found in the laboratory is documented by a free-living fish with the telling name of 'naked minnow' *(Phoxinellus)*. It shows a single row of scales on its lateral line, very reminiscent of the pattern in *Saurichthys*, and has a

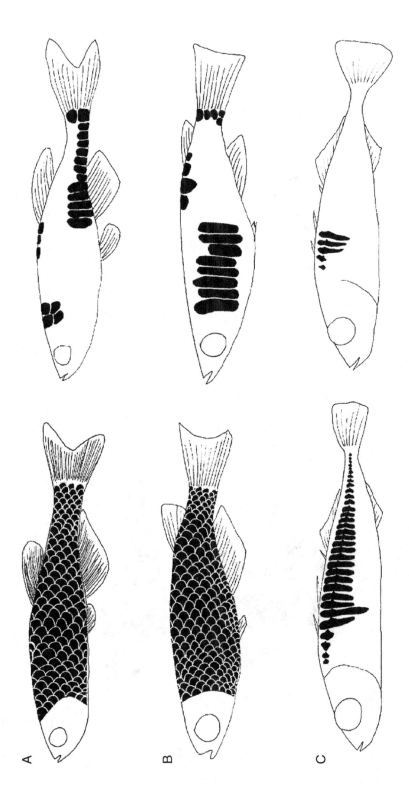

Figure 6.7 Sketches of mutants with mutations in the ectodysplasin pathway (right) and corresponding wildtypes (left). In the mutants (right) the scales (black) are reduced with a tendency to dorsoventral enlargement and alignment in the midlateral, dorsomedian and ventromedian lines. A, zebrafish; B, medaka; C, threespine stickleback. (A and B after Harris *et al.* 2008, C after Schluter *et al.* 2010.)

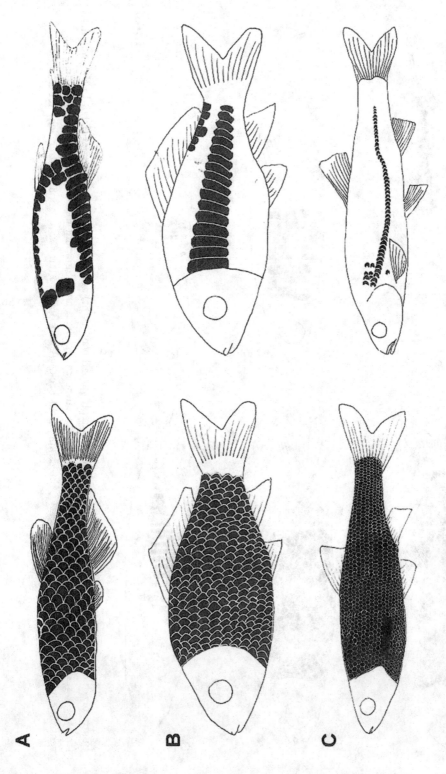

Figure 6.8 Sketches of fishes with mutations in the fibroblast growth factor pathway (right) and corresponding species or wildtypes without this mutation (left) demonstrating reduction and alignment of scales. (A left) Zebrafish wildtype; (A right) zebrafish mutant ('spiegeldanio'); (B left) wildtype carp; (B right) domesticated linear carp; (C left) *Telestes*; (C right) *Phoxinellus*. (A and B after Rohner *et al.* 2009, C after Bogutskaya and Zupancic 2003.)

mutation in the fibroblast growth factor when compared with its fully scaled sister species *Telestes* (Rohner *et al.* 2008) (Figure 6.8 bottom). Moreover, in carps *(Cyprinus carpio)* there exists a morph called linear carp with scales reduced to a single row on its flank. It has been artificially selected for reduced scales. It too has a mutation in the fibroblast growth factor pathway (Rohner *et al.* 2009) (Figure 6.8 middle). Its fitness in the wild might be reduced. But in its niche of artificial ponds it has not only survived but has been productive for food over many centuries.

By contrast, in mutants of zebrafish with reduced scales due to mutations of the ectodysplasin pathway, pleiotropic effects can be observed in fins, cranial bones and the dentition (Harris *et al.* 2008). This is exactly the combination of traits important in the diversification of *Saurichthys*. Indeed, the species in this genus differ, apart from scalation, size and several minor details, in the robustness of the dentition, the number of fin rays and fin ray segments, the number of cranial bones and their proportions (Rieppel 1985, 1992; Mutter *et al.* 2008). By chance the combination of pleiotropic effects could be selectively neutral or even positive. Again a living species proves that a mutation in the ectodysplasin pathway has not always negative pleiotropic effects and is not only viable in lab conditions. Threespine sticklebacks with such a mutation are indeed successful in the wild. They have colonized virtually all freshwater habitats of the northern hemisphere since the last ice age (Schluter *et al.* 2010). The alteration of the shape of their cranial bones is due to pleiotropic effects (or tight linkage) of the mutation in *ectodysplasin* which also causes the reduction of the number of their lateral plates (modified scales) (Albert *et al.* 2008). However, it is not certain if the same is true for the concomitant alterations in fin rays.

Extrapolating from these findings, we previously inferred a loss of function or a regulatory change of either the ectodysplasin pathway or the fibroblast growth factor pathway as a key mechanism behind the differences of scale patterning in *Saurichthys* (Schmid and Sánchez-Villagra 2010). As mentioned above, the presence of the ectodysplasin pathway in *Saurichthys* can be inferred by an extant phylogenetic bracket (Pantalacci *et al.* 2008). The fibroblast growth factor pathway, in contrast, is known to be essential for the development of many different structures and organs in vertebrates and in invertebrates (Wolpert *et al.* 2007) so that its presence in *Saurichthys* can be justifiably assumed. If we suppose an involvement of the fibroblast growth factor for scale patterning in *Saurichthys* we must assume that a gene duplication event had occurred as in teleosts (Kuraku and Meyer 2009) or in the closely related acipenserids famous for their polyploidy (Ludwig *et al.* 2001). If, conversely, the ectodysplasin pathway were to be responsible we must assume that its pleiotropic effects have led to viable morphological diversification.

Conclusion

The elucidation of the genetic underpinnings of morphological diversity of extinct organisms is a great challenge. Here, I have proposed that the reconstruction of molecular mechanisms underlying changes in characters like body shape, jaw elongation and scale reduction for a Triassic fish is feasible under reasonable assumptions. However, the results presented remain open to interpretation. Molecular reconstruction is confronted with many ambiguities and pitfalls. For instance, disparate morphologies can be controlled by similar genetic processes and similar morphologies can be achieved by different genetic mechanisms. Molecular reconstructions have to be carried out on a case by case basis. Rather than testing by falsification only, the more appropriate question is whether or not there is good evidence for the reconstruction proposed. A method yielding good evidence is phylogenetic bracketing by closely related extant (model) organisms, the molecular morphogenetic processes of which are well documented. Evidence is also provided by extrapolation from consistent data of extant organisms approaching phylogenetically the extinct organism of interest.

Present attempts, sometimes based on admittedly scanty evidence, nevertheless provide a basis for constructing specific hypotheses on genetics and development in fossil taxa. Advances in research on genetics and evolutionary development will enhance evidence and permit more detailed and more comprehensive reconstructions of the molecular mechanisms behind the morphological diversity of extinct organisms.

Summary

Saurichthys is a basal actinopterygian fish, best known from Triassic and Early Jurassic deposits worldwide. Among its particular traits are an extremely long snout, reduced scales arranged in few longitudinal lines, and an elongated body with caudally positioned pelvic, anal and dorsal fins. The comparison of these traits with discoveries made in evo-devo research on extant organisms serves as the basis to reconstruct some molecular mechanisms which operated in the fossil.

The reconstruction of molecular underpinnings of traits of fossils would be straightforward if evolution were easy to predict. However, as it depends on contingencies, molecular reconstructions remain hypothetical. Nevertheless, their level of support can be assessed using phylogenetic bracketing, extrapolation, and information derived from developmental studies of extant organisms. Molecular reconstructions take advantage of the growing insight that decisive molecular mechanisms are highly conserved. By contrast, caution is required since evolution of similar traits can be caused by different molecular mechanisms.

The elongation of the jaws and snout in *Saurichthys* can be compared with studies on beak lengths of Darwin's finches and face length of Carnivora, suggesting that expression of calmodulin and/or runt-related transcription factor 2 played a role. Scale reduction, scale elongation and their arrangement in longitudinal lines of *Saurichthys* is paralleled by mutations of ectodysplasin and fibroblast growth factor pathways in zebrafish, medaka and carp, indicating that these molecules were involved in the morphological diversification of the fossil fish. Body elongation and fin position can be compared with investigations on the length and regionalization of the axial skeleton of extant vertebrates. They suggest a shift of the anterior expression boundary of certain *Hox* genes (influenced by retinoic acid, some wingless, fibroblast growth factor and growth differentiation factor) for the characteristic shape and fin position of *Saurichthys*.

Acknowledgements

I am very grateful to my mentor Marcelo R. Sánchez-Villagra for his support and for the revision and improvement of this work. I thank the two anonymous reviewers for their valuable comments which helped considerably to improve the text. Many thanks to Rob Asher for his comments and his careful corrections. I thank also Matthew Cartner for linguistic help.

REFERENCES

Abzhanov, A., Protas, M., Grant, B. R., Grant, P. R. and Tabin, C. J. (2004). Bmp4 and morphological variation of beaks in Darwin's finches. *Science*, **305**(5689), 1462–5.

Abzhanov, A., Kuo, W. P., Hartmann, C., *et al.* (2006). The calmodulin pathway and evolution of elongated beak morphology in Darwin's finches. *Nature*, **442**(7102), 563–7.

Albert, A. Y., Sawaya, S., Vines, T. H., *et al.* (2008). The genetics of adaptive shape shift in stickleback: pleiotropy and effect size. *Evolution*, **62**(1), 76–85.

Albertson, R. C., Streelman, J. T., Kocher, T. D. and Yelick, P. C. (2005). Integration and evolution of the cichlid mandible: the molecular basis of alternate feeding strategies. *Proceedings of the National Academy of Sciences of the United States of America*, **102**(45), 16 287–92.

Alexander, T., Nolte, C. and Krumlauf, R. (2009). Hox genes and segmentation of the hindbrain and axial skeleton. *Annual Review of Cell Developmental Biology*, **25**, 431–56.

Arendt, J. and Reznick, D. (2008). Convergence and parallelism reconsidered: what have we learned about the genetics of adaptation? *Trends in Ecology and Evolution*, **23**(1), 26–32.

Asher, R. J., Lin, K. H., Kardjilov, N. and Hautier, L. (2011). Variability and constraint in the mammalian vertebral column. *Journal of Evolutionary Biology*, **24**(5), 1080–90.

Aulehla, A. and Pourquié, O. (2010). Signaling gradients during paraxial mesoderm development. *Cold Spring Harbor Perspectives in Biology*, **2**, a000869.

Barabási, A. L. and Oltvai, Z. N. (2004). Network biology: understanding the cell's functional organization. *Nature Reviews Genetics*, **5**(2), 101–13.

Benton, M. J. (2009). Paleontology and the history of life. In *Evolution: The First Four Billion Years*, ed. M. Ruse and J. Travis, pp. 80–104. Cambridge, MA: Harvard University Press.

Benton, M. J. (2010). Studying function and behavior in the fossil record. *PLoS Biology*, **8**(3), e1000321.

Bird, N. C. and Mabee, P. M. (2003). Developmental morphology of the axial skeleton of the zebrafish, *Danio rerio* (Ostariophysi: Cyprinidae). *Developmental Dynamics*, **228**(3), 337–57.

Bogutskaya, N. and Zupancic, P. (2003). *Phoxinellus pseudalepidotus* (Teleostei: Cyprinidae), a new species from the Neretva basin with an overview of the morphology of *Phoxinellus* species of Croatia and Bosnia-Herzegovina. *Ichthyological Exploration of Freshwaters*, **14**, 369–83.

Bürgin, T. (1990). Reproduction in Middle Triassic actinopterygians; complex fin structures and evidence of viviparity in fossil fishes. *Zoological Journal of the Linnean Society*, **100**, 379–91.

Bürgin, T. (1999). Middle Triassic marine fish faunas from Switzerland. In *Mesozoic Fishes*, Vol. 2 *Systematics and Fossil Record*, ed. G. H. Arratia and H. P. Schultze, pp. 481–94. München, Germany: Pfeil.

Burke, A. C., Nelson, C. E., Morgan, B. A. and Tabin, C. (1995). Hox genes and the evolution of vertebrate axial morphology. *Development*, **121**, 333–46.

Carroll, S. B. (2008). Evo-devo and an expanding evolutionary synthesis: a genetic theory of morphological evolution. *Cell*, **134**, 25–36.

Cloutier, R. (2010). The fossil record of fish ontogenies: insights into developmental patterns and processes. *Semin Cell Developmental Biology*, **21**(4), 400–13.

Conant, C. G. and Wagner, A. (2003). Convergent evolution of gene circuits. *Nature Genetics*, **34**(3), 264–6.

Conway Morris, S. (2003). *Life's Solution: Inevitable Humans in a Lonely Universe*. Cambridge, UK: Cambridge University Press.

Conway Morris, S. (2010). Evolution: like any other science it is predictable. *Philosophical Transactions of the Royal Society B*, **365**, 133–45.

Darwin, C. (1859). *On the Origin of Species by Means of Natural Selection, or the Preservation of Favoured Races in the Struggle for Life*. London: John Murray.

Davidson, E. H. (2006). *The Regulatory Genome*. Burlington, MA: Academic Press.

Donoghue, P. C. J. (2002). Evolution of development of the vertebrate dermal and oral skeletons: unraveling concepts, regulatory theories, and homologies. *Paleobiology*, **28**, 474–507.

Donoghue, P. C. and Purnell, M. A. (2005). Genome duplication, extinction and vertebrate evolution. *Trends in Ecology and Evolution*, **20**(6), 312–9.

Donoghue, P. C., Graham, A. and Kelsh, R. N. (2008). The origin and evolution of the neural crest. *BioEssays*, **30**(6), 530–41.

Durand, D. and Hoberman, R. (2006). Diagnosing duplications – can it be done? *Trends in Genetics*, **22**(3), 156–64.

Elmer, K. R., Kusche, H., Lehtonen, T. K. and Meyer, A. (2010). Local variation and parallel evolution: morphological and genetic diversity across a species complex of neotropical crater lake cichlid fishes. *Philosophical Transactions of the Royal Society B*, **365**(1547), 1763–82.

Fraser, G. J., Hulsey, C. D., Bloomquist, R. F., *et al.* (2009). An ancient gene network is co-opted for teeth on old and new jaws. *PLoS Biology*, **7**(2), e1000031.

Freitas, R., Zhang, G. and Cohn, M. J. (2006). Evidence that mechanisms of fin development evolved in the midline of early vertebrates. *Nature*, **442**(7106), 1033–7.

Furrer, H. (2003). *Der Monte San Giorgio im Südtessin – vom Berg der Saurier zur Fossil-Lagerstätte internationaler Bedeutung*. Zürich: Naturforschende Gesellschaft.

Gao, X. and Lynch, M. (2009). Ubiquitous internal gene duplication and intron creation in eukaryotes. *Proceedings of the National Academy of Sciences of the United States of America*, **106**(49), 20 818–23.

Gardiner, B. G., Schaeffer, B. and Masserie, J. A. (2005). A review of lower actinopterygian phylogeny. *Zoological Journal of the Linnean Society*, **14**, 511–25.

Gerhart, J. and Kirschner, M. (2007). The theory of facilitated variation. *Proceedings of the National Academy of Sciences of the United States of America*, **104** Suppl 1, 8582–9.

Gibb, S., Maroto, M. and Dale, J. K. (2010). The segmentation clock mechanism moves up a notch. *Trends in Cell Biology*, **20**(10), 593–600.

Gitton, Y., Heude, E., Vieux-Rochas, M., *et al.* (2010). Evolving maps in craniofacial development. *Semin Cell Developmental Biology*, **21**(3), 301–8.

Gomez, C. and Pourquié, O. (2009). Developmental control of segment numbers in vertebrates. *Journal of Experimental Zoology B–Molecular and Developmental Evolution*, **312**(6), 533–44.

Gompel, N. and Prud'homme, B. (2009). The causes of repeated genetic evolution. *Developmental Biology*, **332**, 36–47.

Gould, J. S. (1990). *Wonderful Life*. London: Hutchinson Radius.

Grande, L. and Hilton, E. J. (2006). An exquisitely preserved skeleton representing a primitive sturgeon from the Upper Cretaceous Judith river formation of Montana (Acipenseriformes: Acipenseridae: N.gen and sp.). *Journal of Paleontology*, **80** (suppl 4), 1–39.

Griffith, J. (1962). The Triassic fish *Saurichthys krambergeri* Schlosser. *Palaeontology*, **5**(2), 344–54.

Grimm, H. (1979). Veränderungen in der Variabilität von Populationen des Zahnkarpfens *Aphanius anatoliae* (Leidenfrost 1912) während 30 Jahren: 1943–1974. *Journal of Zoological Systematics and Evolutionary Research*, **17**(4), 272–80.

Hall, B. K. (1975). Evolutionary consequences of skeletal development. *American Zoology*, **15**, 329–50.

Harris, M. P., Rohner, N., Schwarz, H., *et al.* (2008). Zebrafish eda and edar mutants reveal conserved and ancestral roles of ectodysplasin signaling in vertebrates. *PLoS Genetics*, **4**, e1000206.

Helfman, G. S., Collette, B. B., Facey, D. E. and Bowen, B. W. (2009). *The Diversity of Fishes*, 2nd edn. Oxford, UK: Wiley-Blackwell.

Hoekstra, H. E. and Nachman, M. W. (2003). Different genes underlie adaptive melanism in different populations of rock pocket mice. *Molecular Ecology*, **12**(5), 1185–94.

Holley, S. A. (2007). The genetics and embryology of zebrafish metamerism. *Developmental Dynamics*, **236**(6), 1422–49.

Johanson, Z., Sutija, M. and Joss, J. (2005). Regionalization of axial skeleton in the lungfish *Neoceratodus forsteri* (Dipnoi). *Journal of Experimental Zoology B–Molecular and Developmental Evolution*, **304**(3), 229–37.

Johanson, Z., Kearsley, A., den Blaauwen, J., Newman, M. and Smith, M. M. (2010a). No bones about it: an enigmatic Devonian fossil reveals a new skeletal framework – a potential role of loss of gene regulation. *Semin Cell Developmental Biology*, **21**(4), 414–23.

Johanson, Z., Carr, R. and Ritchie, A. (2010b). Fusion, gene misexpression and homeotic transformations in vertebral development of the gnathostome stem group (Placodermi). *International Journal of Developmental Biology*, **54**(1), 71–80.

Kondo, S., Kuwahara, Y., Kondo, M., *et al.* (2001). The medaka rs-3 locus required for scale development encodes ectodysplasin-A receptor. *Current Biology*, **11**, 1202–6.

Kuraku, S. and Meyer, A. (2009). The evolution and maintenance of Hox gene clusters in vertebrates and the teleost-specific genome duplication. *International Journal of Developmental Biology*, **53**, 765–73.

Ludwig, A., Belfiore, N. M., Pitra, C., Svirsky, V. and Jenneckens, I. (2001). Genome duplication events and functional reduction of ploidy levels in sturgeon (*Acipenser, Huso* and *Scaphirhynchus*). *Genetics*, **158**, 1203–15.

Mallo, M., Wellik, D. M. and Deschamps, J. (2010). Hox genes and regional patterning of the vertebrate body plan. *Developmental Biology*, **344**(1), 7–15.

Mercader, N. (2007). Early steps of paired fin development in zebrafish compared with tetrapod limb development. *Development, Growth and Differentiation*, **49**(6), 421–37.

Minoux, M. and Rijli, F. M. (2010). Molecular mechanisms of cranial neural crest cell migration and patterning in craniofacial development. *Development*, **137**(16), 2605–21.

Moyle, P. B. and Cech, J. J. (2000). *Fishes: An Introduction to Ichthyology*, 4th edn. Upper Saddle River, NJ: Prentice Hall.

Müller, J., Scheyer, T. M., Head, J. J., *et al.* (2010). Homeotic effects, somitogenesis and the evolution of vertebral numbers in recent and fossil amniotes. *Proceedings of the National Academy of Sciences of the United States of America*, **107**, 2118–23.

Mundy, N. I. (2009). Conservation and convergence of colour genetics: MC1R mutations in brown cavefish. *PLoS Genetics*, **5**(2), e1000388.

Murata, Y., Tamura, M., Aita, Y., *et al.* (2010). Allometric growth of the trunk leads to the rostral shift of the pelvic fin in teleost fishes. *Developmental Biology*, **347**(1), 236–45.

Mutter, R. J. and Herzog, A. (2004). A new genus of Triassic actinopterygian with an evaluation of deepened flank scales in fusiform fossil fishes. *Journal of Vertebrate Paleontology*, **24**, 794–801.

Mutter, R. J., Cartanyà, J. and Basaraba, A. U. (2008). New evidence of *Saurichthys* from the Lower Triassic with an evaluation of early saurichthyid diversity. In *Mesozoic Fishes*, Vol. 4, ed. G. Arratia, H. P. Schultze and V. H. Wilson, pp. 103–27. München, Germany: Pfeil.

Nelson, J. S. (2006). *Fishes of the World*, 4th edn. Hoboken, NJ: Wiley & Sons.

Nielsen, E. (1949). Studies on Triassic fishes from East Greenland. II. *Australosomus* and *Birgeria*. *Palaeozoologica Groenlandica*, **146**(1), 1–309.

Pantalacci, S., Chaumot, A., Benoît, G., *et al.* (2008). Conserved features and evolutionary shifts of the EDA signaling pathway involved in vertebrate skin appendage development. *Molecular Biology and Evolution*, **25**, 912–28.

Parsons, K. J. and Albertson, R. C. (2009). Roles for Bmp4 and CaM1 in shaping the jaw: evo-devo and beyond. *Annual Reviews in Genetics*, **43**, 369–88.

Popper, K. R. (1963). *Conjectures and Refutations: The Growth of Scientific Knowledge*. London: Paul.

Renesto, R. and Stockar, R. (2009). Exceptional preservation of embryos in the actinopterygian *Saurichthys* from the Middle Triassic of Monte San Giorgio, Switzerland. *Swiss Journal of Geoscience*, **102**, 323–30.

Rieppel, O. (1980). Additional specimens of *Saurichthys madagascariensis* Piveteau, from the Eotrias of Madagascar. *Neues Jahrb für Geologie und Paläeontologie–Monatshefte*, **1980**, 43–51.

Rieppel, O. (1985). Die Triasfauna der Tessiner Kalkalpen. XXV. Die Gattung *Saurichthys* (Pisces, Actinopterygii) aus der mittleren Trias des Monte San Giorgio, Kanton Tessin. *Schweizerische Paläontologische Abhandlungen*, **108**, 1–103.

Rieppel, O. (1992). A new species of the genus *Saurichthys* (Pisces: Actinopterygii) from the Middle Triassic of Monte San Giorgio (Switzerland), with comments on the phylogenetic interrelationships of the genus. *Palaeontographica Abteilung A*, **221**, 63–94.

Rohner, N., Harris, M., Bercsényi, M., Orban, L. and Nüsslein-Volhard, C. (2008). Fibroblast growth factor signaling in skeletal evolution. *Developmental Biology*, **319**, 499.

Rohner, N., Bercsényi, M., Orban, L., *et al.* (2009). Duplication of fgfr1 permits Fgf signaling to serve as a target for selection during domestication. *Current Biology*, **19**, 1642–7.

Salzburger, W. and Meyer, A. (2004). The species flocks of East African cichlid fishes: recent advances in molecular phylogenetics and population genetics. *Naturwissenschaften*, **91**(6), 277–90.

Schluter, D., Marchinko, K. B., Barrett, R. D. and Rogers, S. M. (2010). Natural selection and the genetics of adaptation in threespine stickleback. *Philosophical Transactions of the Royal Society B–Biological Sciences*, **365**(1552), 2479–86.

Schmid, L. and Sánchez-Villagra, M. R. (2010). Potential genetic bases of morphological evolution in the Triassic fish *Saurichthys*. *Journal of Experimental Zoology B–Molecular and Developmental Evolution*, **314B**(7), 519–26.

Schröter, C. and Oates, A. C. (2010). Segment number and axial identity in a segmentation clock period mutant. *Current Biology*, **20**(14), 1254–8.

Schwarz, W. (1970). Die Triasfauna der Tessiner Kalkalpen. XX. *Birgeria stensiöi* Aldinger. *Schweizerische Paläontologische Abhandlungen*, **89**, 1–93.

Scotland, R. W. (2010). Deep homology: a view from systematics. *BioEssays*, **32**, 438–49.

Sears, K. E., Goswami, A., Flynn, J. J. and Niswander, L. A. (2007). The correlated evolution of Runx2 tandem repeats, transcriptional activity, and facial length in Carnivora. *Evolution and Development*, **9**(6), 555–65.

Shapiro, M. D., Bell, M. A. and Kingsley, D. M. (2006). Parallel genetic origins of pelvic reduction in vertebrates. *Proceedings of the National Academy of Sciences of the United States of America*, **103**(37), 13 753–8.

Shubin, N., Tabin, C. and Carroll, S. (1997). Fossils, genes and the evolution of animal limbs. *Nature*, **388**(6643), 639–48.

Shubin, N., Tabin, C. and Carroll, S. (2009). Deep homology and the origins of evolutionary novelty. *Nature*, **457**, 818–23.

Sire, J. Y. and Akimenko, M. A. (2004). Scale development in fish: a review, with description of sonic hedgehog (shh) expression in the zebrafish (*Danio rerio*). *International Journal of Developmental Biology*, **48**(2–3), 233–47.

Stanley, S. M. (2005). *Earth System History*. New York: Freeman.

Steiner, C. C., Römpler, H., Boettger, L. M., Schöneberg, T. and Hoekstra, H. E. (2009). The genetic basis of phenotypic convergence in beach mice: similar pigment patterns but different genes. *Molecular Biology and Evolution*, **26**(1), 35–45.

Stern, D. L. and Orgogozo, V. (2009). Is genetic evolution predictable? *Science*, **323**, 746–51.

Suzuki, D., Brandley, M. C. and Tokita, M. (2010). The mitochondrial phylogeny of an ancient lineage of ray-finned fishes (Polypteridae) with implications for the evolution of body elongation, pelvic fin loss, and craniofacial morphology in Osteichthyes. *BMC Evolutionary Biology*, **10**, 209.

Van Fraassen, B. C. (1980). *The Scientific Image*. Oxford, UK: Clarendon.

Ward, A. B. and Brainerd, E. L. (2007). Evolution of axial patterning in elongate fishes. *Biological Journal of the Linnean Society*, **90**, 97–116.

Ward, A. B. and Mehta, R. S. (2010). Axial elongation in fishes: using morphological approaches to elucidate developmental mechanisms in studying body shape. *Integrative and Comparative Biology*, **50**, 1106–19, doi:10.1093/icb/icq029.

Weinreich, D. M., Delaney, N. F., DePristo, M. A. and Hartl, D. L. (2006). Darwinian evolution can follow only very few mutational paths to fitter proteins. *Science*, **312**(5770), 111–14.

Wellik, D. M. (2007). Hox patterning of the vertebrate axial skeleton. *Developmental Dynamics*, **236**(9), 2454–63.

Wildekamp, R. H., Küçük, F., Ünlüsayin, M. and Neer, W. V. (1999). Species and subspecies in the genus *Aphanius* Nardo 1897 (Pisces Cyprinodontidae) in Turkey. *Turkish Journal of Zoology*, **23**, 23–44.

Witmer, L. M. (1995). The extant phylogenetic bracket and the importance of reconstructing soft tissues in fossils. In *Functional Morphology in Paleontology*, ed. J. Thomason, pp. 19–33. Cambridge, UK: Cambridge University Press.

Wolpert, L., Jessel, T., Lawrence, P., *et al.* (2007). *Principles of Development*, 3rd edn. Oxford, UK: Oxford University Press.

Woltering, J. M. and Duboule, D. (2010). The origin of digits: expression patterns versus regulatory mechanisms. *Developmental Cell*, **18**(4), 526–32.

Wood, T. E., Burke, J. M. and Rieseberg, L. H. (2005). Parallel genotypic adaptation: when evolution repeats itself. *Genetica*, **123**, 157–70.

Wray, G. A. (2007). The evolutionary significance of cis-regulatory mutations. *Nature Reviews Genetics*, **8**, 206–16.

Wu, F., Sun, Y., Hao, W., *et al.* (2009). New species of *Saurichthys* (Actinopterygii: Saurichthyidae) from middle Triassic (Anisian) of Yunnan province, China. *Acta Geologica Sinica*, **83**, 440–50.

7

A molecular guide to regulation of morphological pattern in the vertebrate dentition and the evolution of dental development

MOYA SMITH AND ZERINA JOHANSON

Introduction

The emphasis in this chapter is on the evolution of the gnathostome dentition, with developmentally reiterated units (individual tooth germs) organized in a specific temporal and spatial pattern in their initiation and for their replacement. The genetic and molecular controls of this patterning process are much less well known than those of the morphogenetic events of single tooth production that is well documented (Thesleff *et al.* 2007; Bei 2009). The individual morphogenetic units develop to become functional teeth in the phenotype and are organized along the jaws or pharyngeal arches for shape, size, spacing and timing of replacement, specific for each tooth in their proximal–distal position. Research has begun to identify the genes involved in patterning for tooth site and tooth replacement via *in situ* hybridization probes to active genes and by generic antibodies to the gene products. In particular, studies of mechanisms for regulation of tooth replacement have benefited by those extended beyond mammals to fishes (Fraser *et al.* 2004, 2006a, 2006b, 2008, 2009) and reptiles such as snakes and lizards (Buchtova *et al.* 2008; Vonk *et al.* 2008; Handrigan *et al.* 2010; Handrigan and Richman 2010a, 2010b).

As in all structures derived from epithelium, including feathers, scales and hair, teeth are formed initially by co-operative interaction between the epithelium and underlying mesenchyme, but for teeth the contribution of cranial neural crest to the process is critical, as reviewed by Smith and Hall (1993) and recently in mammals (Miletich and Sharpe 2004). Initially, in tooth formation the neural-crest-derived mesenchymal condensation is related to a thickened epithelial band, restricting the regions that can make teeth to this odontogenic

From Clone to Bone: The Synergy of Morphological and Molecular Tools in Palaeobiology,
ed. Robert J. Asher and Johannes Müller. Published by Cambridge University Press.
© Cambridge University Press 2012.

band, as an early morphogenetic process. Both cell types are also identified by different localized, intensified gene expression at equivalent times to the morphological stages (examples are *Otlx2*, *Pitx2*, *Fgf8*, *Bmp4* in the epithelium and *Pax9*, *Barx1*, *Msx1*, *Msx2* in the mesenchyme). This topic, of the localization of the odontogenic cells and early signalling patterns before tooth formation, has been reviewed from comparison between mouse and both osteichthyan and chondrichthyan fish, to suggest that there is much conservation (Smith *et al.* 2009b; Fraser *et al.* 2010). Interestingly, in the toothless birds the early epithelial odontogenic signalling pathways have been conserved (Chen *et al.* 2000) whereas those of the mesenchyme are lost. Teeth were restored in chick embryos after mouse neural crest transplantations (Mitsiadis *et al.* 2003) and chick *Pitx2* expressed in the epithelium with mouse *Msx1* and *Barx1* expressed in the neural crest cells of the dental mesenchyme. This is a good example from experimental data to show that the complementary molecular information can inform morphogenetic studies. In this example, molecular data illustrate the importance of cranial neural-crest-derived mesenchyme and its timing of migration, to the odontogenic process and the essential reciprocal interaction of gene products from both cell layers.

As dentitions comprise developmentally reiterated modular units (individual tooth germs), these also exhibit at each tooth site reciprocal gene expression activities progressively in developmental time, as demonstrated in the mammalian dentition where 'the same signaling cascade is used reiteratively throughout tooth development' (Jernvall and Thesleff 2000). The same unit structure (tooth germ) and interactive molecular interactions can also be shown in osteichthyan examples (Figure 7.1) as the units are said to develop as semi-autonomous genetic modules, where each can undergo phenotypic evolution (Stock 2001), as will be discussed later (see 'The first dentition'). As Stock (2001) discusses, there may be both correlated and independent evolution of groups of teeth, to explain region-specific tooth loss as a common trend in vertebrate evolution. It is important to note that Stock regards the dentition itself as a module of the dermal exoskeleton (in evolutionary terms, not developmental), but with long-acquired independent genetic control. This is a topic central to our debate in the following sections, as we consider an alternative evolutionary acquisition, not from the external dermal skeleton but from the pharyngeal endoskeleton (Smith and Coates 1998).

Determining how dentitions evolved during diversification of vertebrates from the early Palaeozoic onwards continues to be controversial and is still much debated (Smith and Coates 2000; Smith 2003; Huysseune *et al.* 2009; Fraser *et al.* 2010; Ohazama *et al.* 2010). Of particular interest, and to which we attach greater significance, is the evolution of patterning mechanisms

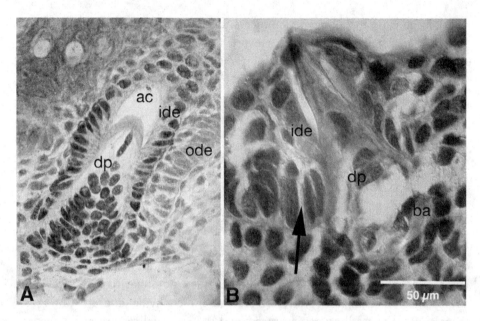

Figure 7.1 (A) *Oncorhynchus mykiss* day 10: Section through developing oral tooth with generic antibody for Pax gene product: this enhances the active inner dental epithelium (ide), and ectomesenchyme of the papilla (dp) (contra the outer dental epithelium (ode)). This shows the co-operative interaction between the two during histogenesis and after the acrodin cap (ac) has formed. (B) *Neoceratodus forsteri* hatchling with mineralized lower jaw tooth at completed stage. Because of the very small size there is only a single column of odontoblasts (dp, compared with A). (Adapted from Smith *et al.* 2009c.) All tissues are present (excluding acrodin, never present in dipnoan teeth) but including bone of attachment (ba) and site of next tooth germ (arrow). (See also colour plate.)

associated with the dentition. In this review we will consider which gene networks control tooth patterning with their intrinsic signalling pathways, and how these may have evolved. Adopting the principle that evolution progresses by co-option of an existing system, did these gene networks evolve by co-option of those associated with external odontodes in the skin (e.g. denticle ornament on dermal scales), or from internal visceral skeletal structures such as denticles attached to the branchial arches?

Of course, molecular information should advance theories derived from structural data, especially those restricted to the fossil record, to provide a potentially evolving developmental mechanism. Could a comparison of gene networks allow us to determine whether unitary teeth evolved from the co-option of mechanisms and networks associated with a skin odontode-based system or a pharyngeal denticle-based system? Both of these concepts embrace the mechanism that heterotropy must occur to ensure teeth are positioned at the margins of the jaws, either from the skin ectoderm, or from the pharyngeal endoderm. On phylogenetic criteria (odontodes present in the conodont pharynx but not in the skin), Donoghue and Sansom (2002) favoured the endoskeletal origin of odontodes with an early shift to the skin of armoured agnathans, but this raised an important issue, as it questions one of the axioms of skeletal evolution, namely, given that the dermal and endoskeletal skeletons have distinct developmental and phylogenetic origins (Patterson 1977), the transfer of endodermal odontodes to the skin is contrary to this generally accepted hypothesis (Donoghue and Sansom 2002). Of significance is the proposal that endoderm has more appropriate gene networks to be recruited for evolution of tooth patterning (Smith 2003). Can this theory be tested and have these alternative networks (skin versus pharynx) been characterized to the degree that we can make this distinction? Experiments have recently been reported in mammals (Ohazama *et al.* 2010) that test these theories, comparing the molecular characteristics of pharyn-geal/foregut endoderm with the molecular characteristics of oral ectoderm during mouse development. They suggest that as teeth develop from ectoderm in the mouse but from endoderm in fish this indicates that teeth can develop and possibly evolve via different mechanisms (see 'The first dentition' below). They take this further by accepting that the two different theories may be correct and can account for 'two different mechanisms of tooth develop-ment that may have provided the developmental and genetic diversity on which evolution has acted to produce heterodont dentitions in mammals' (Ohazama *et al.* 2010). We would draw attention to the known heterodont dentitions in chondrichthyan (Port Jackson shark) and osteichthyan fish (even cichlid pharyngeal teeth), an unexplored area of genetic control. Also an important mechanism in this debate, recently developed, is how evolution-ary change may have occurred, such as 'tinkering' with the genes, where minor changes in gene regulation can produce notable developmental or morphological changes (Tummers and Thesleff 2009) (see 'The first denti-tion' below).

In this chapter we will also discuss how recent advances in understanding the molecular basis of developmental mechanisms involved in living vertebrate dentitions can be used to interpret the development of the dentition in early vertebrates, dominated by fossil taxa such as the Conodonta, Thelodonti, Anaspida, Osteostraci and Placodermi (Figure 7.2). Three of these taxa, conodonts, thelodonts and placoderms, are resolved to key points on the phylogeny and will feature in our review. Can patterns of gene expression and

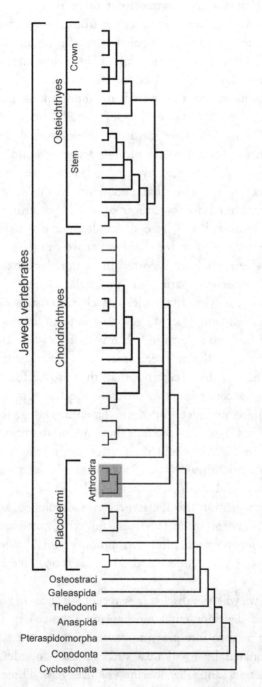

Figure 7.2 Cladogram showing relationships of the Vertebrata, including jawed and jawless vertebrates. (Compiled from Brazeau 2009 and Donoghue and Purnell 2005.) Note paraphyletic Placodermi, following Brazeau (2009).

mechanisms associated with the development of teeth in living vertebrates be extrapolated to fossil taxa? Can molecules and morphology be combined in fossils to illuminate evolution of vertebrate dentitions? When early vertebrate taxa are considered, evolution of major characters such as dentitions are seen to evolve in a stepwise manner with macroevolution, as characters are acquired at successive nodes.

Naturally, much of the new molecular work has focused on laboratory animals such as the mouse (Jernvall and Thesleff 2000), but other mammals such as the shrew and ferret have been used and also more phylogenetically basal extant vertebrates such as python, lungfish, trout and cichlids (Fraser *et al.* 2004, 2006a, 2006b, 2008, 2009, 2010; Buchtova *et al.* 2008; Handrigan and Richman 2010a, 2010b). Comparative molecular data that demonstrate differences in the timing and position of gene activity within development and growth of the dentition in extant forms can illuminate the developmental mechanisms that produce dental diversity. These methods use *in situ* hybridization with cDNA probes for genes that are active at the time of initiation of tooth production and regulation of each tooth unit into a spatial and temporal pattern. One example is from the work on dental diversity in cichlid species with their very distinct tooth morphologies, row spacing and multiple tooth rows (Streelman *et al.* 2003; Fraser *et al.* 2008; Figure 7.3).

As well, targeted gene knockouts in model animals such as the mouse, chick and also the zebrafish can create modified dentition morphologies. Many alter tooth number and also tooth shape and these changes in structural features can replicate morphologies characterizing more primitive taxa, and changes that occur in wild populations (Line 2003). All these data illuminate how the molecular mechanisms of development might have been employed in the evolution of morphological diversity. It has been noted that genetic variations within a population can modulate evolutionary changes in dental patterning and may promote morphological diversity. Using single and double knockout mice, Line (2003) showed that gene mutations affect tooth development and by comparison with phenotypic variation in sciurids and mice, these may be translated into macroevolutionary changes in morphology.

We examine these questions in the context of two current theories as to the origin and evolution of dentitions: the 'inside-out' theory versus the 'outside-in' theory. Most recently, Huysseune *et al.* (2009) and Fraser *et al.* (2010) have proposed modifications to these theories. Importantly, Ohazama *et al.* (2010) have suggested that both theories with their different mechanisms have evolved to produce the mammalian heterodont dentition with different genetic parameters in each region.

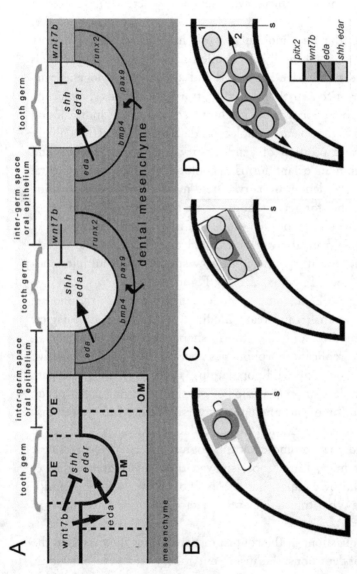

Figure 7.3 (A–D) Gene expression and tooth patterning in Malawi cichlids (figure and caption adapted from Fraser *et al.* 2008). (A) first tooth germ forms in region of restricted *shh* expressing oral epithelium (OE, in which the odontogenic band has formed DE) through interaction with dental mesenchyme (DM, genes in both, *shh, edar, eda, wnt7b,* + *bmp4, pax9, runx2*). (B–C) Oral views of subsequent primary tooth germs (1) show involvement of six genes, some in the zone of inhibition (*wnt7b, eda, shh*) establishing spacing of sequential tooth location in odontogenic band, through interaction with dental mesenchyme (DM, *eda, bmp4, pax9, runx2*); OM, oral mesenchyme without this gene expression. (D) Sequentially added teeth (2) to extend primary row (1) and form new rows with spacing regulated within the *pitx2*-positive field. Each new tooth row employs a 'copy-and-paste' mechanism using the same genes from the previous tooth row to pattern the dentition, until this mechanism fails to restrict tooth number, or row number. Line (s) is the jaw symphysis.

Historical concepts questioned: with molecular perspectives on structural pattern

Old theories

In the first instance, the validity of the historical concept that external skin denticles gave rise to teeth on the jaws has to be addressed, known by comparison with newer theories as the 'outside-in' theory of dental evolution. The dentition of jawed vertebrates is assumed to have evolved after modification of the mandibular arch to form jaws. Teeth individually were considered to derive from the non-growing, scattered skin denticles in sharks (Reif 1982), by their migration into the mouth, hence creating a new site for tooth production. However, if this theory is accepted, then how did the distinctive pattern differences arise between the ordered dentition and superficial and unordered placoid denticles in extant chondrichthyans, as discussed by Fraser and Smith (2011; Figure 7.4)? Was it simply through evolution of a new structure, the dental lamina (Reif 1982), as exemplified in sharks (Figure 7.5)? Many papers have been published since Reif's that illustrate how the debate is still unresolved (Smith and Coates 1998, 2000, 2001; Smith 2003; Fraser *et al.* 2006b, 2010; Huysseune *et al.* 2009; Fraser and Smith 2011). Molecular details from data on gene regulation show how the pattern can be achieved and is the goal of both theories as discussed below.

Figure 7.4 (A–C) *Scyliorhinus caniculus*, MicroX-ray tomogram of juvenile (BL- 27.5 cm). Contrast between dentition patterned into vertical rows developed from a dental lamina, and placoid scales with shape polarity but without space, size and time order. (A) Labial view of anterior jaw margin showing ordered tooth sets of functional and replacement sets, including smaller size of symphyseal and parasymphyseal (sy, p.sy). External dermal denticles (Sk.d) at oral margin packed into available space on the jaw, flattened non-cusped crown. (B, C) Placoid scales in the dorsal skin of same specimen, anterior is to the right. B is in external view and C in visceral view at same magnification. Note irregular spacing of their arrangement and size differences of the crown and base. Varying times of formation are shown both by small (oldest), or wide pulp openings (youngest) and from white (oldest) to grey (youngest) for level of mineralization. (Adapted from Fraser and Smith 2011.)

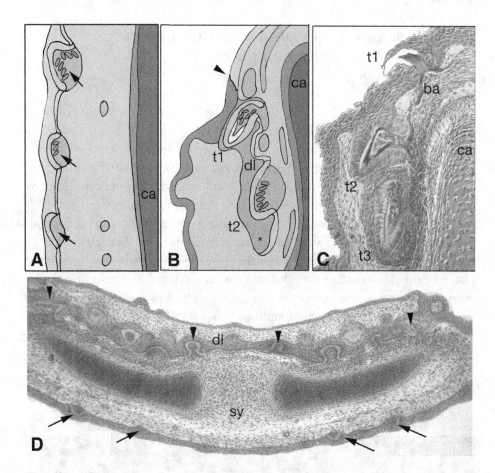

Figure 7.5 (A–D) Comparison of external skin denticle development (placoid scales), with teeth from early ontogenetic stages of *Scyliorhinus caniculus* (Chondrichthyes) embryos in sectioned views. (A) Schematic of three early epithelial buds for denticles (arrows) formed through interaction between the skin epithelium with dermal papillae of the external surface of the jaw (ca, cartilage); this is superficial development and they do not form from an invaginated epithelium. (B) Schematic of early stage of tooth development, lingual aspect of lower jaw, with tooth germs t1 and t2 formed within a dental lamina (dl), equivalent to those in (C) at eruption stage; oral epithelium boundary with skin epithelium (arrow head). Primordial tooth bud (asterix) develops within the expanded extension of the dental lamina (dl). (C) A photomicrograph of 5-day post-hatchling at start of tooth eruption, oldest tooth (t1) to newest (t3) developing in a dental lamina, with all tissues formed, including enameloid (white space) on tooth crown of t1, and bone of attachment (ba) (ca, jaw cartilage). (D) Coronal section of embryo to show continuous dental lamina (dl) across lower jaw symphysis (sy) and 14 tooth buds (seven each side – between arrow heads). These are well advanced whereas skin denticles are only just beginning to develop (arrows). These are superficial, not on a dental lamina, in the ventral ectoderm of the jaw and develop as shown in A. (A–C adapted from Smith *et al.* 2009a.)

Figure 7.6 *Loganellia scotica* (Thelodonti), pharyngeal denticle whorls. (A) Pharyngeal denticle whorls *in situ*, showing common orientation of denticles within the whorl. (B) Individual whorl from the pharyngeal region and one from the oral region with larger separate skin denticle on the outside and smaller joined ones on the inside. (Used with permission, from Smith and Coates 2001: figs. 14.1B, G, H.)

Each specific dentition pattern involves the spatiotemporal order of tooth induction, referred to as the 'dental pattern' and gives rise to the order and number of teeth in the functional dentition. This property was assumed to arise in the innovative subepithelial dental lamina that originated when skin denticles migrated into the mouth to create a meristic series of teeth (Reif 1982). In order to discuss the origins of a dental lamina and evolution of vertebrate dental patterns (Smith *et al.* 2009b), it is important to understand the spatial and temporal parameters of these dental patterns because they make each higher-level vertebrate clade distinctively different with respect to their dentition (Smith 2003).

Secondly, following the same principle of using an existing structure to convert into another (e.g. skin denticles into teeth), an alternative hypothesis based on new fossil data was proposed, suggesting that iterative sets of denticle whorls in the pharynx of the jawless fossil taxon Thelodonti (Van der Bruggen 1993) could have been co-opted for a tooth succession system ('inside-out' hypothesis). Here, denticles were ordered in time and space, forming whorls with denticles showing a common orientation (spatial patterning; Figure 7.6). Denticle bases were thicker with closed pulp canals in older denticles, and a thinner base and more open canals in newly added denticles, indicating a temporal patterning with respect to addition. Hence it was suggested that the spatiotemporal patterning characterizing the vertebrate dentition evolved from these ordered sets on the pharyngeal arches (Smith and Coates 2001). These whorls are homologous with those of pharyngeal arches in sharks, and could provide a tooth succession mechanism in a way that skin denticles did not (Smith and Coates 1998, 2001). The developmental mechanisms and genetic patterning information for pharyngeal denticle sets (sequential, joined tooth units) could be co-opted, via heterotopy, from inside the pharynx to the margins of the oral jaws to pattern successive sets for teeth. This challenged the 'outside-in' theory, as skin denticles did not have similar time-and-space linked sets for renewal (Smith and Coates 2001; Figure 7.4).

This idea has since been controversial and much debated (Johanson and Smith 2005; Huysseune *et al.* 2009; Fraser *et al.* 2010), and two modifications of these theories have recently been proposed. These incorporate aspects of both theories and can be described as 'inside and outside' theories, incorporating elements of both (Huysseune *et al.* 2009; Fraser *et al.* 2010).

Alternative hypotheses

The 'inside-out hypothesis' has been criticized by those who support the evolution of teeth from external dermal elements such as placoid scales. For example, Hecht *et al.* (2008) noted similarity in expression of the *Runx* gene family between scales and teeth in the shark *Scyliorhinus canicula*, and suggested that teeth and the placoid scales of sharks evolved from a homologous developmental module, supporting the 'outside-in' hypothesis (Hecht *et al.* 2008) but *Runt* is an invertebrate orthologue of the *Runx* gene, hence is primitive for animals and the major gene involved in skeletogenesis. Also, homology has been generally accepted for the dermal epithelial appendages representing morphogenetic units (Schaeffer 1977), so that neither inform the question of from where the order of the sequential tooth units came. Similarly, *ephA4* is expressed in shark dermal denticles and mouse teeth (Freitas and Cohn 2004; Luukko *et al.* 2005). However, *Runx2* is also present in the pharyngeal dentition of cichlids (Fraser *et al.* 2009), which contradicts Hecht *et al.* (2008) and leaves the question still open. We emphasize that these gene similarities may simply be due to homology of their origin as individual developmental units (Schaeffer 1977), rather than any mechanism responsible for their time order and patterning.

Huysseune *et al.* (2009) presented a modified 'outside-in' hypothesis, where external dermal denticles migrated through the gill slits into the pharyngeal cavity. They explained that initiation of denticles internally was only possible for structural reasons dependent on presence or absence of gill slits, because ectoderm could extend deeply into the slits to encounter pharyngeal endoderm, whereas in tetrapods and other taxa where gill slits were absent there was an absence of pharyngeal denticles. This seemed not to consider that the evolutionary loss of pharyngeal denticles could occur for other reasons and failed to address from where dentition patterning is derived, because as noted this patterning is largely absent from dermal denticles, particularly in chondrichthyans. Although placoid scales in adult fish clearly show patterning, this pattern is absent from early ontogenetic stages (Reif 1982); see also new data on absence of pattern in growth of denticles in shark skin (Fraser and Smith 2011; Figure 7.4). Huysseune *et al.* (2009) dismissed the importance of identifying the source of patterning in dentitions, noting that some teeth are not organized, in some fish. This, however, does not provide an explanation for pattern that clearly does exist in the majority of dentitions. Moreover, Huysseune *et al.* (2009) attempt to explain observed patterning of zebrafish pharyngeal dentitions into whorls as due to space constraints on the pharyngeal arches, forcing the dentition into a whorl shape. Thelodont pharyngeal whorls are said to be due to a heterochronic shift of patterning of the crowns of skin denticles, implying that instead of forming as separate, spaced units, a shift in developmental timing again forced the denticles into a whorl shape. Also absent from the revised theory of Huysseune *et al.* (2009) is an explanation of tooth replacement mechanisms as the dermal denticles are non-growing (see Reif 1982), and scales are only replaced after they have been damaged or removed.

Fraser *et al.* (2009, 2010) suggested a compromise between the two hypotheses, as did Ohazama *et al.* (2010). In the former, emphasis is on the primary role of neural-crest-derived mesenchyme in the interaction with the tooth-inducing potential of both ectoderm and endoderm, as the co-operative epithelial cell layer in the mouth. This puts on one side the classic experiments that show epithelial cells to be the first pattern-inducing cells in this co-operative event (Graham *et al.* 1988). The later work on the mouse (Ohazama *et al.* 2010) does recognize that the epithelium from which the molar teeth come is different from that in which the incisor teeth develop, due to endodermal origin versus ectodermal. They further suggest that oral teeth in mammals have a dual molecular origin contributing to the different morphology of incisors and molars; these suggestions provide the future direction for comparative experiments in fish, amphibians and reptiles. The theory of Fraser *et al.* (2010) was largely stimulated by new data on developing teeth of the axolotl in which transgenic and fate-mapping methods have been used (Soukup *et al.* 2008). These confirm the original studies that oral teeth can be formed from either ectoderm, endoderm, or both together, but Fraser *et al.* (2010) emphasize that the key to this innovation is the odontogenic capacity of neural crest cells. The linking of the two cell types is seen as permitting the collaboration between two different pre-existing gene co-expression groups (with neural-crest-derived mesenchyme) to form separately in evolution both skin (denticles) and oral teeth (Fraser *et al.* 2010). However, data for neural-crest-derived origin of skin mesenchyme (ectomesenchyme) in fish is currently lacking and only one study, in the lungfish *Neoceratodus forsteri*, has demonstrated with vital dye labelling that ectomesenchyme is present in oral teeth of non-tetrapod vertebrates (Kundrat *et al.* 2008). The review by Fraser *et al.* (2010) represents an appreciation of basal homology among odontodes using the presumptive odontode gene regulatory network (oGRN) as the homologous unit. Importantly, it does acknowledge as significant 'the idea of co-option of patterning information, as stated by the "inside-out" theory'. The dissection of genes in the network that are used in spatiotemporal dentition pattern, rather than those of the oGRN has not yet been completed, although Fraser *et al.* (2008) have identified the involvement of *wnt7b* (epithelial) and *eda* (mesenchymal) in spatial patterning. These genes express between teeth and between tooth rows, relative to *shh*, responsible for location of tooth initiation. Fraser *et al.* (2010) also note that it still needs to be determined whether skin denticles share a more substantial gene expression with teeth.

Skin appendage patterning

As described by Donoghue and Sansom (2002), patterning of external skin appendages has been examined in the context of a variety of models that discuss the response of cells to gradients resulting from the interaction of various molecules within the skin. Patterning has been most intensively studied for feathers, where feather buds appear originally in the lumbar region, and develop along the midline both anteriorly and posteriorly, regulated both spatially (in lines) and temporally (Mou *et al.* 2006; Baker *et al.* 2009). An early study by Jung *et al.* (1998) showed that 'the feather tract is initiated by a continuous stripe of Shh, Fgf-4, and Ptc expression in the epithelium, which then segregates into discrete feather primordia that are strongly *Shh* and *Fgf-4* positive'. From these observations Jung and colleagues proposed a model that 'involves (1) homogeneously distributed global activators that define the field, (2) a position-dependent activator of competence that propagates across the field, and (3) local activators and inhibitors triggered in sites of individual primordia that act in a reaction-diffusion mechanism'. These buds (primordia) and the interbud region develop via the interaction of the promotors/activators of cell accumulation (FGFs, follastatin) and inhibitors (BMPs). Similarly, patterning of mammalian hair relies on the Edar ectodyplasin receptor as a promotor, and again, BMP as an inhibitor (Mou *et al.* 2006).

Various genes have been found to express during fish scale development, either by *in situ* hybridization or in gene mutants. These include *shh*, involved in scale development but not origination (Sire and Akimenko 2004), *ephA4* (Freitas and Cohn 2004), and three genes of the *Runx* family expressing in different parts of the shark placoid scale (Hecht *et al.* 2007). *Eda, Edar* (necessary for scale placode development (Kondo *et al.* 2001; Sire and Akimenko 2004), *bmp2b* and *shh* are expressed in the zebrafish scale, with these genes acting to organize cells into signalling centres (Kondo *et al.* 2001; Harris *et al.* 2008). Besides this recent work on zebrafish scales, observations on scale patterning have been limited. For example, Donoghue (Donoghue and Sansom 2002; Donoghue *et al.* 2006) reviewed work by Reif on shark scale patterning and compares this to other early vertebrates and feather patterning. Donoghue noted that while Reif observed that the first scale generation in shark developed almost simultaneously and without any organization (also Johanson *et al.* 2007, 2008), the first scales in a variety of fish taxa originate in rows or discrete areas, or are associated with the lateral line. This is similar to the original lines of feather buds in birds.

While Donoghue and Sansom (2002) suggested that the chondrichthyan condition might be unique for the group, in terms of evaluating from where dentition pattern may have originated, chondrichthyans are more relevant phylogenetically than osteichthyan (bony) fishes such as the zebrafish and the medaka. Fraser and Smith have described the pattern of scale development in sharks as non-ordered in *Scyliorhinus canicula* (Fraser and Smith 2011; Figure 7.4). Close spacing only developed with time (Reif 1982). Also, Fraser and Smith did conclude that regulation of the distance between the denticles was by some biological patterning constraint (nearest neighbour hypothesis) and they were not initially evenly spaced in the skin (Fraser and Smith 2011). Possibly this is a requirement to be near other denticles to initiate denticle formation for the involvement, or recruitment, of stem cells.

As noted above, patterning is also relevant in terms of how skin appendages are replaced through time. Teeth, feathers and hair are replaced from a stem cell population associated with the base of the pre-existing appendage. In teleost fish, when scales are lost, cells from the 'scale pocket' are recruited to form a new scale; these have also been described as stem cells (Sire and Akimenko 2004). However, it is important to note replacement scales are generated in response to loss or damage (noted for chondrichthyans; Reif 1980), and are not continuous or cyclical in nature. This is an important point of difference between pharyngeal dental elements and external elements such as scales.

We emphasize that with respect to structural pattern for spacing and replacement, there is no iterative mechanism for organization of external epithelial denticles, for example (as first proposed in chondrichthyans), that is suitable to be co-opted and converted into a pattern that evolves into 'tooth sets' (Johanson *et al.* 2008; Fraser *et al.* 2010; Fraser and Smith 2011) The first developmental appearance of external patterning in chondrichthyan ontogeny occurs at the tip of the caudal fin, a significant distance from the pharyngeal slits and mouth, and it is also established as rows in a caudal–rostral direction (Johanson *et al.* 2007, 2008). By comparison, a 'tooth whorl pattern' is observed in the oropharyngeal cavity of living and fossil sharks. Although molecular data are currently lacking, an organized pattern is seen in toothed pads on the 3rd–6th pharyngeal arches of the Carcharinidae as denticles 'lined up in rows' (Nelson 1969, 1970). Although these are derived taxa, the ability to form patterned rows as pre-whorls may have been present early on the phylogeny, reflected by differences in the dental gene regulatory network and that associated with external skin appendages such as scales, and by morphological observations in fossil taxa. For example, denticle whorls occur in Carboniferous sharks such as *Akmonistion*, although a regionalization of these in the oropharyngeal cavity, comparable to that described in extant sharks by Nelson, is absent (Coates and Sequeira 2001).

Molecular data for tooth pattern induction

How can molecular data inform these questions regarding structural pattern? To understand the molecules controlling patterned modular development, a tentative start has been made, via gene expression studies. More specifically, genes are now known that orchestrate the intricate and precise patterns in extant basal vertebrate dentitions, with teeth regulated in time and space (Figure 7.3). The ability to clone genes for tooth induction in mice and to view their expression patterns in time-staged embryos is a powerful tool to establish the meristic control for tooth number and spacing in fish and reptile dentitions (Buchtova *et al.* 2008; Fraser *et al.* 2008; Vonk *et al.* 2008). This, combined with morphological data to correlate the site of gene expression with early dental placodes, identifies both the timing and spacing of the first teeth (Fraser *et al.* 2004, 2006a). By extending the number of genes used in this way to include most of those identified in the dental network, both an ancient and a core gene network for evolution and development of individual teeth has been proposed, including *bmp2*, *bmp4*, *dlx2*, *eda*, *edar*, *pax9*, *pitx2*, *runx2*, *shh*, and *wnt7b* (Fraser *et al.* 2009). Some of these gene pathways will be involved in spatiotemporal patterning (Fraser *et al.* 2009), as noted above, but those for morphogenesis of the individual teeth are currently better known. Data for specific dentition patterns of fish and tetrapods were acquired in the 1970s through longitudinal studies of their development and growth (Osborn 1971; Berkovitz 1975, 1977, 1978; Berkovitz and Moore 1974, 1975). Much of this involves the sequential and continuous addition of successional teeth that act as replacements for the first formed sets along the jaw margins, and are often but not always formed from a dental lamina. The genes involved in these processes, generating successional teeth ordered in time and space, are only beginning to be understood (see below) (Jarvinen *et al.* 2006, 2009; Buchtova *et al.* 2008).

Dental lamina reconsidered

We propose that the criterion for identifying teeth, the presence of a dental lamina (Reif 1982), should be re-examined (Smith *et al.* 2009b). Both the old and new concepts of a developmental character for 'true teeth' can be informed by structural observations from fossil taxa. These combined with molecular developmental data from extant forms, can be used to evaluate the earliest node in the gnathostome phylogeny where dentitions evolved (Smith and Johanson 2003). New molecular data on gene expression in the epithelial cells of the dental lamina during the production of both initial and successional teeth can establish the real site of tooth initiation and function of the lamina (Jarvinen *et al.* 2009; Smith *et al.* 2009b). Mammalian tooth renewal seems to be supported by the beta-catenin pathway as an activator of the enamel knot (sites of additional tooth formation) and Wnt signalling may also be implicated. Many of the molecular studies show that true patterned teeth as in teleosts can form without the presence of a classic structural dental lamina (Fraser *et al.* 2006a; Huysseune and Witten 2006). These studies also show that genes used for setting up the replacement tooth series, even without a dental lamina (Fraser *et al.* 2006a), are different from the first tooth inductive genes: that is, *pitx2* together with *bmp4*, but without *shh*. Similarly, by comparison, studies on reptilian replacement teeth show that although *shh* is expressed in the odontogenic band, it is not expressed in the cells producing the successional tooth (Buchtova *et al.* 2008). Genes that are translated into this secondary tooth replacement pattern may be used as a general mechanism for patterning the vertebrate dentition in renewal for life.

If we use the continuously growing mouse incisor as a model for tooth replacement and a pseudo-dental lamina at the base of the open root, then genes are known that localize to putative stem cells within the *Notch*-expressing epithelium together with *Lunatic fringe* expression (Harada *et al.* 1999) such as expression of *Fgf-3* and *Fgf-10* restricted to the mesenchyme underlying these basal epithelial cells, as cells expressing their receptors *Fgfr1b* and *Fgfr2b*. Complex studies like this have not yet been applied to non-mammalian dentitions.

Originally, in fossil jawed vertebrates, a dentition comprising many separate tooth whorls at the jaw margins (as in some acanthodians) was cited as evidence for developmental successive tooth formation, but from separate dental laminae (Reif 1982). Placoderms, in contrast, lacked tooth whorls and were deemed not to have true teeth defined as those produced by a dental lamina (Reif 1982). With regard to chondrichthyans, all recent sharks have a continuous dental lamina, and Reif (1982) noted that the discontinuous lamina of *Chlamydoselache* in the adult specimens had developed as a specialization from a continuous lamina in the embryonic specimens. In the adult, separate dental laminae for each partially joined tooth set are seen as the epithelial pockets separated by connective tissue and by skin denticles from outside the mouth.

The dentition of the earliest articulated shark, the Lower Devonian *Doliodus*, possesses small tooth families, where each is separate along the jaws (Maisey *et al.* 2009). Maisey *et al.* (2009) commented that there was no evidence that tooth families attached to the ethmoid region in *Doliodus* were formed by a separate dental lamina relative to those articulating to the palatoquadrates. Instead, the dental lamina would have extended between the palatoquadrates and onto the ethmoid region. *Akmonistion* shows widely separated tooth families, each in a pocket in the jaw cartilage, and also had sets of denticle whorls on the pharyngeal bars, organized in a time sequence (Smith and Coates 2001). These tooth families were thought to have developed from separate dental laminae (Smith and Coates 2001), as proposed for acanthodians. Although, living sharks possess a well-developed dental lamina (the continuous type; Figure 7.5D) we cannot yet determine the primitive condition for the group, as we have too few, and contradictory, data from early sharks. As noted, teleost teeth may develop without the contribution of a dental lamina, and as described further below, this developmental model for teleosts could also be applied to placoderms where in each statodont tooth row, new teeth are added to the side of the last teeth to have formed (Smith *et al.* 2009a).

Principles of co-option of structural pattern with molecular mechanisms

Co-option models

The evolutionary mechanism of co-option could be applied to any modular system (Raff 1996), in particular the dentition including cartilages of the branchial arch modules and replacing tooth set modules. In this concept, a new functional structure arises by co-option and modification of a system evolved earlier in the phylogeny, utilizing the gene networks of those developmental modules for morphogenetic programs. One example given was derivation of dorsoventral articulated jaws from the agnathan condition of non-jointed branchial arches. Together with a new function as predatory feeding, anteroposterior (A/P) Hox patterned cranial neural crest was recruited from patterns already present in the lamprey, with the main gnathostome novelty being the evolution of jointed arches (Cerny *et al.* 2010). Additionally, one can also propose recruitment of pattern information of the denticle sets to the jaws.

Developmental modules are those that can undergo temporal transformation through development and can also undergo evolutionary transformation in geological time, as set out by Raff (1996). He discusses the evolution of feeding as a co-option event at both the morphological and gene level. For the jaws and dentition this co-option may be of a serial homologue as are the pharyngeal arches, but there are also superimposed on each of the arches, separate serial homologues in the ordered sets of denticles (Nelson 1969). The genes that accompany this co-option of one system to function as another are modified in development.

The classic theory of the origin of teeth for dentitions is one that suggests co-option of external denticles as modules into the mouth and recruitment of their molecular mechanisms for morphogenesis. However, as noted above, revision of this theory was considered necessary because of new structural fossil evidence in jawless fish that challenged the theory of the co-option of teeth from skin denticles. The critical observation was that *Loganellia* (a member of the Thelodonti), an agnathan fossil, had an internal skeleton as well as external denticles, with the former manifested as sets of patterned denticles on the pharyngeal arches (Figure 7.6).

Previously unrecognized was the fact that in many jawed vertebrates, both chondrichthyans and osteichthyans, there are iterative sets of denticles or teeth on all the arches of the common oropharyngeal cavity, and proposed to be homologous (Nelson 1970). Because in agnathans they were patterned as an iterative series, this spatiotemporal program for pharyngeal denticles could be co-opted to the oral jaw margin in gnathostomes and function as a patterned dentition, linked with the regulatory gene networks. As well, in the Conodonta, a basal vertebrate group (Figure 7.2), a functionally organized oropharyngeal dentition is present, but external scales are not (Donoghue and Purnell 1999). These observations provide a solid phylogenetic basis for determining the type and site of denticles for co-option as developmental modules for tooth sets in the dentition, together with their operational genes (e.g. those identified for the cichlid pharyngeal dentition; Fraser *et al.* 2009). Also, new molecular evidence on the Hox environment in which teeth will develop, both pharyngeal and oral, suggests that this event is restricted to Hox-negative regions (i.e. Hox expression absent), including the vertebrate jaw and homologous elements in the lamprey (Takio *et al.* 2004, 2007).

Molecular mechanisms

The idea that a Hox-negative environment was necessary for teeth to be generated on the mandibular arch was tested by experimental manipulation studies in the mouse (James *et al.* 2002) where it was found that development of these mammalian teeth was unaffected even in an induced Hox-positive environment. They concluded that teeth develop independently of Hox gene expression whether negative or positive, and are 'subject to local interactions and not part of the axial patterning process'. Also, in Malawi cichlids, observations on comparative *hox* expression data between oral and pharyngeal tooth regions showed that homologous tooth sets do form in a Hox-positive environment from the undisputed endodermal tissues of the pharyngeal arches (Fraser *et al.* 2009). However, this union has since become decoupled only in the first arch, where oral teeth develop without any associated Hox genes (Hox-negative regions). These experiments, therefore, do not deny the 'inside-out' theory that teeth likely originated in the endoderm-dominated pharyngeal environment of early vertebrates ahead of oral jaws as they can develop independently of Hox expression.

Part of this theory focuses on the developmental layer through which the pattern was generated, ectoderm, endoderm or from the region of union of the two tissue layers (Johanson and Smith 2005; Huysseune *et al.* 2009). The role of ectoderm or endoderm is discussed in recent reviews of this topic (Soukup *et al.* 2008; Fraser *et al.* 2010). If the serial addition pattern mechanism (SAM; Smith 2003) had been co-opted from the pharyngeal denticle sets instead of from the skin then a putative primitive tooth set would express a collection of genes from deep within the pharyngeal endoderm, but also found for teeth associated with oral regions. This could be tested against an alternative origin from the ectodermal genes as experimental data from the mammalian dentition show, in which molecular characteristics of pharyngeal foregut endoderm are compared with oral ectoderm in mouse development, seeking to find if this contributes to heterodont dentitions (Ohazama *et al.* 2010).

We assume that the core networks of genes in relation to morphogenetic development of a tooth are conserved from fish through to mammals (Buchtova *et al.* 2008; Fraser *et al.* 2009; Handrigan and Richman 2010a, 2010b) especially as the tooth module is considered to be homologous throughout jawed vertebrates, and with skin denticles, as odontodes (Ørvig 1977). These core networks include genes for tooth morphogenesis, although vertebrate dentitions show a great morphological diversity. However, core networks also include genes that operate in establishing the spatiotemporal pattern of the teeth and their replacements with respect to the order of timing and position (Fraser *et al.* 2008). Can a core gene network actually produce the wide variety of vertebrate dentitions, and is there only one way or one time to use an operational gene? An example focusing on tooth shape modification through a molar series, taken from the primates, uses repeated expression of genes in a developmental module to explain cusp shape diversity and homoplasy with respect to more primitive dentitions (Jernvall and Thesleff 2000; Streelman *et al.* 2003; Streelman and Albertson 2006). They suggested that 'later-developing cusps may be evolvable but also more homoplastic'. Streelman *et al.* (2003) adopted this model to explain tooth shape diversity in cichlid fish, both intraspecific and interspecific. They found that there was a positive correlation between number of teeth in the first row and tooth shape and this was maintained through generations of natural cichlid populations. The molecular basis of this model was the iterative deployment of the signalling molecules as used in mammalian teeth (Streelman *et al.* 2003). Later, Streelman and Albertson (2006) used natural mutants of cichlids to show patterns of *bmp4* expression in early odontogenesis to address and refine predictions of models linking tooth shape and tooth number. They were able to show that 'mutations in the Bmp4 cistron do not control tooth shape in this mapping cross'.

The complexity of the regulatory network in all aspects of mammalian tooth development was reviewed by Tummers and Thesleff (2009), who concluded that a restricted number of major signalling pathways were used, including TGFß, BMP, Wnt, FGF, Eda and Hedgehog. The complexity of tooth shape, orientation and number was achieved by modifying existing interactive genetic pathways (tinkering), leading to the intricate system of complex signalling compartments. As noted, 'the number of teeth as well as molar cusp patterns can be modified by tinkering with several different signal pathways' (Tummers and Thesleff 2009).

We have attempted to use expression data for genes known to be required for determining tooth position in mammals (such as *shh* and *pitx2*), to interpret developmental stages of fish representing major clades in the gnathostome phylogeny (Actinopterygii, Sarcopterygii). These genes are precise markers of timing and spatial organization. Within some actinopterygian species two sets of functional biting jaws can be compared, those on the oral jaws representing the first pharyngeal arch and those on the last pharyngeal arch, with teeth arranged in time and space in both. We have questioned whether patterning genes and the gene networks used are the same at the two sites, and whether pharyngeal networks could have been co-opted to the oral jaw. Gene expression studies in cichlid fish have provided new data and a hypothesis that there is an archaic or core gene network that is shared between the two sites, including *bmp2*, *bmp4*, *dlx2*, *eda*, *edar*, *pax9*, *pitx2*, *runx2*, *shh*, and *wnt7b* (Fraser *et al.* 2009), and that this is significant for our understanding of evolution of dentitions. Other studies on the zebrafish and medaka have explored the gene expression differences between the oral and pharyngeal jaws, the former without oral teeth, and used these to explain tooth loss as an evolutionary novelty (Stock 2001; Stock *et al.* 2006; Debiais-Thibaud *et al.* 2008).

The genetic toolkit for building a dentition

Fish dentitions compared with reptiles and mammals

Genes involved in sequential stages of tooth development have become increasingly well understood. Over 200 genes are known to be active in tooth development and can be seen as gene expression profiles at a website generated in Helsinki (http://bite-it.helsinki.fi; Nieminen *et al.* 1998, from Cobourne and Sharpe 2003). Apart from the mouse, such expression profiles are now known in reptiles and various fishes as well, spanning a wide range of the vertebrate phylogeny (Fraser *et al.* 2004, 2006a, 2009; Vonk *et al.* 2008; Handrigan *et al.* 2010; Handrigan and Richman 2010a, 2010b). As frequently stated in mammalian studies 'although both new fossil and molecular data can be expected to resolve many of the phylogenetic incongruences, there is a continuing debate about the best way to use dental evidence in evolutionary taxonomy' (Kangas *et al.* 2004). After a study of changes in a single gene (ectodysplasin) in a natural mutant mouse and a transgenic one, they concluded that, through tooth-specific cis-regulation, 'extreme reduction of dental features in mammalian lineages such as seals and whales may have required only a simple developmental change' (Kangas *et al.* 2004).

Unlike these two mammalian piscivorous homodont dentitions, teeth of reptiles and fish are especially noted for the ability to be replaced continually, such that information derived from the mouse dentition (complex cusp morphology, teeth not replaced) was not always applicable to broader discussions of earlier vertebrate tooth development. Much less is known about the control of overall numbers in dentitions and where this occurs. A central question is how gene mutations affect development and are translated into macroevolutionary changes in morphology. In one very valuable study, to understand how changes at the molecular level affect the distinct stages of tooth development, Line (2003) looked at the genetics of variation in tooth number, where phenotypic variation occurs in populations of sciurid and murid rodents with the arrest of tooth production in single and double knockout mice. This was first explored in human conditions of hypodontia (Line 2001) to explore new hypotheses linking development with the patterning of the dentition during mammalian evolution. Furthermore, the developmental changes can be linked to changes in the molecular morphogenetic field of specific genes. Line (2003) has shown that phenotypic variations in tooth number relate to mutations in the homeobox genes *Msx1* and *Pax9* by the inactivation of one allele. As previously shown, tooth type appears to be determined by epithelial signals and to involve differential activation of homeobox genes in the mesenchyme. Jernval and Thesleff (2000) were able to establish that molecular details of signalling networks, particularly in the families BMP, FGF, Hh and Wnt, were significant. The same signalling cascade is used reiteratively throughout tooth development, not just in production of the serial homologues along the jaw: 'The successional determination of tooth region, tooth type, tooth crown base and individual cusps involves signals that regulate tissue growth and differentiation' (Jernvall and Thesleff 2000).

Also, in this debate distinction should be made between successional teeth for replacement and supernumerary teeth added to the primary tooth set. Cobourne and Sharpe (2010) described a number of fundamental mechanisms that facilitated supernumerary tooth formation. Interestingly, they show that disrupting pattern along the buccolingual aspect of the jaws could produce extra teeth directly from the oral epithelium. Teeth form an extra row by co-ordinated restriction of *Bmp4* activity through loss of *Osr2* (odd-skipped related) in mutant mice (Mikkola 2009; Zhang *et al.* 2009; Figure 7.7): *Osr2*-deficient mice required *Msx1*, a feedback activator of *Bmp4* expression. These findings suggest that the *Bmp4*–*Msx1* pathway propagates mesenchymal activation for sequential and repeated tooth induction. We would not be able to understand the pattern of tooth presence/absence along the jaw axis without these mouse knockout models that allow us to propose the molecular mechanisms that regulate spatial tooth order. Zhang *et al.* (2009) discovered that levels of *BMP4* are critical so that in $Osr2^{-/-}$ with $Msx^{-/-}$ mice first molars will form but second ones do not, because these are successive teeth as part of the same molar family.

Kavanagh *et al.* (2007) proposed an inhibitory cascade model in which initiation of posterior molars depended on a balance between intermolar inhibition and mesenchymal activation, as it was shown experimentally by organ culture that mouse mandibular first molar tooth germs inhibited second molar development. Kangas *et al.* (2004) observed that wild type mice have weak *Shh* expression activity in the same location at which, in the transgenic mice (K14-Eda), an extra tooth develops. The detection of transient gene expression for *Shh* in the tooth germ (placode or bud stage) can confirm the

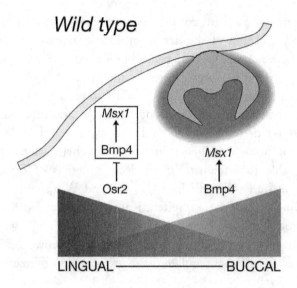

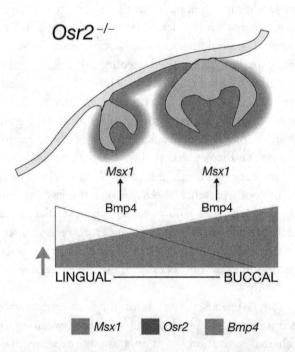

Figure 7.7 In the mouse, misexpression of the gene *Osr2*⁻/⁻ results in expression of *Bmp4* and *Msx1* lingual to the pre-existing tooth resulting in development of a new tooth. This occurs because of the lack of the Osr gene product. This misexpression mimics the condition seen in fishes, where multiple sequential secondary teeth (replacing teeth) are present. (Adapted from Cobourne and Sharpe 2010). (See also colour plate.)

structural information that vestigial teeth may develop but not continue to form the phenotypic adult tooth, as in the diastema region of mice, seen as the 'phylogenetic memory' for premolars now lost (Peterkova *et al.* 2006).

One intriguing aspect of the adaptive feature of continuous tooth replacement is the inherent ability to repeat an embryonic process – that of initiating a tooth. This involves sequestration of a stem cell population and ensures regular control in terms of space and time (Harada *et al.* 1999; Tummers and Thesleff 2003; Wang *et al.* 2007). The debate concerns the structural location of where the stem cell niche is located more than the gene activity that may identify these cells. Studies (Huysseune and Thesleff 2004) on the zebrafish, rainbow trout and lizards suggested that the sequestration of epithelial stem cells occurs only with the initiation of replacement teeth (Fraser *et al.* 2006a; Handrigan *et al.* 2010). This may relate to a different genetic control of the initiation of replacement teeth from that of the primary teeth.

With regard to patterning genes it has been shown that there are conserved gene networks involving *bmp2, bmp4, eda, edar, fgf8, pax9, pitx2, runx2, shh* and *wnt7b*, known to pattern iterative structures and teeth in other vertebrates (Fraser *et al.* 2008, 2009; Vonk *et al.* 2008; Handrigan *et al.* 2010; Handrigan and Richman 2010a, b). To specify this pattern further in order to establish the spacing of tooth rows and tooth positions in the rows it was shown that epithelial *wnt7b* and mesenchymal *eda* are expressed in the intergerm and inter-row regions; these likely regulate the spacing of the *shh*-positive units. Whereas, combinatorial epithelial expression of *pitx2* and *shh* appears to govern the position of the competence of these initial tooth sites and also future tooth rows where the loci for tooth units will later appear.

Fraser and colleagues (Fraser *et al.*, 2004) identified several genes involved in the dentition of the rainbow trout, *Oncorhynchus mykiss*. *Shh* and *pitx2* were expressed in the dental epithelium corresponding with *bmp4* in the mesenchyme, with *shh* and *pitx2* expressed first, in a restricted region of the epithelium called the odontogenic band. This band is a common feature of all dentitions so far described from gene expression studies, and although sometimes called a dental lamina, it is only a thickening of the epithelium, which may later grow deeply into the connective tissue to form either continuous or separate dental laminae. It is equivalent to a dental field of tooth-forming ability, restricted to the margins of the mouth, and/or the pharyngeal surfaces of the branchial bars. In a recent review Fraser and Smith (2011) commented that most teleosts do not have a deep dental lamina, and the primary epithelial odontogenic band is restricted to a superficial region. If placoderms show the teleost condition (see below), this could be modified through evolution into a deeper dental lamina that occurs in chondrichthyans and tetrapods. Expression at restricted points in the overlying odontogenic band, coincident with local thickenings, are focal spots of *shh* expression and *Bmp4* follows in the mesenchyme, representing loci of tooth-forming potential. In all these developmental stages, expression of *shh* and *pitx2* is restricted to the sites of odontogenic potential, such as continuation of the odontogenic band for successive teeth. With morphogenesis of the tooth, *shh* becomes restricted to the sides of the developing tooth (inner dental epithelium), and then *pitx2* on the lingual side only of the outer dental epithelium (Fraser *et al.* 2004). Importantly, the timing and spatial expression of these genes also characterizes the mouse and reptile dentition (although studies on the latter are less advanced). The trout and many other fish also develop a dentition on the fifth branchial arch and the hyoid for a tongue; the expression of *shh* and *bmp4* is the same as described for the teeth on the jaws (also palatal bones), while *pitx2* expression becomes downregulated.

Regarding studies on tooth initiation and loss of the gene for *bmp* in fish, that of Wise and Stock (2006) is very informative as they compared two teleost species, zebrafish and medaka, respectively without and with oral jaw teeth. They showed from comparative expression patterns that at the tooth initiation stage *bmp4* is missing from the oral region in the zebrafish but present in the medaka, such that the zebrafish morphology is due to regional loss of *bmp4* expression. Previous studies (Stock *et al.* 2006) had shown that *dlx2* expression, the downstream target of *bmp4*, is also absent from the oral region. One study (Borday-Birraux *et al.* 2006) compared expression of all the *dlx* genes in zebrafish with mouse to discover that *dlx1a* and *dlx6b* were absent from the tooth germs in the zebrafish. This was put into an evolutionary context with two alternative hypotheses (Borday-Birraux *et al.* 2006), either that this was due to a loss of function in the actinopterygian lineage leading to teleosts, or a difference between oral and pharyngeal dentitions; both require more functional genetic data to provide an answer. Stock *et al.* (2006) explored the developmental genetic mechanisms in cypriniform fishes, with tooth presence or absence in these two regions, and suggested that loss of fgf signalling to oral epithelium accounted for evolutionary tooth loss.

Molecular modification of gene networks

Regulatory tinkering

Tummers and Thesleff and colleagues also review the signal pathways involved in various aspects of tooth development (Thesleff *et al.* 2007; Tummers and Thesleff 2009). Perhaps somewhat surprisingly, given the diversity of teeth and dentitions, genes in these signal pathways belong to a small number of conserved families, including TGFß, Wnt, FGF and Hedgehog.

Chan *et al.* (2009) stated that there were strong evolutionary constraints on tissue specific gene expression, implying limited value in elucidating vertebrate evolution. However, Tummers and Thesleff (2009) suggest that the evolution of new tooth morphologies is not dependent on the evolution of new pathways, but the 'tinkering' of these existing ones. For example, the BMP (Zhang *et al.* 2009) signalling pathway regulates the patterning of tooth cusps. Misexpression of genes associated with this pathway, such as ectodysplasin (KT14-Eda) and ectodin can turn the cusp pattern in mouse molars to that of a kangaroo and hippopotamus, respectively (Tummers and Thesleff 2009). The same group investigated induction of mouse incisors by deletion of *Sostdc1* (ectodin) found in the mesenchyme and a putative antagonist of BMP signalling in the induction of teeth morphogenesis (Munne *et al.* 2009; Ahn *et al.* 2010). This resulted in extra incisors next to the normal single pair and they concluded that *Sostdc1* might act to limit supernumerary tooth induction. They further tested this potential role of the dental mesenchyme by reducing it in incisor tooth germs, which they then cultured and found the same phenotype as the *in vivo* results, extra incisors in the explant. This correlated with the early development of the dentition in *Sostdc1*-deficient mice when *Shh* expression loci for the incisors were much enlarged. This gene has an inhibitory role for successional teeth, at least in the mouse dentition restricted to one set of teeth. Munne *et al.* (2009) commented that the role of *Sostdc1* may be to modulate Bmp and Wnt signalling in limiting tooth replacement (also Thesleff and Tummers 2009; Ahn *et al.* 2010). Also, this signalling pathway may be part of the developmental programme limiting induction of teeth through the inhibitory role of the dental mesenchyme via both BMP and Wnt signalling.

The complexity of the pathways, with feedback loops and various activators and inhibitors, offers unlimited possibilities for tinkering or fine-scale modulation of the development process.

Misexpression of genes and atavistic dental morphologies

'Reverse tinkering' via gene misexpression studies in the mouse can reproduce many additional sets of teeth and identify genes that are relevant in dental evolution. For example, stimulation of Wnt signalling in the mouse oral ectoderm results in repeated tooth development where all shapes are simple cone morphologies (Jarvinen *et al.* 2006; Tummers and Thesleff 2009). Mouse molars normally have complex cusp morphologies and are not replaced; this Wnt signalling reproduced a more fish-like developmental pattern (simple morphology, tooth replacement) in the mouse. Also, $Osr2^{-/-}$ mice have a lingually expanded odontogenic field (Mikkola 2009; Zhang *et al.* 2009), again resembling a more fish-like morphology seen, for example, in cichlids (Fraser *et al.* 2008, 2009) and phylogenetically basal fossil osteichthyan taxa such as *Lophosteus* and *Andreolepis* (Botella *et al.* 2007).

The original dentition and structural constraints

The first dentition

One approach to the question of where does the first dentition for jawed vertebrates appear, and from which system did it evolve, of necessity will involve fossil taxa and use the criteria that we have established in the earlier sections to define a dentition. This will determine a developmental model that might explain that comparing oral and pharyngeal dentitions is relevant to the question of how the first dentition evolved. In both oral and pharyngeal dentitions, there are spatially discrete tooth sets on the jaw (first pharyngeal arch) and on one or more of the more posterior pharyngeal arches. The two dentitions show similarity with respect to developmental timing of the jaw and arch replacement tooth sets and the core dental gene (CDG) network, the majority of which (*pax9* being an exception) showed similar expression in both sets of cichlid jaws (Fraser *et al.* 2008, 2009). This similarity suggests that the two dentitions may be linked evolutionarily through co-option of common gene networks (Figure 7.8).

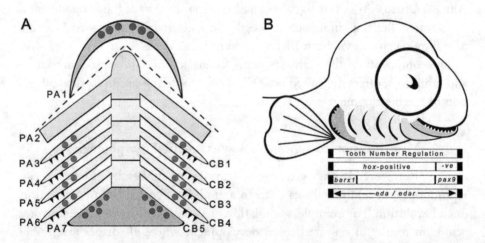

Figure 7.8 (A) Teleost fish with teeth (black spots) as well as gill rakers on the pharyngeal arches (CB1–CB5) and on the jaws. Both share expression of genes from a 'core gene network'. (B) Pharyngeal teeth develop in a Hox+ve environment, while oral teeth develop in a Hox−ve environment, associated with the oral jaw. (Adapted from Fraser *et al.* 2009.)

As noted above ('Old theories'), considerable controversy continues to surround the question of how dentitions evolved, whether from external body scales or pharyngeal teeth. The latter hypothesis not only derives from the observed similarity in oral and pharyngeal dental networks, it also has support from the fossil record, where phylogenetically basal vertebrate taxa such as the Conodonta, Thelodonta, Placodermi and Chondrichthyes show evidence of dentition patterning in the pharyngeal region (Smith and Coates 2000, 2001; Figures 7.4, 7.6). In the conodonts, this takes the form of opposing dental elements with matched occlusion (Donoghue and Purnell 1999), and in thelodonts, with the presence of whorls of dental elements (Smith and Coates 1998), occurring in both groups in the pharyngeal region. Although replacement has not been clearly established in the conodont dentition, the thelodont whorls show a time patterning of elements within the whorl, suggesting that new elements are forming as sequentially added units. The basal jawed vertebrate group 'Placodermi' (currently thought to be a paraphyletic clade; Brazeau 2009) also presents evidence of denticle pattern within the pharyngeal region (Johanson and Smith 2003, 2005), as they cover the part of the bony trunk-shield forming the rear wall of the pharyngeal cavity (the post-branchial lamina, discussed next).

Dentitions at the origin of jaws in the Placodermi

Placoderms represent the most basal jawed vertebrate group to possess a patterned oral dentition. This patterned dentition occurs only within certain placoderm groups, for example within the Arthrodira (Johanson and Smith 2003, 2005; Smith and Johanson 2003). Here, in various arthrodiran taxa, upper and lower dentitions possess teeth arranged in two or three opposing rows, with a clear size gradation within the row, indicating the addition of new teeth (largest, unworn, at the end of the row) and differential wear of older teeth (near the front of the row) relative to younger. Notably, more basal placoderms lack such distinctly patterned dentitions, including the Acanthoracida (e.g. *Romundina*) and the Antiarchi (e.g. *Bothriolepis*; Smith and Johanson 2003; Goujet and Young 2004). Smith and Johanson (2003; also Janvier 1996) mapped this character distribution onto pre-existing placoderm and jawed vertebrate phylogenies (Goujet and Young 1995; Janvier 1996) to suggest that such a dentition evolved separately in derived placoderms such as the Arthrodira, and in other jawed vertebrates (the early history of chondrichthyans is dominated by scales, with a dentition first appearing millions of years later; this suggests teeth may have also evolved independently in chondrichthyans; Wilson *et al.* 1999). However, new phylogenies of jawed vertebrates resolve the Placodermi as a paraphyletic group with respect to other taxa (crown group Gnathostomata; Brazeau 2009). Here, derived arthrodires are more closely related to the crown group (Figure 7.2), suggesting that teeth may have a single origin, at this node (but see comments on chondrichthyan dentitions just above).

With respect to the origins of this dentition, and the two hypotheses of dental evolution described above, placoderms add solid data to the hypothesis that teeth formed first internally, as the 'origin of an ancient co-ordinated regulatory system' (Smith *et al.* 2009b). This is supported by the presence of patterned denticles within the pharyngeal region, covering the post-branchial lamina in all taxa except *Weejasperaspis* (Smith and Coates 2001; Johanson and Smith 2003, 2005).

Pharyngeal pattern

The post-branchial lamina is a medially directed flange associated with the bones supporting the pectoral fin in jawed vertebrates. This includes the cleithrum and clavicle in bony fishes, and their putative homologues in the Placodermi, including the anterior trunkshield plates (Zhu *et al.* 2001). The post-branchial lamina is particularly well developed in the Placodermi, with the lamina effectively forming a posterior wall to the pharyngeal chamber, just posterior to the branchial arches. Johanson and Smith (2005) surveyed the post-branchial lamina across the Placodermi, and noted that almost without exception (see below), this lamina was covered in rows of denticles, including phylogenetically basal groups such as the Acanthothoraci and Antiarchi (basal antiarchs such as *Yunnanolepis*; the post-branchial lamina is lost in more derived antiarchs; Zhang 1980). These denticles differed morphologically from those covering the head and trunkshield and rows originated repeatedly along the dorsal margins of the lamina. Johanson and Smith (2005) suggested that the patterning represented by these ordered rows could have been co-opted to the dentitions within the Placodermi, in the derived Arthrodira. The morphological differences suggested the influence of a different epithelium from that which produced external shield denticles, as the change occurred coincident with the junction of skin with pharynx. Therefore, Johanson and Smith (2005) claimed that the post-branchial lamina denticles could have developed under the influence of endoderm. Moreover, with respect to these patterned denticles it appears that these were present, phylogenetically, before a patterned dentition evolved on the placoderm jaws, in the Arthrodira. Along with conodonts and thelodonts, the post-branchial lamina in placoderms provides evidence for pharyngeal patterning existing prior to patterned oral jaws. This provides an evolutionary sequence involving this internal patterning, biting jaws, and co-option of patterning to the dentition of more derived placoderms and subsequently, other crown-group gnathostomes.

Husseyeune *et al.* (2009) suggested that patterned placoderm post-branchial denticles represented support for their hypothesis that teeth evolved as competent ectoderm migrated into the gill arches, with the space between the placoderm headshield and trunkshield being another point of entry for the ectoderm. However, the placoderm acanthothoracid taxon, *Weejasperaspis*, almost uniquely among the placoderms, lacks patterned, morphologically distinct denticles on its post-branchial lamina (Johanson and Smith 2005). We suggest this resulted from the invasion of external ectoderm and accompanying denticles, and shows that these external denticles carry no patterning information that could have been subsequently co-opted to the dentition. As noted, the different morphology and patterning of the post-branchial denticles in other placoderm groups suggested the influence of endoderm in this region. Organized denticle rows on the post-branchial lamina are retained in basal osteichthyan fishes such as *Amia* and *Lepisosteus* (Johanson and Smith 2003), and in fossil taxa such as *Guiya*, *Psarolepis*, *Achoania* and *Youngolepis* (Chang 1991; Zhu *et al.* 2009).

Developmental data applied to fossils

Observations from the teleost developmental data can be applied to the placoderm dentitions described above, particularly for the derived arthrodiran taxa. As noted, in arthrodires, new teeth are added to the ends of pre-existing rows, along the biting margin, and at the symphyseal margin. One criticism of the identification of teeth in the Placodermi was that a tooth-producing dental lamina, comparable to that in sharks, was absent (e.g. Goujet and Young 2004). However, we suggest that the spatiotemporally patterned teeth in the arthrodires could be derived in development from the teleost model where new teeth are regulated not from a deep dental lamina but from the epithelium of the previous tooth (see Smith *et al.* 2009b). This would involve a small pocket of competent cells held within inner and outer dental epithelia, at the base of the pre-existing tooth (Harada *et al.* 1999; Smith *et al.* 2009b, see also 'The original dentition' section below). Phylogenetically, this would imply that this type of tooth replacement was primitive for jawed vertebrates, being present in placoderms and retained in osteichthyan dentitions including sarcopterygians (e.g. lungfish) and actinopterygians.

Notably, development of teeth from a small epithelial pocket of competent cells also characterizes the pharyngeal dentition, as investigated in zebrafish (lacking teeth on the oral jaws; Van der Heyden and Huysseune 2000; Van der Heyden *et al.* 2000). In zebrafish, these cells form a 'strand' connecting the functional tooth to the newly developing tooth beneath. Teeth are continuously replaced on both the oral and pharyngeal jaws, prior to functional need. This differs from external dermal appendages such as feathers, scales and hair. Replacement appendages also develop from a pocket of stem cells (feathers: Yue *et al.* 2005; scales: Sire and Akimenko 2004), but these only develop when the functional unit has been lost (moulting, damage). The capacity for continual replacement is an important patterning difference of internal denticles and teeth, not shared with external dermal elements such as scales.

Lungfish dentitions and modifications of the stereotypical osteichthyan pattern

Constraint and conservation from diversity

We have examined conservation and constraint in the use of genes in the development of dentitions to suggest how they might have evolved. Lungfish tooth plates appear to have been a unique morphological innovation in which all teeth of the adult dentition have been retained, never shed but organized into growth rows (smallest to largest) radiating from the first growth centre in a medial to lateral direction (reviewed by Smith and Johanson 2010). One question is how were they transformed from the stereotypic osteichthyan pattern? Although there are numerous fossil forms only three extant genera remain that can be used to study development of the dentition (*Lepidosiren*, *Neoceratodus*, *Protopterus*). *Neoceratodus forsteri* is the most primitive of these and it retains a marginal row of teeth, notable for only being present in the larval to hatchling stages and then lost (Smith *et al.* 2009c). In the Devonian, lungfish taxa were numerous and dentitions diverse, but once the dentition pattern became restricted to the typical palatal and pre-articular toothplates of post-Devonian lungfish, as in taxa of Devonian age (e.g. *Diabolepis*, *Andreyevichthys*), it was retained in the extant forms (Ahlberg *et al.* 2006).

Potential gene loss

From gene expression studies we can formulate how the genes may have been regulated in the early, Devonian members of the group to produce distinctive differences from the osteichthyan pattern of tooth development. Smith *et al.* (2009c) reported from an expression study in *Neoceratodus forsteri*, that *shh* was one gene required for tooth bud initiation, and that the pattern order (timing and jaw location) for the marginal tooth row (studied in the lower jaw, labial to the pre-articular toothplates) was the same as that for teleost osteichthyans. In this way *Neoceratodus*, through conservation of the developmental genes for location of the first set of separate teeth, has retained these at the margins of the lower jaw. However, those genes that regulate initiation of timing for a second replacement set have been lost from the jaw or have failed to be expressed during lungfish evolution. Subsequent dental development occurs on the pre-articular and palate, where the process of resorption and loss of teeth has been decoupled from the additive programme, so that all are retained into the adult stages (a statodont dentition) and growth is by tooth row addition (Smith *et al.* 2002). Only the rudimentary first marginal jaw teeth are shed in *Neoceratodus*, as part of a remodelling process, and no replacement teeth are added. Typical osteichthyan resorption of teeth, prior to eruption of the newly developed tooth, is lost from all lungfish palatal and pre-articular dentitions, with a remodelling type of resorption instead occurring along a broad front across the tooth plate (e.g. *Holodipterus*; Ahlberg *et al.* 2006). Further studies using a more comprehensive set of genes would yield more detailed information as to how gene loss or altered pathways in development (via the tinkering and gene misexpression described above) could have contributed to evolution of this highly specialized dentition.

Conclusions and prospects

The origin and evolution of the vertebrate dentition remains controversial. Although much of the work on the molecular mechanisms responsible for tooth development has focused on lab animals such as mouse (with a restricted dentition) and chick (teeth are evolutionarily missing), there is now a reasonable body of information examining homologous mechanisms in more basal vertebrates such as the reptiles, lungfish, salmon and cichlids. As these investigations of gene activity in developmental processes are extended more basally in the vertebrate phylogeny, they can be applied to the fossil taxa resolved to significant nodes in the phylogeny. Thus, Fraser *et al.* (2009) identified ancient and core gene networks in the cichlid oral and pharyngeal dentitions, including *bmp2*, *bmp4*, *dlx2*, *eda*, *edar*, *pax9*, *pitx2*, *runx2*, *shh*, and *wnt7b*, with the pharyngeal dentition differing in the absence of *pax9* expression. Fraser *et al.* (2008) also identified genes involved in tooth patterning, such as *wnt7b* and *eda*; research is also beginning to examine gene expression in the continuously replaced dentitions of snakes and lizards (Buchtova *et al.* 2008; Handrigan and Richman 2010a, 2010b; Handrigan *et al.* 2010); future research can expand upon these early results and make comparisons to gene networks identified by Fraser and colleagues. Included in these networks are genes responsible for dentition origins including spatiotemporal patterning and replacement. One should take cogniscence of the fact that external denticles (odontodes) lack early patterning and are non-growing scales in chondrichthyans, reviewed by Fraser and Smith (2011), only being replaced in response to loss or damage. Some revised hypotheses suggest that these external denticles entered the gill slits and so transferred patterning mechanisms to the pharyngeal arches (Van der Heyden and Huysseune 2000; Van der Heyden *et al.* 2000; Huysseune *et al.* 2009). However, iterative scale patterning (from caudal to rostral) early in chondrichthyan ontogeny only occurs at the extreme tip of the caudal fin, and is so far removed from the gill openings as to be excluded (Johanson *et al.* 2007, 2008). The post-branchial lamina of the placoderm *Weejasperaspis* is interpreted as being covered in denticles derived from the external ectodermal epithelium, with these denticles showing no pattern, contra all other post-branchial laminae (Johanson and Smith 2003).

In contrast, internal oropharyngeal denticles can be organized into spatiotemporal groups and distributed on all, or a few, of the pharyngeal arches (including the rear wall of the pharyngeal cavity in placoderms). It is proposed that gene networks for tooth addition are co-opted from the pharyngeal region of conodonts, thelodonts and basal placoderms, and subsequently co-opted to the oral dentitions of derived placoderms and other jawed vertebrates (Fraser and Smith 2011). Notably, these characteristics evolved stepwise through vertebrate phylogeny, emphasizing the modular nature of the vertebrate dentitions. This reflects the fact that the dental gene network is an oropharyngeal specific mechanism (Fraser and Smith 2011). As Fraser and Smith emphasized, Reif predicted many years ago that the external dermal skeleton of denticles is a separate system from that of the internal visceral skeleton of denticles in the oropharynx. In this context recent research on molecular expression confirms topographic structural isolation. However, our hypothesis can be tested by further determining the genetic profile of patterning these two sets of denticles in the skin and oropharynx (particularly with respect to the continual replacement of oropharyngeal teeth). This includes the genetic profile of the patterned chondrichthyan skin denticles at the tail (Silvan *et al.* 2010) and the rest of the body, and in phylogenetically basal actinopterygians such as *Polypterus*.

The modified inside and out hypothesis presented by Fraser *et al.* (2010) stated that denticles could develop internally or externally, wherever competent epithelium met neural-crest-derived mesenchyme, and moreover, that denticles evolved when genes associated with this mesenchyme and epithelium were combined, 'in embryological time and space'. One way to test this hypothesis would be to examine neural-crest contribution to external scale development and the gene networks associated with this (presumably) trunk neural crest, as opposed to the cranial neural crest associated with the oropharyngeal cavity.

Summary

This chapter focuses on evolution of the gnathostome dentition, with emphasis on morphological patterns in the initiation and replacement of all teeth in the dentition. Whereas the genetic and molecular controls for morphogenesis of single tooth production are well documented, those that regulate shape change, size, spacing and timing of the first tooth rows and those that replace them are not as well understood nor documented. These features are the characteristic of the majority of fish dentitions, with notable morphological diversity related to functional adaptations. We review all the known morphological and molecular data for this group and that of reptilian dentitions, which are also endowed with multiple sets of replacement teeth. Importantly, accepting the axiom that evolution may progress through co-option of existing developmental modules together with their functional gene networks, through modification of these for a different function, we compare pattern data for denticles in the external skin with those covering the internal visceral skeleton. In this context recent research on spatial and temporal molecular expression patterns supports the morphological data that the external dermal skeleton of denticles is a separate system from that of the internal visceral skeleton of the oropharynx, confirming their topographical structural isolation. Those of the oropharynx, organized as spatiotemporal groups of morphogenetic units, provide an evolutionary model for co-option into the dentition on the margins of the jaws from deeper in the pharyngeal cavity, rather than from those of the external skin. Based on this concept, their genetic profiles can be compared to assess the extent of conservation in regulation of the patterning between pharyngeal and oral teeth, and can be tested by comparing similar molecular data for denticles of the skin, which at present is scarce; ideally this should be obtained from a basal actinopterygian such as *Polypterus*.

Two papers relevant to the issues discussed in this chapter have recently been published: Fraser *et al.* (2012) and Smith *et al.* (2013).

Acknowledgements

We would like to thank Rob Asher and Johannes Müller for inviting us to participate in this volume, and we also thank the two reviewers for their constructive critical comments. We are indebted to all colleagues with whom we have worked over the years, especially Gareth Fraser, Jean Joss and Anthony Graham.

REFERENCES

Ahlberg, P. E., Smith, M. M. and Johanson, Z. (2006). Developmental plasticity and disparity in early dipnoan (lungfish) dentitions. *Evolution and Development*, **8**, 331–49.

Ahn, Y., Sanderson, B. W., Klein, O. D. and Krumlauf, R. (2010). Inhibition of Wnt signaling by Wise (Sostdc1) and negative feedback from Shh controls tooth number and patterning. *Development*, **137**, 3221–31.

Baker, R. E., Schnell, S. and Maini, P. K. (2009). Waves and patterning in developmental biology: vertebrate segmentation and feather bud formation as case studies. *International Journal of Developmental Biology*, **53**, 783–94.

Bei, M. (2009). Molecular genetics of tooth development. *Current Opinion in Genetics and Development*, **19**, 504–10.

Berkovitz, B. K. (1975). Observations on tooth replacement in piranhas (Characidae). *Archives of Oral Biology*, **20**, 53–6.

Berkovitz, B. K. (1977). The order of tooth development and eruption in the Rainbow trout (*Salmo gairdneri*). *Journal of Experimental Zoology*, **201**, 221–6.

Berkovitz, B. K. (1978). Tooth ontogeny in the upper jaw and tongue of the rainbow trout (*Salmo gairdneri*). *Journal de Biologie Buccale*, **6**, 205–15.

Berkovitz, B. K. and Moore, M. H. (1974). A longitudinal study of replacement patterns of teeth on the lower jaw and tongue in the rainbow trout *Salmo gairdneri*. *Archives of Oral Biology*, **19**, 1111–19.

Berkovitz, B. K. and Moore, M. H. (1975). Tooth replacement in the upper jaw of the rainbow trout (*Salmo gairdneri*). *Journal of Experimental Zoology*, **193**, 221–34.

Borday-Birraux, V., Van der Heyden, C., Debiais-Thibaud, M., *et al.* (2006). Expression of Dlx genes during the development of the zebrafish pharyngeal dentition: evolutionary implications. *Evolution and Development*, **8**, 130–41.

Botella, H., Blom, H., Dorka, M., Ahlberg, P. E. and Janvier, P. (2007). Jaws and teeth of the earliest bony fishes. *Nature*, **448**, 583–6.

Brazeau, M. D. (2009). The braincase and jaws of a Devonian 'acanthodian' and modern gnathostome origins. *Nature*, **457**, 305–8.

Buchtova, M., Handrigan, G. R., Tucker, A. S., *et al.* (2008). Initiation and patterning of the snake dentition are dependent on sonic hedgehog signaling. *Developmental Biology*, **319**, 132–45.

Cerny, R., Cattell, M., Sauka-Spengler, T., *et al.* (2010). Evidence for the prepattern/cooption model of vertebrate jaw evolution. *Proceedings of the National Academy of Sciences of the United States of America*, **107**, 17 262–7.

Chan, E. T., Quon, G. T., Chua, G., *et al.* (2009). Conservation of core gene expression in vertebrate tissues. *Journal of Biology*, **8**, 33.

Chang, M.-M. (1991). Head exoskeleton and shoulder girdle of *Youngolepis*. In *Early Vertebrates and Related Problems of Evolutionary Biology*, ed. M.-M. Chang, Y. H. Liu, and G. R.Zhang. Beijing: Science Press.

Chen, Y., Zhang, Y., Jiang, T. X., *et al.* (2000). Conservation of early odontogenic signaling pathways in Aves. *Proceedings of the National Academy of Sciences of the United States of America*, **97**, 10 044–9.

Coates, M. I. and Sequeira, S. E. K. (2001). A new stethacanthid chondrichthyan from the Lower Carboniferous of Bearsden, Scotland. *Journal of Vertebrate Paleontology*, **21**, 438–59.

Cobourne, M. T. and Sharpe, P. T. (2003). Tooth and jaw: molecular mechanisms of patterning in the first branchial arch. *Archives of Oral Biology*, **48**, 1–14.

Cobourne, M. T. and Sharpe, P. T. (2010). Making up the numbers: the molecular control of mammalian dental formula. *Seminars in Cell and Developmental Biology*, **21**, 314–24.

Debiais-Thibaud, M., Germon, I., Laurenti, P., Casane, D. and Borday-Birraux, V. (2008). Low divergence in Dlx gene expression between dentitions of the medaka (*Oryzias latipes*) versus high level of expression shuffling in osteichtyans. *Evoution and Development*, **10**, 464–76.

Donoghue, P. C. J. and Purnell, M. A. (1999). Mammal-like occlusion in conodonts. *Paleobiology*, **25**, 58–74.

Donoghue, P. C. J. and Purnell, M. A. (2005). Genome duplication, extinction and vertebrate evolution. *Trends in Ecology and Evolution*, **20**, 312–19.

Donoghue, P. C. and Sansom, I. J. (2002). Origin and early evolution of vertebrate skeletonization. *Microscopy Research and Technique*, **59**, 352–72.

Donoghue, P. C. J., Sansom, I. J. and Downs, J. P. (2006). Early evolution of vertebrate skeletal tissues and cellular interactions, and the canalization of skeletal development. *Journal of Experimental Zoology–Molecular and Developmental Evolution*, **306**, 278–94.

Fraser, G. J. and Smith, M. M. (2011). Evolution of development for patterning vertebrate dentitions: an oro-pharyngeal specific mechanism. *Journal of Experimental Zoology–Molecular and Developmental Evolution*, **316B**, 99–112.

Fraser, G. J., Graham, A. and Smith, M. M. (2004). Conserved deployment of genes during odontogenesis across osteichthyans. *Proceedings of Royal Society B–Biological Sciences*, **271**, 2311–17.

Fraser, G. J., Berkovitz, B. K., Graham, A. and Smith, M. M. (2006a). Gene deployment for tooth replacement in the rainbow trout (*Oncorhynchus mykiss*): a developmental model for evolution of the osteichthyan dentition. *Evolution and Development*, **8**, 446–57.

Fraser, G. J., Graham, A. and Smith, M. M. (2006b). Developmental and evolutionary origins of the vertebrate dentition: molecular controls for spatio-temporal organisation of tooth sites in osteichthyans. *Journal of Experimental Zoology–Molecular and Developmental Evolution*, **306**, 183–203.

Fraser, G. J., Bloomquist, R. F. and Streelman, J. T. (2008). A periodic pattern generator for dental diversity. *BMC Biology*, **6**, 32.

Fraser, G. J., Hulsey, C. D., Bloomquist, R. F., *et al.* (2009). An ancient gene network is co-opted for teeth on old and new jaws. *PLoS Biology*, **7**, e31.

Fraser, G. J., Cerny, R., Soukup, V., *et al.* (2010). The odontode explosion: the origin of tooth-like structures in vertebrates. *BioEssays*, **32**, 808–17.

Fraser, G. J., Britz, R., Hall, A., Johanson, Z. and Smith, M. M. (2012). Replacing the first generation dentition in pufferfish with a unique beak. *Proceedings of the National Academy of Sciences of the United States of America*, **109**, 8179–84.

Freitas, R. and Cohn, M. J. (2004). Analysis of *EphA4* in the lesser spotted catshark identifies a primitive gnathostome expression pattern and reveals co-option during evolution of shark-specific morphology. *Development Genes and Evolution*, **214**, 466–72.

Goujet, D. and Young, G. C. (1995). Interrelationships of placoderms revisited. *Geobios Mémoire Spécial*, **19**, 89–96.

Goujet, D. and Young, G. C. (2004). Placoderm anatomy and phylogeny: new insights. In *Recent Advances in the Origin and Early Radiation of Vertebrates*, ed. G. Arratia, M. V. H. Wilson, and R. Cloutier. Munchen, Germany: Verlag Dr. Friedrich Pfeil.

Graham, A., Holland, P. W., Lumsden, A., Krumlauf, R. and Hogan, B. L. (1988). Expression of the homeobox genes Hox 2.1 and 2.6 during mouse development. *Current Topics in Microbiology and Immunology*, **137**, 87–93.

Handrigan, G. R. and Richman, J. M. (2010a). A network of Wnt, hedgehog and BMP signaling pathways regulates tooth replacement in snakes. *Developmental Biology*, **348**, 130–41.

Handrigan, G. R. and Richman, J. M. (2010b). Autocrine and paracrine *Shh* signaling are necessary for tooth morphogenesis, but not tooth replacement in snakes and lizards (Squamata). *Developmental Biology*, **337**, 171–86.

Handrigan, G. R., Leung, K. J. and Richman, J. M. (2010). Identification of putative dental epithelial stem cells in a lizard with life-long tooth replacement. *Development*, **137**, 3545–9.

Harada, H., Kettunen, P., Jung, H. S., *et al.* (1999). Localization of putative stem cells in dental epithelium and their association with Notch and FGF signaling. *Journal of Cell Biology*, **147**, 1051–20.

Harris, M. P., Rohner, N., Schwarz, H., *et al.* (2008). Zebrafish eda and edar mutants reveal conserved and ancestral roles of ectodysplasin signaling in vertebrates. *PLoS Genetics*, **4**, e1000206.

Hecht, J., Seitz, V., Urban, M., *et al.* (2007). Detection of novel skeletogenesis target genes by comprehensive analysis of a *Runx2(−/)* mouse model. *Gene Expression Patterns*, **7**, 102–12.

Hecht, J., Stricker, S., Wiecha, U., *et al.* (2008). Evolution of a core gene network for skeletogenesis in chordates. *PLoS Genetics*, **4**, e1000025.

Huysseune, A. and Thesleff, I. (2004). Continuous tooth replacement: the possible involvement of epithelial stem cells. *BioEssays*, **26**, 665–71.

Huysseune, A. and Witten, P. E. (2006). Developmental mechanisms underlying tooth patterning in continuously replacing osteichthyan dentitions. *Journal of Experimental Zoology–Molecular and Developmental Evolution*, **306**, 204–15.

Huysseune, A., Sire, J. Y. and Witten, P. E. (2009). Evolutionary and developmental origins of the vertebrate dentition. *Journal of Anatomy*, **214**, 465–76.

James, C. T., Ohazama, A., Tucker, A. S. and Sharpe, P. T. (2002). Tooth development is independent of a Hox patterning programme. *Developmental Dynamics*, **225**, 332–5.

Janvier, P. (1996). *Early Vertebrates*. Oxford, UK: Oxford University Press.

Jarvinen, E., Salazar-Ciudad, I., Birchmeier, W., *et al.* (2006). Continuous tooth generation in mouse is induced by activated epithelial Wnt/beta-catenin signaling. *Proceedings of the National Academy of Science of the United States of America*, **103**, 18 627–32.

Jarvinen, E., Tummers, M. and Thesleff, I. (2009). The role of the dental lamina in mammalian tooth replacement. *Journal of Experimental Zoology–Molecular and Developmental Evolution*, **312B**, 281–91.

Jernvall, J. and Thesleff, I. (2000). Reiterative signaling and patterning during mammalian tooth morphogenesis. *Mechanisms of Development*, **92**, 19–29.

Johanson, Z. and Smith, M. M. (2003). Placoderm fishes, pharyngeal denticles, and the vertebrate dentition. *Journal of Morphology*, **257**, 289–307.

Johanson, Z. and Smith, M. M. (2005). Origin and evolution of gnathostome dentitions: a question of teeth and pharyngeal denticles in placoderms. *Biological Reviews*, **80**, 303–45.

Johanson, Z., Smith, M. M. and Joss, J. (2007). Early scale development in *Heterodontus* (Heterodontiformes; Chondrichthyes): a novel chondrichthyan scale pattern. *Acta Zoologica*, **88**, 249–56.

Johanson, Z., Tanaka, M., Chaplin, N. and Smith, M. (2008). Early Palaeozoic dentine and patterned scales in the embryonic catshark tail. *Biology Letters*, **4**, 87–90.

Jung, H. S., Francis-West, P. H., Widelitz, R. B., *et al.* (1998). Local inhibitory action of BMPs and their relationships with activators in feather formation: implications for periodic patterning. *Developmental Biology*, **196**, 11–23.

Kangas, A. T., Evans, A. R., Thesleff, I. and Jernvall, J. (2004). Nonindependence of mammalian dental characters. *Nature*, **432**, 211–14.

Kavanagh, K. D., Evans, A. R. and Jernvall, J. (2007). Predicting evolutionary patterns of mammalian teeth from development. *Nature*, **449**, 427–32.

Kondo, S., Kuwahara, Y., Kondo, M., *et al.* (2001). The medaka rs-3 locus required for scale development encodes ectodysplasin-A receptor. *Current Biology*, **11**, 1202–6.

Kundrat, M., Joss, J. M. and Smith, M. M. (2008). Fate mapping in embryos of *Neoceratodus forsteri* reveals cranial neural crest participation in tooth development is conserved from lungfish to tetrapods. *Evolution and Development*, **10**, 531–6.

Line, S. R. (2001). Molecular morphogenetic fields in the development of human dentition. *Journal of Theoretical Biology*, **211**, 67–75.

Line, S. R. (2003). Variation of tooth number in mammalian dentition: connecting genetics, development, and evolution. *Evolution and Development*, **5**, 295–304.

Luukko, K., Løes, S., Kvinnsland, I. H. and Kettunen, P. (2005). Expression of ephrin-A ligands and EphA receptors in the developing mouse tooth and its supporting tissues. *Cell and Tissue Research*, **319**, 143–52.

Maisey, J., Miller, R. and Turner, S. (2009). The braincase of the chondrichthyan *Doliodus* from the Lower Devonian Campbellton Formation of New Brunswick, Canada. *Acta Zoologica*, **90**, 109–22.

Mikkola, M. L. (2009). Controlling the number of tooth rows. *Science Signalling*, **2**, pe53.

Miletich, I. and Sharpe, P. T. (2004). Neural crest contribution to mammalian tooth formation. *Birth Defects Research Part C: Embryo Today*, **72**, 200–12.

Mitsiadis, T. A., Cheraud, Y., Sharpe, P. and Fontaine-Perus, J. (2003). Development of teeth in chick embryos after mouse neural crest transplantations. *Proceedings of the National Academy of Sciences of the United States of America*, **100**, 6541–5.

Mou, C., Jackson, B., Schneider, P., Overbeek, P. A. and Headon, D. J. (2006). Generation of the primary hair follicle pattern. *Proceedings of the National Academy of Sciences of the United States of America*, **103**, 9075–80.

Munne, P. M., Tummers, M., Jarvinen, E., Thesleff, I. and Jernvall, J. (2009). Tinkering with the inductive mesenchyme: Sostdc1 uncovers the role of dental mesenchyme in limiting tooth induction. *Development*, **136**, 393–402.

Nelson, G. J. (1969). Gill arches and the phylogeny of fishes, with notes on the classification of vertebrates. *Bulletin of the American Museum of Natural History*, **141**, 479–552.

Nelson, G. J. (1970). Pharyngeal denticles (placoid scales) of sharks, with notes on the dermal skeleton of vertebrates. *American Museum Novitates*, **2415**, 1–26.

Ohazama, A., Haworth, K. E., Ota, M. S., Khonsari, R. H. and Sharpe, P. T. (2010). Ectoderm, endoderm, and the evolution of heterodont dentitions. *Genesis*, **48**, 382–9.

Osborn, J. W. (1971). The ontogeny of tooth succession in *Lacerta vivipara* Jacquin (1787). *Proceedings of the Royal Society of London B*, **179**, 261–89.

Ørvig, T. (1977). A survey of odontodes ('dermal teeth') from developmental, structural, functional, and phyletic points of view. In *Problems in Vertebrate Evolution*, ed. S. M. Andrews, R. S. Miles, and A. D. Walker. London: Academic Press.

Patterson, C. (1977). Cartilage bones, dermal bones and membrane bones, or the exoskeleton versus the endoskeleton. In *Problems in Vertebrate Evolution*, ed. S. M. Andrews, R. S. Miles and A. D. Walker. London: Academic Press.

Peterkova, R., Lesot, H. and Peterka, M. (2006). Phylogenetic memory of developing mammalian dentition. *Journal of Experimental Zoology–Molecular and Developmental Evolution*, **396**, 234–50.

Raff, R. A. (1996). *The Shape of Life: Genes, Development, and the Evolution of Animal Form.* Chicago, IL: University of Chicago Press.

Reif, W.-E. (1980). Development of dentition and dermal skeleton in embryonic *Scyliorhinus canicula*. *Journal of Morphology*, **166**, 275–88.

Reif, W.-E. (1982). Evolution of dermal skeleton and dentition in vertebrates: the odontode-regulation theory. *Evolutionary Biology*, **15**, 287–368.

Schaeffer, B. (1977). The dermal skeleton in fishes. In *Problems in Vertebrate Evolution*, ed. S. M. Andrews, R. S. Miles and A. D. Walker. London: Academic Press

Silvan, O., Germon, I., Seydi, T., *et al.* (2010). A survey of the expression of key genes during tooth and dermal denticle development in dogfish. Paper presented at Euro Evo Devo Conference, Paris, July 6–9, 2010, p. 157.

Sire, J.-Y. and Akimenko, M.-A. (2004). Scale development in fish: a review, with description of sonic hedgehog (shh) expression in the zebrafish (*Danio rerio*). *International Journal of Developmental Biology*, **48**, 233–47.

Smith, M. M. (2003). Vertebrate dentitions at the origin of jaws: when and how pattern evolved. *Evolution and Development*, **5**, 394–413.

Smith, M. M. and Coates, M. I. (1998). Evolutionary origins of the vertebrate dentition: phylogenetic patterns and developmental evolution. *European Journal of Oral Sciences*, **106** Supplement 1, 482–500.

Smith, M. M. and Coates, M. I. (2000). Evolutionary origins of teeth and jaws: developmental models and phylogenetic patterns. In *Development, Function and Evolution of Teeth*, ed. M. F. Teaford, M. M. Smith, and M. W. J. Ferguson. Cambridge, UK: Cambridge University Press.

Smith, M. M. and Coates, M. I. (2001). The evolution of vertebrate dentitions: phylogenetic pattern and developmental models. In *Major Events in Early Vertebrate Evolution*, ed. P. Ahlberg. London: Taylor and Francis.

Smith, M. M. and Hall, B. K. (1993). A developmental model for evolution of the vertebrate exoskeleton and teeth: the role of cranial and trunk neural crest. *Evolutionary Biology*, **27**, 387–448.

Smith, M. M. and Johanson, Z. (2003). Separate evolutionary origins of teeth from evidence in fossil jawed vertebrates. *Science*, **299**, 1235–6.

Smith, M. M. and Johanson, Z. (2010). The dipnoan dentition: a unique adaptation with a longstanding evolutionary record. In *The Biology of Lungfishes*, ed. J. M Jorgensen and J. Joss. London: CRC Press.

Smith, M. M., Krupina, N. I. and Joss, J. (2002). Developmental constraints conserve evolutionary pattern in an osteichthyan dentition. *Connective Tissue Research*, **43**, 113–19.

Smith, M. M., Fraser, G. J., Chaplin, N., Hobbs, C. and Graham, A. (2009a). Reiterative pattern of sonic hedgehog expression in the catshark dentition reveals a phylogenetic template for jawed vertebrates. *Proceedings of the Royal Society B–Biological Sciences*, **276**, 1225–33.

Smith, M. M., Fraser, G. J. and Mitsiadis, T. A. (2009b). Dental lamina as source of odontogenic stem cells: evolutionary origins and developmental control of tooth generation in gnathostomes. *Journal of Experimental Zoology–Molecular and Developmental Evolution*, **312B**, 260–80.

Smith, M. M., Okabe, M. and Joss, J. (2009c). Spatial and temporal pattern for the dentition in the Australian lungfish revealed with sonic hedgehog expression profile. *Proceedings of the Royal Society B–Biological Sciences*, **276**, 623–31.

Smith, M. M., Johanson, Z., Underwood, C. and Diekwisch, T. (2013). Pattern formation in development of chondrichthyan dentitions: a review of an evolutionary model. *Historical Biology*, doi: 10.1080/08912963.2012.662228.

Soukup, V., Epperlein, H. H., Horacek, I. and Cerny, R. (2008). Dual epithelial origin of vertebrate oral teeth. *Nature*, **455**, 795–8.

Stock, D. W. (2001). The genetic basis of modularity in the development and evolution of the vertebrate dentition. *Philosophical Transactions of the Royal Society London B–Biological Sciences*, **356**, 1633–53.

Stock, D. W., Jackman, W. R. and Trapani, J. (2006). Developmental genetic mechanisms of evolutionary tooth loss in cypriniform fishes. *Development*, **133**, 3127–37.

Streelman, J. T. and Albertson, R. C. (2006). Evolution of novelty in the Cichlid dentition. *Journal of Experimental Zoology–Molecular and Developmental Evolution*, **306B**, 216–26.

Streelman, J. T., Webb, J. F., Albertson, R. C. and Kocher, T. D. (2003). The cusp of evolution and development: a model of cichlid tooth shape diversity. *Evolution and Development*, **5**, 600–8.

Takio, Y., Pasqualetti, M., Kuraku, S., *et al.* (2004). Evolutionary biology: lamprey Hox genes and the evolution of jaws. *Nature*, **429**, 262–3.

Takio, Y., Kuraku, S., Murakami, Y., *et al.* (2007). Hox gene expression patterns in *Lethenteron japonicum* embryos – insights into the evolution of the vertebrate Hox code. *Developmental Biology*, **308**, 606–20.

Thesleff, I. and Tummers, M. (2009). Tooth organogenesis and regeneration (January 31, 2009). The Stem Cell Research Community, ed. StemBook, doi/10.3824/stembook.1.37.1.

Thesleff, I., Jarvinen, E. and Suomalainen, M. (2007). Affecting tooth morphology and renewal by fine-tuning the signals mediating cell and tissue interactions. *Novartis Foundation Symposium*, **284**, 142–53; discussion 153–63.

Tummers, M. and Thesleff, I. (2003). Root or crown: a developmental choice orchestrated by the differential regulation of the epithelial stem cell niche in the tooth of two rodent species. *Development*, **130**, 1049–57.

Tummers, M. and Thesleff, I. (2009). The importance of signal pathway modulation in all aspects of tooth development. *Journal of Experimental Zoology–Molecular and Developmental Evolution*, **312B**, 309–19.

Van der Bruggen, W. A. J. P. (1993). Denticles in thelodonts. *Nature*, **364**, 107.

Van der Heyden, C. and Huysseune, A. (2000). Dynamics of tooth formation and replacement in the zebrafish (*Danio rerio*) (Teleostei, Cyprinidae). *Developmental Dynamics*, **219**, 486–96.

Van der Heyden, C., Huysseune, A. and Sire, J. Y. (2000). Development and fine structure of pharyngeal replacement teeth in juvenile zebrafish (*Danio rerio*) (Teleostei, Cyprinidae). *Cell and Tissue Research*, **302**, 205–19.

Vonk, F. J., Admiraal, J. F., Jackson, K., *et al.* (2008). Evolutionary origin and development of snake fangs. *Nature*, **454**, 630–3.

Wang, X. P., Suomalainen, M., Felszeghy, S., *et al.* (2007). An integrated gene regulatory network controls stem cell proliferation in teeth. *PLoS Biology*, **5**, e159.

Wilson, M. V. H., Hanke, G. F. and Sahney, S. (1999). Teeth, jaws, and ears in some early gnathostomes. *Journal of Vertebrate Paleontology*, **19** (supplement to No. 3), 85A.

Wise, S. B. and Stock, D. W. (2006). Conservation and divergence of *bmp2a*, *bmp2b*, and *bmp4* expression patterns within and between dentitions of teleost fishes. *Evolution and Development*, **8**, 511–23.

Yue, Z., Jiang, T. X., Widelitz, R. B. and Chuong, C. M. (2005). Mapping stem cell activities in the feather follicle. *Nature*, **438**, 1026–9.

Zhang, M. M. (1980). Preliminary note on a Lower Devonian antiarch from Yunnan, China. *Vertebrata PalAsiatica*, **28**, 179–90.

Zhang, Z., Lan, Y., Chai, Y. and Jiang, R. (2009). Antagonistic actions of *Msx1* and *Osr2* pattern mammalian teeth into a single row. *Science*, **323**, 1232–4.

Zhu, M., Yu, X. and Ahlberg, P. E. (2001). A primitive sarcopterygian fish with an eyestalk. *Nature*, **410**, 81–4.

Zhu, M., Zhao, W., Jia, L., *et al.* (2009). The oldest articulated osteichthyan reveals mosaic gnathostome characters. *Nature*, **458**, 469–74.

8

Molecular biology of the mammalian dentary: insights into how complex skeletal elements can be shaped during development and evolution

NEAL ANTHWAL AND ABIGAIL S. TUCKER

Introduction

The mammalian dentary

One of the defining features of mammals is the manner in which the lower and upper jaws articulate. In most vertebrates with jaws (gnathostomes), the jaw joint is formed by the articulation of the quadrate bone in the cranial base and the articular, the proximal element of the compound tetrapod mandible. These elements are both derived from the proximal part of Meckel's cartilage, and are initially part of a single cartilaginous condensation, which later subdivides, hypertrophies and ossifies (Wilson and Tucker 2004). In the mammals, the jaw joint is formed by the articulation of two previously unopposed elements, the squamosal bone in the cranial base and the dentary bone, which in mammals forms the entire mandible. In humans this joint is known as the temporomandibular joint or TMJ. The squamosal and dentary are derived directly from the mesenchyme of the first pharyngeal arch via membranous ossification, and therefore have no cartilage template.

The German embryologist and anatomist Karl Reichert suggested in 1837 that the mammalian homologues of the tetrapod jaw articulation are to be found in the ossicles of the mammalian middle ear (1837, as reviewed in English by de Beer 1937), with the homologue of the articular bone being the malleus, while the quadrate's homologue is the incus. Other homologous components include the tympanic ring, which is a homologue of the angular, and gonial, the homologue of the pre-articular. This anatomical homology has recently been supported by genetic evidence, from studies in mice, chick and zebrafish (Miller *et al.* 2003; Tucker *et al.* 2004; Wilson and Tucker 2004).

From Clone to Bone: The Synergy of Morphological and Molecular Tools in Palaeobiology, ed. Robert J. Asher and Johannes Müller. Published by Cambridge University Press.

A transformation series, moving from a quadro-articular articulation to a squamosal-dentary articulation, can be observed in the fossil record, as the post-dentary bones reduce in size and the dentary enlarges (Luo 2007). Several premammalian cynodonts and mammaliaforms of the Late Triassic and Early Jurassic, such as *Morganucodon*, were in possession of both quadro-articulate and squamosal-dentary joints, where the quadro-articular had become somewhat more specialized in hearing, but has not completely lost its function as a jaw articulation (Kermack *et al.* 1981; Luo 2007).

This transformation of jaw articulation is also observed in neonatal marsupials, where the jaw articulation is formed initially at birth by the quadro-articular/malleo-incudal joint rather than the squamosal-dentary. At this stage the neonate must suckle in the mother's pouch. However, the membraneous bones that make the jaw joint, the squamosal and dentary, have not fully developed and are unable to make a functional joint. The primary jaw joint between the malleus and incus therefore is utilized to support the upper and lower jaws while suckling. Once the membraneous bones have formed fully in the pouch, the jaw articulation shifts and the malleo-incodal joint becomes internalized within the cranial base and is dedicated exclusively to auditory function (Filan 1991; Clark and Smith 1993; Smith 2006).

The forces driving the evolution of the squamosal-dentary joint appear likely to be the increased masticatory efficiency afforded by the evolution of the masseter muscle and the coronoid process, and the relocation of the attachment site of the muscles of mastication from the post-dentary bones to the dentary. In extant reptiles, the feeding apparatus serves to hold food. Chunks may be torn if the piece of food is too large, but by and large food is swallowed whole and not chewed. Chewing requires both vertical movements, such as those allowed by the reptilian jaw articulation, and horizontal movements, which is the great innovation afforded by the new mammalian musculature and joint (Kemp 2005). There is a close functional correlation between the presence of a squamosal-dentary joint and the evolution of unilateral occlusion, with fossil mammaliaform reptiles showing no evidence of significant molar occlusion (Crompton and Hylander 1986). The advent of the squamosal-dentary joint therefore opened up new avenues for evolution of the dentition.

In addition, the increasing importance of hearing to basal synapsids resulted in the recruitment of the post-dentary bones to the transmission of sound to the inner ear, allowing for their reduction in size and eventual detachment from the jaw articulation (Allin 1975). Expansion of the neocortex has also been implicated in the separation of the two jaw joints, allowing for functional specialization (one to the jaw articulation, the other to the ear; Rowe 1996).

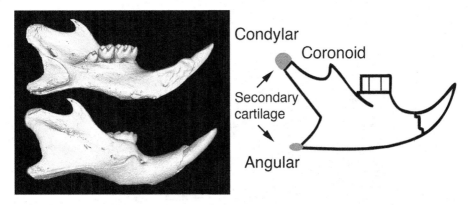

Figure 8.1 MicroCT (computerized tomography) of a six-week-old mouse showing the dentary with its three proximal processes. Top image shows buccal side, bottom image lingual side. Right-hand side, schematic of a dentary with condylar, coronoid and angular processes.

Although all mammals have a squamosal-dentary joint, the shape of the dentary and its elements varies considerably. It is this variation that we will investigate in this chapter. The mandible of mammals is made up of two dentary bones often fused at the midline. The most distal portion is made up of a rostral process, whilst the proximal portion exhibits three processes: a condylar process that forms the articulation with the skull flanked superiorly by the coronoid process and inferiorly by the angular process (Figure 8.1). The coronoid and angular processes serve as muscle attachment sites for the jaw opening and closing muscles respectively. These posterior processes of the mandible are capped with secondary cartilages, the function of which is to facilitate growth and to enable the articulation of the dentary with the cranial base at the squamosal (or squamosal portion of the temporal) bone (Frommer 1964; Beresford 1981; Depew *et al.* 2002b).

Modularity

Biological modularity allows for discrete integrated units, or modules, that interact with, and are dependent on, other units to contribute to a biological system. Such modularity allows for complex changes to occur to the development of organism without the need for system-wide changes, and so represents the framework upon which evolution can act within the context of a viable organism. These units can be, amongst other things, spatial, evolutionary or morphological (Goswami 2007), and can exist at almost all orders of magnitude, from the modular nature of the genome, to the various anatomical components of a whole organism, or even to the various specialized individual

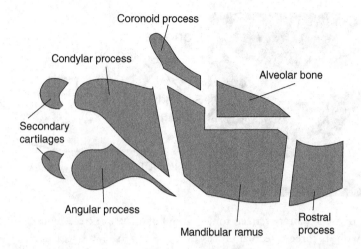

Figure 8.2 The dentary can be thought of as consisting of a number of distinct but interconnecting modules.

animals that make up a colony of social insects. Within the context of the gross anatomy of an individual organism, morphological units are structures, or group of structures, that exhibit less variation within the traits of the units than with traits outside of the unit (Klingenberg *et al.* 2003). Once again, such units can exist in several orders of magnitude. For example, the head of a tetrapod is a morphological unit, heads of different individuals are less different than the head of one individual compared with the upper limb of another. The head in turn is made up of a number of different units: the mandible, maxilla, calvarium and so forth, and each of these is made up of yet more units.

The mammalian dentary, itself a component of the lower jaw arcade, is made up of a number of different morphological units, commonly described as the mandibular body, alveolar bones, the mental/rostral process and the proximal coronoid, condylar and angular processes, and, when present, their secondary cartilages (Atchley *et al.* 1985; Hall and Hallgrímsson 2008) (Figure 8.2).

In this chapter we will focus on the three proximal processes (condylar, coronoid and angular) and their secondary cartilages.

Morphological variation in the dentary

The proximal processes of the mandible act as a set of functionally integrated units; they are involved in jaw articulation and jaw opening and closing. Although integrated, the processes are morphologically distinct units, and variation can be seen across species. This variation depends on function,

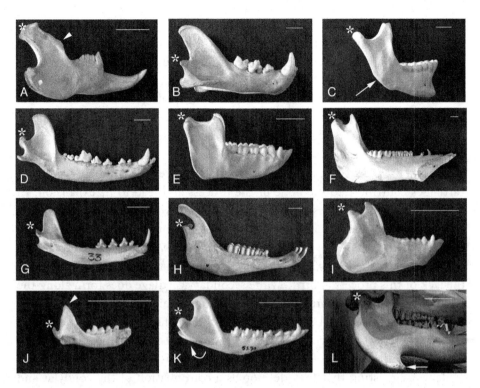

Figure 8.3 Morphological variation of the proximal processes of the dentary seen throughout the mammalian clade. Asterisk indicates the condylar process; arrowheads indicate examples of coronoid process variation; arrows indicate variation in the angular process. Specimens are (A) leporid (rabbit), (B) felid (lion), (C) human, (D) canid (wolf), (E) old world monkey, (F) suid (wild boar), (G) canid, (H) ruminant (deer), (I) new world monkey, (J) mustelid (weasel), (K) *Dasyurus* (quoll), (L) hippo. (All specimens are part of the Department of Craniofacial Development teaching collection at Guy's Hospital, London.) Scale bar in (A–K) represents 2 cm. Scale bar in (L) represents 10 cm.

with the process size and shape dependent on the diet of the species, and on the inherited characteristics of the processes. The processes responsible for this variation are, as yet, unknown. It is likely that a number of mechanisms lie at the heart of variation seen in the dentary, including variation in developmental signals and processes through heterochrony or allometry (Klingenberg 1998). In many species, the adult dentary is missing one or both of the angular and coronoid processes, or one or other of the processes is greatly enlarged. For example, in the guinea pig and rabbit (Figure 8.3A), the coronoid process is greatly diminished compared with that of the mouse, whist in the weasel (Figure 8.3J) the coronoid process is greatly enlarged. Common shrews have a long angular process, whereas the human mandible lacks an angular process (Figure 8.3C). The relative position of the proximal elements also varies; for example, among the morphological differences between placental and marsupial mammals is the position of the angular process (Figure 8.3K) which is medially inflected in almost all marsupials (Sánchez-Villagra and Smith 1997).

The distribution of secondary cartilages is also very varied. In mice, the coronoid process does not develop a secondary cartilage during normal development whereas the angular and condylar processes do. These secondary cartilages contribute to the bone of the process and lead to its expansion. In rats the coronoid process also develops an embryonic secondary cartilage, which is small, appears relatively late in development and does not contribute to the growth of the process (Vinkka 1982; Tomo et al. 1997). In humans no secondary cartilage forms in the region of the angular process, in keeping with the lack of this process at birth, while secondary cartilage does form along the coronoid process. The presence, size and timing of the secondary cartilage therefore plays a central role in shaping the dentary during development, and may present a mechanism whereby process size can be manipulated during evolution. How secondary cartilages form is still under debate, with evidence suggesting that they develop as a continuation of the condylar and angular processes, or as a sesamoid, apparently distinct from the developing bone (Vinkka 1982; Vinkka-Puhakka and Thesleff 1993; Anthwal et al. 2008). The secondary cartilages in mammals differ from those formed in birds, as they can initiate without the need for mechanical stimulation (Vinkka-Puhakka and Thesleff 1993; Anthwal et al. 2008). Therefore, mechanical forces acting upon the dentary do not dictate the initial position of secondary cartilages.

The morphological variations seen in mammalian dentaries reflect the adaptations of the mandible to distinct ecologies and feeding strategies that require different mechanical properties. An extreme example of this is shown in the Western Australian honey possum (*Tarsipes rostratus*), where the dentary is reduced to a slender rod with no angular and coronoid process, as its honey-eating diet requires limited muscle force (Hall and Hallgrímsson 2008). A more subtle example can be seen in old world rats and mice (Muridae), where omnivorous species tend to have a smaller angular process compared with herbivorous species (Michaux et al. 2007). Compared with omnivores, herbivores require a stronger lateral force in their jaw closure action in order to grind down the highly cellulous material that they eat. Consequently, those muscles responsible for this action are relatively larger and more powerful, and the process to which they attach is correspondingly enlarged. Beyond the Muridae, the powerful jaw closure muscles and large gape of the hippopot-amus, which is a large herbivore, results in a large angular process with a large surface area for the attachment of the masseter and internal pterygoid muscles (Figure 8.3L), while humans, an omnivore with a relatively weak bite, have no angular process.

Whilst these variations in morphology reflect variation in the mechanical forces acting on the dentary, and these epigenetic forces have an effect on the shape of the mandible, mandible morphology is also controlled by many other influences (Atchley *et al.* 1985; Atchley and Hall 1991; Klingenberg *et al.* 2003; Michaux *et al.* 2007). An example of this is shown by the medial inflection of the angular, which is a feature of marsupials but not placentals. Variation in the shape of the inflected angular (shelf-like or rod-like) within marsupials can be correlated with diet; however, the presence of the inflection itself cannot be explained adequately by chewing mechanism or muscle morphology (Sánchez-Villagra and Smith 1997). An explanation may lie in the fact that during marsupial development the angular maintains a close relationship to the tympanic ring (ectotympanic). The explanation for the medial inflection may therefore be linked to ontogeny and the changing role of the jaw in marsupial pouch young (Maier 1987, 1990).

Embryology: changes to proximal elements of the dentary during development

The development of the dentary in the mouse

The house mouse, *Mus musculus*, is the model organism for the study of mammalian development, in particular for analysis of molecular biology. In the adult mouse, the basic morphology of the mandible is faithfully represented (Figure 8.1). By looking at the development of the dentary in the mouse embryo, the way by which this basic shape is laid down can be assessed. During development, the dentary bone is first detectable by alizarin red staining at embryonic day (e) 14.5 (Figure 8.4). One day earlier, at e13.5, expression of *Runx2*, a transcription factor key to the early osteogenic pro-gramme, can be detected in the mouse dentary anlage and already marks out the coronoid, condylar and angular processes. The basic three-processed pattern of the proximal dentary is therefore patterned before ossification (Anthwal *et al.* 2008). Secondary cartilages are first observed at the condylar process by alcian blue staining at e15.5, followed by the angular cartilage at e16.5 (Livne and Silbermann 1990; Miyake *et al.* 1997; Anthwal *et al.* 2008; Figure 8.4).

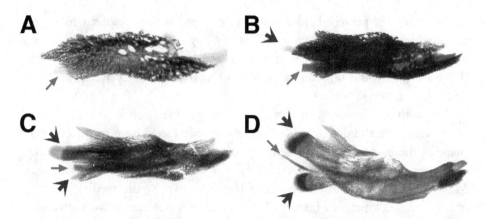

Figure 8.4 Development of the mouse dentary and its secondary cartilages from ossification at e14.5 to birth (P0). (A) e14.5. (B) e15.5. (C) e16.5. (D) P0. Cartilage shows up as blue (alcian blue stain), bone as red (alizarin red stain). The black arrows mark the three proximal processes visible at e14.5. Red arrows indicate Meckel's cartilage. The blue arrows indicate the formation of secondary cartilage at the condylar and then at the angular. (See also colour plate.)

The development of the dentary in other species

As previously discussed, mammals with different diets and lifestyles have differently shaped dentary bones. To identify when these differences become apparent we can turn to embryonic development to ask whether dentary bones are initiated in different patterns, or whether they begin with a similar template and then are modified.

Evidence from the guinea pig

Herbivorous guinea pigs have a larger angular process and a greatly reduced coronoid process (similar to that shown in Figure 8.3A). During the initial stages of ossification, before secondary cartilage formation, both the mouse (e14.5) and guinea pig (e37) dentary exhibit a condylar, coronoid and angular process (compare Figures 8.4 and 8.5). A coronoid process therefore is present in the embryo but is lost by birth. This suggests that the guinea pig dentary is under similar patterning signals to that of the mouse early in its ontogeny, but as the embryo develops, other factors cause the loss of the process. The early angular process of the guinea pig appears relatively larger compared with that of the mouse, although not to the same proportion seen in the adult. Since the mandibles of the mouse and the guinea pig are broadly similar early in their development, it seems reasonable to suggest that much of their differences, in respect to the proximal processes, are a

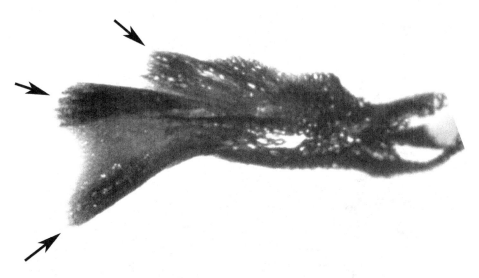

Figure 8.5 Alizarin red staining of an e37 guinea pig dentary. At this stage
the proximal dentary of the guinea pig possesses a coronoid, condylar
and angular process, unlike in the adult, which does not possess a distinct
coronoid process.

consequence of differential growth of the processes once the initial pattern
has been set up. The reduction of the coronoid process in the guinea pig
may be an adaptation to intrauterine jaw movements in preparation for
postnatal use. The plasticity of the adult guinea pig residual coronoid
process and its relationship with the action of the jaw musculature can be
demonstrated by experimental cutting of the temporalis muscle, which
results in a change in the size and shape of the residual process (Boyd
et al. 1967). These differences, whilst reflecting a mechanical necessity, also
no doubt have a genetic basis.

Evidence from marsupials

We have already discussed the inflected angular of marsupials. Some
marsupials, however, such as the koala (*Phascolarctos cinereus*) and striped
possum (*Dactylopsila trivirgata*), do not have an inflected angular in the adult
(Sánchez-Villagra and Smith 1997). Importantly, however, when the pouch
young are analysed an inflected angular is present, showing that this marsu-
pial-specific dentary morphology is preserved during development and then
modified later in some species (Sánchez-Villagra and Smith 1997). This fact
again stresses an important ontogenic reason for the presence of an inflected
angular in marsupials.

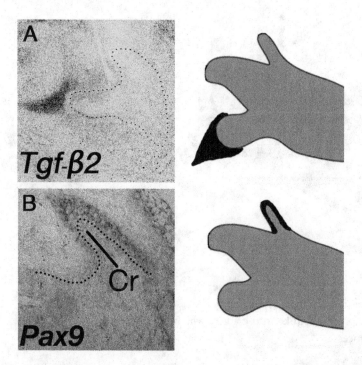

Figure 8.6 (A) Expression of *Tgf-β2* around the developing angular process at e14.5 by radioactive *in situ* hybridization. (B) Expression of *Pax9* in the osteogenic front of the coronoid process at e14.5 by dig-labelled *in situ* hybridization.

Molecular data: linking genes with anatomy

Gene expression in the mouse dentary

Using the mouse as a model, the genes that may play a role in shaping the dentary can be analysed. The first step is the identification of candidate genes that show isolated expression patterns associated with the proximal processes of the dentary and then the manipulation of these genes to investigate the effect on dentary patterning. A number of genes are expressed in the developing dentary, associated with the formation of bone and cartilage (*Runx2*, Type II collagen). There are, however, in addition, a number of genes with more specific patterns. For example, *Tgf-β2* (transforming growth factor beta 2) is expressed in the tissue surrounding the developing angular process of the dentary, but is not expressed around the equivalent region of the condylar or coronoid (Anthwal *et al.* 2008; Figure 8.6). In contrast, *Pax9*, a paired homeobox transcription factor, is expressed within the developing coronoid, at the osteogenic front, but is not associated with the other proximal processes (Figure 8.6). These expression domains fit with the proposed modular nature of the dentary.

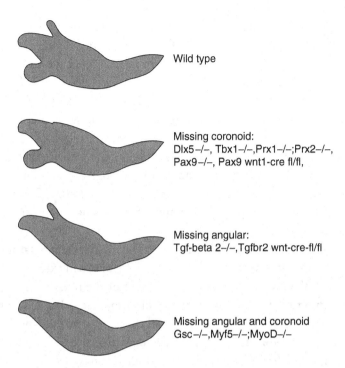

Figure 8.7 Summary of the effects of a selection of targeted gene deletions on the proximal processes of the mouse dentary bone.

Mouse mutants

A large number of mouse full and conditional knockouts have been generated, which provide a wealth of molecular insight into the mechanisms involved in embryonic processes. A number of these phenotypes differentially affect the processes and secondary cartilages of the proximal dentary, shedding light on the signals necessary to pattern and influence its growth. A summary of knockout mice with dentary defects is shown in Figure 8.7.

The angular process

Given the expression of *Tgf-β2* around the developing angular process it would be predicted that loss of this gene would affect this process. In mouse knockouts of *Tgf-β2* the angular process is completely missing (Sanford *et al.* 1997). The coronoid and condylar process, and whole mandible, are also smaller, indicating the important role this gene plays in chondrogenesis and osteogenesis (Alvarez *et al.* 2002; Janssens *et al.* 2005; Mukherjee *et al.* 2005). The loss of the angular is specific to *Tgf-β2* compared with the other isoforms of Tgf-β. Targeted deletion of *Tgfbr2*, the common type II receptor for all three isoforms of *Tgf-β*, in the neural crest, using the cre-recombinase system (*Tgfbr2 wnt1-cre fl/fl*) produces a similar phenotype in the dentary, although in addition the secondary cartilage on the angular fails to form and the condylar cartilage fails to develop mature chondrocytes or undergo ossification (Ito *et al.* 2003; Oka *et al.* 2007). The *Tgf-β*s have been shown to control expression of the cartilage marker *Sox9* and play an important role in secondary cartilage initiation (Anthwal *et al.* 2008). In the *Tgfβ2* knockout the angular fails to form but the secondary cartilage does appear to develop, indicating that loss of secondary cartilage is not enough to lead to loss of the angular. In keeping with this, loss of the type I Bmp receptor gene *Alk2* results in failure of the secondary cartilages to develop, whilst maintaining a small angular process (Dudas *et al.* 2004).

Interestingly, in the *Tgfbr2* conditional mouse knockout, the angular initially forms but fails to grow out (Anthwal *et al.* 2008). The failure in extension of the angular process is linked with a defect in attachment of the muscles around the angular process, due to localized loss of the tendon marker *Scleraxis* (Anthwal *et al.* 2008; Oka *et al.* 2008). Loss of the tendon therefore prevents mechanical force from acting on the forming angular. The angular process also undergoes accelerated osteoblast differentiation, with heightened levels of *Dlx5* (Oka *et al.* 2008). Loss of *Tgf-β* signalling at the angular therefore disrupts formation of secondary cartilages, osetoblast differentiation and muscle attachment, which when combined lead to a failure in the angular process to grow.

In keeping with the need for mechanical force in shaping the dentary, the angular process is also lost in *Myf5;MyoD* double knockouts. These two genes are muscle differentiation factors and the resultant null embryos lack the majority of skeletal muscle (Rot-Nikcevic *et al.* 2006). Staining of the skeletal tissue reveals that secondary cartilage develops on the condylar process, but not at the angular process. In these mutants, in addition to loss of the angular process, there is no apparent coronoid process, indicating that the coronoid, like the angular, is regulated by mechanical force. *Tgf-β* signalling, however, is not involved in co-ordinating the formation of the mechanical force for the coronoid process (Rot-Nikcevic *et al.* 2007).

The coronoid process

The coronoid process is affected in a number of mouse knockouts, providing insight into how it is patterned and regulated. Given the expression of *Pax9* within the developing coronoid it would be predicted that this structure would be defective in *Pax9* mutant mice. The phenotype of the *Pax9* null mutant mouse includes defects in a number of tissues in the head such as clefting of the palate, arresting of dental development at the bud stage, failure of the thymus and parathyroid glands to develop, ear defects such as a deformed tympanic ring, and, most interestingly for this discussion, the absence of the coronoid process (Peters *et al.* 1998). Conditional inactivation of *Pax9* in *Wnt1* expressing neural crest cells results in a similar phenotype in the neural-crest-derived elements of the head. For example, there is no tooth development beyond the bud stage, the palate is cleft and the coronoid process of the mandible is missing (Kist *et al.* 2007). The loss of the coronoid in *Pax9* mutant mice is therefore associated with an innate defect in the neural crest cells that form the bone and cartilage of the dentary, rather than a muscle defect.

Prx1 null (homozygous null designated by the symbol −/−) mice have a reduced coronoid process, which is lost in the $Prx1^{-/-}$; $Prx2^{-/-}$ double knock-out (ten Berge *et al.* 1998). The $Prx1^{-/-}$; $Prx^{-/-}$ mouse has a reduced lower jaw, the dentaries are shortened, most of Meckel's cartilage is deleted and the incisors are either absent or fused (ten Berge *et al.* 1998). Of particular interest is the finding that *Pax9* expression is downregulated, indicating that the origin of the coronoid defect may be due to loss of this transcription factor.

Loss of the coronoid is also observed in mice with mutations in the T-box transcription factor *Tbx1*. *Tbx1* is a candidate gene for DiGeorge syndrome, which is caused by the deletion of chromosome 22q11 (Jerome and Papaioannou 2001). DiGeorge syndrome, also known as velocardialfacial syndrome, has a spectrum of symptoms including cleft palate and other craniofacial dysmorphias including micrognathia, or a small mandible (Kobrynski and Sullivan 2007). *Tbx1* is involved in the proper development of neural-crest-derived cells, yet does so indirectly as it is not expressed in neural crest cells, but is expressed in the mesoderm-derived component of the first pharyngeal arch. Loss of the coronoid is also observed in conditional knockouts, where loss of *Tbx1* is driven only in mesodermally derived cells in the head (Aggarwal *et al.* 2010). The loss of the coronoid process may be related to a perturbation in the development of the musculature of the jaw, since the branchiomeric muscles are derived from the pharyngeal arch mesoderm, and loss of *Tbx1* expression results in the loss of the expression of *MyoD* in the first and second pharyngeal arches. In a similar manner to the *Myo;Myf5* double knockout, disruption of the muscular development of the mandible therefore results in the abnormal morphology of the coronoid.

Finally, loss of the coronoid is also associated with loss of *Dlx* genes. The *Dlx* family members are expressed in the first pharyngeal arch in a nested pattern in the mouse embryo, and the loss of each family member affects the development of the maxilla and/or the mandible depending on where the genes are expressed. Loss of *Dlx5* results in a hypoplastic mandible, including reduced condylar and angular processes, and a complete loss of the coronoid process, while overexpression of *Dlx5* is also linked with the angular defect in *Tgfbr2* conditional mutants (Depew *et al.* 2002a; Oka *et al.* 2008). In the *Dlx5* mutant there is a massive reduction in the expression of the homeobox gene *Goosecoid* (*Gsc*) throughout the pharyngeal arches (Depew *et al.* 1999, 2002a). In keeping with this, knockout of *Gsc* leads to a similar dentary phenotype with a reduced coronoid process, and a reduced angular process with secondary cartilage, although the condylar process is unaffected (Rivera-Perez *et al.* 1995). Mice chimeric for *Gsc* exhibit a range of phenotypes dependent on the proportion of *Gsc*-null cells (Rivera-Perez *et al.* 1999).

The above examples demonstrate that the non-articular angular and coronoid processes are often affected by gene knockouts of a number of different signalling pathways and transcription factors, with the coronoid process appearing particularly vulnerable. The condylar process is, however, more robust, although it is affected and hypoplastic in a number of gene knockouts. This suggests that, compared with the other proximal processes, the condylar process is under the control of additional signals that can act in a redundant manner to preserve the condylar. This preservation may be, in part at least, due to the crucial nature of the condylar process in articulating with the cranial base.

Human syndromes

In addition to studies in the mouse we can also learn about the genes involved in patterning the dentary by studying human craniofacial syndromes. The dentary is hypoplastic in a number of craniofacial syndromes where we have a good understanding of the underlying genetic mutations. For example, in patients with Treacher Collins syndrome, caused by mutations in *TCOF1*, the dentary is small and the condylar process defective (Posnik *et al.* 2004). Subtle dentary defects are also shown in patients with X-linked Hypohidrotic ectodermal dysplasia, which is caused by mutants in *EDA* (Clauss *et al.* 2008). In patients with cleidocranial dysplasia, which is caused by mutations in *Runx2*, the position and angle of the coronoid process is disrupted, along with a number of other skeletal defects (Jensen and Kreiborg 1993). We can therefore take this information and move to animal models of these syndromes to further understand the role of these genes in shaping the jaw.

The patterning of the different proximal processes is dependent on different morphogenetic and mechanical signals, and utilizes different transcription factors

The numerous knockout mice with proximal dentary phenotypes demonstrate how the processes of the dentary are under the control of varying developmental signals and processes, including a number of genetic signals that are able to control the development of each process individually, such as *Tgf-β2* and *Pax9*. It is also clear that the proximal dentary processes require mechanical action from the muscles that attach to them for their proper development, and so the patterning of the jaw musculature plays a key role in the patterning of the dentary bone. The dentary therefore does not form in isolation but is part of a larger unit, with the skeletal elements intimately interconnected with the muscles and tendons. As such, genetic changes that affect the forces acting upon the dentary during development may be as crucial in explaining the variation in mandibular morphologies seen throughout mammals as those genetic changes that directly alter bone and cartilage.

Genes that affect the positioning, size and timing of the secondary cartilages may also play important roles in shaping the dentary. The mouse does not have secondary cartilage associated with its coronoid process. However, in *Parathyroid hormone related protein (PTHrP)* null mice, the coronoid process develops a secondary cartilage, similar to that observed in the rat, indicating that in normal development *PTHrP* may play a role in suppressing the formation of secondary cartilage on this process (Shibata *et al.* 2003). This is rather intriguing as it indicates that secondary cartilages may form as a default state on all proximal processes, stimulated by *Tgf-β* signalling, but that other genes are involved in creating the pattern of secondary cartilages by actively inhibiting their development at specific positions.

The coronoid process is a separate developmental unit, and exhibits a reduced degree of integration with the other processes possibly due to its evolutionary and developmental origins

The evolution of the coronoid process of the dentary was one of the key events in proto- and early mammal dentary evolution. In reptiles, the action of jaw closure muscles simply brings the lower jaw into occlusion with the upper jaw, the force acting in an upward direction (Kemp 2005). Mammalian mandibles, however, thanks to the new musculature and attachment sites afforded by the increased complexity of the dentary bone, enable two vectors of the force while closing the jaw: the upward and forward force of the masseter, and an upward and backward force of the temporalis attached to the coronoid process (Crompton 1963; Crompton and Hylander 1986). These forces working together allowed for increased efficiency in chewing, and reduced the loading burden on the quadro-articular joint therefore allowing the post-dentary bones to decrease in size (Crompton and Hylander 1986). Variation in the size of the coronoid has also allowed for adjustment of the opening and closing action of the jaw, with an increased coronoid, and therefore a higher and more posterior coronoid, resulting in an increase in the ability to raise and lower the jaw, and consequently an increased ability to chew (Demar and Barghuse 1972).

The evolutionary novelty of the coronoid process appears to have utilized transcription factors such as *Dlx5*, *Tbx1* and *Pax9*, which are known to have roles in tooth development, a tissue that juxtaposes the developing coronoid, and as such are topographically well situated to contribute to the development of a new mandibular process. The persistence of the condylar and angular processes in the mutants for these transcription factors highlights the independence of the coronoid process module within the mandible. Loss of the angular process in contrast results in a perturbation of the other processes implying a more general bone or cartilage defect (as in *Tgf-β* signalling mutants). It can therefore be inferred that the coronoid process is less integrated with the condylar and angular processes than those processes are with each other. The coronoid is a more ancient and more conserved feature of cynodonts and mammaliaforms, compared with the condylar and angular processes (Kemp 2005), and it is therefore perhaps logical that these more derived processes are more integrated with each other than either is with the coronoid. The independence of the coronoid is perhaps reflected in the huge variation in size and shapes of this element across mammals.

The angular and condylar processes, and the secondary cartilages of these processes, are separate but integrated developmental units

The dependence of the angular and condylar process on secondary cartilage development in the mouse demonstrates that in each case the process and its cartilage are highly integrated. This is not at all surprising, since secondary endochondral ossification results in the secondary cartilages contributing to the bone of the processes. This, one may argue, means that a process and its secondary cartilage do not represent separate developmental units. However, the processes of the membranous dentary appear to be patterned at the very initial stages of ossification long before the initiation of the secondary cartilages. Additionally, it seems that the secondary cartilages of the processes develop from a common, but distinct, population of skeletoblasts. It can therefore be suggested that initially the cartilage and the ossified process are a highly integrated, but separate, pair of developmental units. As development proceeds to the postnatal period, this integration becomes so great that they effectively function as one unit.

Independence of functional units allows for variations in dentary morphology across species. For example, as previously mentioned the dentary of the guinea pig and that of humans can be contrasted in the elongation or absence of the coronoid and angular processes, and this variation has been enabled by the differential growth of this process. In the angular process this could hypothetically be a consequence of different levels of *Tgf-β2* signalling in the developing embryo, much as the variation in the coronoid process may be due to different levels of *Tbx1* or *Pax9* activity.

An evo-devo approach to mammalian dentary development

The laboratory mouse enables the researcher to investigate the potential developmental, molecular and genetic basis for the adaptations in the mammalian dentary seen throughout evolution. Targeted deletions of key genes involved in process development, either directly through osteo- and chondrogenesis, or indirectly through processes such as the development of the jaw musculature and muscle attachment sites, can help identify the basis of the modularity and plasticity of the dentary.

Insights from molecular data

Alterations in one process can be independent of another process, particularly in the case of the coronoid process. Thus, changes in the coronoid do not impact on the condylar, allowing jaw dynamics to change without influencing the essential jaw joint. This would have allowed for almost complete loss of the coronoid, as observed in rabbits and guinea pigs, or large extension of the process, as observed in many carnivores, without altering articulation.

Different genes play a role in patterning different parts of the dentary. For example, muscle attachment and mechanical force are equally important for the correct development of the coronoid and angular process, but different genes are involved in creating these attachments.

The condylar is the most conserved of the proximal processes, reflecting the importance of this structure in jaw articulation.

It is clear that changes in gene expression alter not only formation of a process but also influence the mechanical load working on that process.

In addition changes in the pattern of secondary cartilage induction may reflect inhibition of cartilage formation at specific locations, allowing alterations in the size of specific processes.

As yet, we are not fully able to understand how these genetic processes account for the differences seen in dentary morphologies, although we now are armed with a number of avenues and candidates for investigation. We can predict changes in gene expression that may play a role in shaping the dentary. For example, reduction of *Pax9* in the coronoid and an increase in *Tgf-β2* around the angular would be predicted to lead to a residual coronoid process and an increased mechanical pull on the angular, creating a dentary that would look similar to that of a guinea pig. Likewise localized inhibition of *PTHrP* might stimulate formation of a coronoid secondary cartilage, leading to increased growth of this process.

To test these ideas, the next step is to investigate expression of these key genes during development of animals with diverse dentary morphology to identify whether any of our predictions from the mouse data are backed up. General trends in gene expression changes would then provide a solid platform from which evolutionary change could be extrapolated.

The expansion in the repertoire of model organisms available to the developmental biologist is beginning to allow for such investigations to be made. The grey short-tailed opossum, *Monodelphis domestica*, is the first marsupial to have had its genome sequenced (Mikkelsen *et al.* 2007), and already represents a useful and usable model marsupial (Smith and van Nievelt 1997; Smith 2001, 2006; Sears 2011; Sears *et al.*, this volume), whilst the development of the shrew dentition is increasingly being studied due to the presence of premolars, which are absent in the mouse (Jarvinen *et al.* 2008). Non-avian diapsids (i.e. 'reptiles') may offer a more basal and suitable model for outgroup comparison of dentary development compared with other non-mammalian models such as chicken or quail, and as a reflection of this reptile colonies for the study of jaw and tooth development have recently been set up in London, Toronto, Prague and Kansas City.

Conclusions

The development of non-traditional animal models, such as *Monodelphis*, represents an exciting opportunity to enhance our understanding of evolution and development of anatomy. However, these models are not yet able to compete with *Mus musculus* in terms of our understanding of its development and our technical ability in the control of gene expression. It is for this reason and others outlined in this chapter that mouse genetics represents a powerful tool for the investigation of the evolution of key anatomical structures. Studies in mice open avenues of investigation into the molecular biology underpinning interspecific and intraspecific morphological variation, and are able to inform and direct research on non-traditional models. This approach is beginning to be used to elucidate our understanding of the development of the jaw, limbs, and beyond (see chapters in this volume by Kuratani and Nagashima; Mitgutsch *et al.*; Richardson; Schmid; Sears *et al.*; and Smith and Johanson). For this reason, mouse genetics remains a vital part of the evo-devo arsenal.

Summary

Molecular biology, in conjunction with developmental biology, can shed light on the mechanisms that may have been active during evolution. Conserved gene expression can highlight homologies between structures, supporting or casting doubt on homologies based on anatomical features alone, while phenotypes generated after loss of gene function can help illuminate how changes in gene expression may have led to changes observed in the fossil record and in extant vertebrates.

In this chapter we examine the mammalian dentary (or mandible) as an example of a skeletal element with large variations in shape and size, looking at how molecular biology can indicate mechanisms of morphological evolution. We first describe the mammalian dentary, and the different patterns of the proximal part of this bone that play a role in the jaw articulation. We then move to the mouse as a model organism, and discuss the expression of genes and impact of mutations in these genes on formation of the dentary. We use this molecular data to hypothesize mechanisms important in dentary evolution.

Acknowledgements

The authors would like to thank the BBSRC for funding.

REFERENCES

Aggarwal, V. S., Carpenter, C., Freyer, L., *et al.* (2010). Mesodermal Tbx1 is required for patterning the proximal mandible in mice. *Developmental Biology*, **344**, 669–81.

Allin, E. F. (1975). Evolution of the mammalian middle ear. *Journal of Morphology*, **147**, 403–37.

Alvarez, J., Sohn, P., Zeng, X., *et al.* (2002). TGFbeta2 mediates the effects of hedgehog on hypertrophic differentiation and PTHrP expression. *Development*, **129**, 1913–24.

Anthwal, N., Chai, Y. and Tucker, A. S. (2008). The role of transforming growth factor-beta signalling in the patterning of the proximal processes of the murine dentary. *Developmental Dynamics*, **237**, 1604–13.

Atchley, W. R. and Hall, B. K. (1991). A model for development and evolution of complex morphological structures. *Biological Reviews of the Cambridge Philosophical Society*, **66**, 101–57.

Atchley, W. R., Plummer, A. A. and Riska, B. (1985). Genetics of mandible form in the mouse. *Genetics*, **111**, 555–77.

Beresford, W. A. (1981). *Chondroid Bone, Secondary Cartilage and Metaplasia*. Baltimore, MD: Urban & Schwarzenberg.

Boyd, T. G., Castelli, W. A. and Huelke, D. F. (1967). Removal of the temporalis muscle from its origin: effects on the size and shape of the coronoid process. *Journal of Dental Research*, **46**, 997–1001.

Clark, C. T. and Smith, K. K. (1993). Cranial osteogenesis in *Monodelphis domestica* (Didelphidae) and *Macropus eugenii* (Macropodidae). *Journal of Morphology*, **215**, 119–49.

Clauss, F., Maniere, M. C., Obry, F., *et al.* (2008). Dento-craniofacial phenotypes and underlying molecular mechanisms in hypohidrotic ectodermal dysplasia (HED): a review. *Journal of Dental Research*, **87**, 1089–99.

Crompton, A. W. (1963). Evolution of mammalian jaw. *Evolution*, **17**, 431–9.

Crompton, A. W. and Hylander, W. L. (1986). Changes in mandibular function following the acquisition of a dentary-squamosal joint. In *The Ecology and Biology of Mammal-like Reptiles*, ed. N. Hotton, J. MacLean, J. Roth and E. C. Roth. Washington, D.C.: Smithsonian Institution Press, pp. 263–82.

de Beer, G. (1937). *The Development of the Vertebrate Skull*. Oxford, UK: Oxford University Press.

Demar, R. and Barghuse, H. (1972). Mechanics and evolution of synapsid jaw. *Evolution*, **26**, 622–37.

Depew, M. J., Liu, J. K., Long, J. E., *et al.* (1999). Dlx5 regulates regional development of the branchial arches and sensory capsules. *Development*, **126**, 3831–46.

Depew, M. J., Lufkin, T. and Rubenstein, J. L. (2002a). Specification of jaw subdivisions by Dlx genes. *Science*, **298**, 381–5.

Depew, M. J., Tucker, A. S. and Sharpe, P. T. (2002b). Craniofacial Development. In *Mouse Development: Patterning, Morphogenesis and Organogenesis*, ed. J. Rossant and P. Tam. London: Academic Press.

Dudas, M., Sridurongrit, S., Nagy, A., Okazaki, K. and Kaartinen, V. (2004). Craniofacial defects in mice lacking BMP type I receptor Alk2 in neural crest cells. *Mechanisms of Development*, **121**, 173–82.

Filan, S. L. (1991). Development of the middle ear region in *Monodelphis domestica* (Marsupialia, Didelphidae): marsupial solutions to an early birth. *Journal of Zoology*, **225**, 577–88.

Frommer, J. (1964). Prenatal development of the mandibular joint in mice. *Anatomical Record*, **150**, 449–61.

Goswami, A. (2007). Cranial modularity and sequence heterochrony in mammals. *Evolution and Development*, **9**, 290–8.

Hall, B. and Hallgrímsson, B. (2008). *Strickberger's Evolution*, 4th edn. Sudbury, MA: Jones and Bartlett Publishers.

Ito, Y., Yeo, J. Y., Chytil, A., *et al.* (2003). Conditional inactivation of Tgfbr2 in cranial neural crest causes cleft palate and calvaria defects. *Development*, **130**, 5269–80.

Janssens, K., Ten Dijke, P., Janssens, S. and Van Hul, W. (2005). Transforming growth factor-{beta}1 to the bone. *Endocrine Reviews*, **26**, 743–74.

Jarvinen, E., Valimaki, K., Pummila, M., Thesleff, I. and Jernvall, J. (2008). The taming of the shrew milk teeth. *Evolution and Development*, **10**, 477–86.

Jensen, B. L. and Kreiborg, S. (1993). Craniofacial abnormalities in 52 school-age and adult patients with cleidocranial dysplasia. *Journal of Craniofacial Genetics and Developmental Biology*, **13**, 98–108.

Jerome, L. A., and Papaioannou, V. E. (2001). DiGeorge syndrome phenotype in mice mutant for the T-box gene, Tbx1. *Nature Genetics*, **27**, 286–91.

Kemp, T. S. (2005). *The Origin and Evolution of Mammals*. Oxford, UK: Oxford University Press.

Kermack, K. A., Mussett, A. F. and Rigney, H. W. (1981). The skull of *Morganucodon*. *Zoological Journal of the Linnean Society*, **71**, 1–158.

Kist, R., Greally, E. and Peters, H. (2007). Derivation of a mouse model for conditional inactivation of Pax9. *Genesis*, **45**, 460–4.

Klingenberg, C. P. (1998). Heterochrony and allometry: the analysis of evolutionary change in ontogeny. *Biological Reviews of the Cambridge Philosophical Society*, **73**, 79–123.

Klingenberg, C. P., Mebus, K. and Auffray, J. C. (2003). Developmental integration in a complex morphological structure: how distinct are the modules in the mouse mandible? *Evolution and Development*, **5**, 522–31.

Kobrynski, L. J. and Sullivan, K. E. (2007). Velocardiofacial syndrome, DiGeorge syndrome: the chromosome 22q11.2 deletion syndromes. *Lancet*, **370**, 1443–52.

Livne, E. and Silbermann, M. (1990). The mouse mandibular condyle: an investigative model in developmental biology. *Journal of Craniofacial Genetics and Developmental Biology*, **10**, 95–8.

Luo, Z. X. (2007). Transformation and diversification in early mammal evolution. *Nature*, **450**, 1011–19.

Maier, W. (1987). Der Processus angularis bei *Monodelphis domestica* und seine Beziehungen zum Mittelohr: Eine ontogenetische und evolutionsmorphologische Untersuchung. *Gegenbaurs morphologisches Jahrbuch*, **133**, 123–61.

Maier, W. (1990). Physiology and ontogeny of mammalian middle ear structures. *Netherlands Journal of Zoology*, **40**, 55–74.

Michaux, J., Chevret, P. and Renaud, S. (2007). Morphological diversity of Old World rats and mice (Rodentia, Muridae) mandible in relation with phylogeny and adaptation. *Journal of Zoological Systematics and Evolutionary Research*, **45**, 263–79.

Mikkelsen, T. S., Wakefield, M. J., Aken, B., *et al.* (2007). Genome of the marsupial *Monodelphis domestica* reveals innovation in non-coding sequences. *Nature*, **447**, 167–77.

Miller, C. T., Yelon, D., Stainier, D. Y. R. and Kimmel, C. B. (2003). Two endothelin 1 effectors, hand2 and bapx1, pattern ventral pharyngeal cartilage and the jaw joint. *Development*, **130**, 1353–65.

Miyake, T., Cameron, A. M. and Hall, B. K. (1997). Stage-specific expression patterns of alkaline phosphatase during development of the first arch skeleton in inbred C57BL/6 mouse embryos. *Journal of Anatomy*, **190** (Pt 2), 239–60.

Mukherjee, A., Dong, S. S., Clemens, T., Alvarez, J. and Serra, R. (2005). Co-ordination of TGF-beta and FGF signaling pathways in bone organ cultures. *Mechanisms of Development*, **122**, 557–71.

Oka, K., Oka, S., Sasaki, T., *et al.* (2007). The role of TGF-[beta] signaling in regulating chondrogenesis and osteogenesis during mandibular development. *Developmental Biology*, **303**, 391–404.

Oka, K., Oka, S., Hosokawa, R., *et al.* (2008). TGF-beta mediated Dlx5 signaling plays a crucial role in osteo-chondroprogenitor cell lineage determination during mandible development. *Developmental Biology*, **321**, 303–9.

Peters, H., Neubuser, A., Kratochwil, K. and Balling, R. (1998). Pax9-deficient mice lack pharyngeal pouch derivatives and teeth and exhibit craniofacial and limb abnormalities. *Genes and Development*, **12**, 2735–47.

Posnik, J. C., Tiwana P. S. and Costello, B. J. (2004). Treacher Collins syndrome: comprehensive evaluation and treatment. *Oral and Maxillofacial Surgery Clinics of North America*, **16**, 503–23.

Reichert, C. (1837). Über die Visceralbögen der Wirbeltiere im allgemeinen und deren Metamorphose bei den Vögeln und Säugetieren. *Archiv füur Anatomie, Physiologie*, 120–222.

Rivera-Perez, J. A., Mallo, M., Gendron-Maguire, M., Gridley, T. and Behringer, R. R. (1995). Goosecoid is not an essential component of the mouse gastrula organizer but is required for craniofacial and rib development. *Development*, **121**, 3005–12.

Rivera-Perez, J. A., Wakamiya, M. and Behringer, R. R. (1999). Goosecoid acts cell autonomously in mesenchyme-derived tissues during craniofacial development. *Development*, **126**, 3811–21.

Rot-Nikcevic, I., Reddy, T., Downing, K. J., *et al.* (2006). Myf5$^{-/-}$:MyoD$^{-/-}$ amyogenic fetuses reveal the importance of early contraction and static loading by striated muscle in mouse skeletogenesis. *Development Genes and Evolution*, **216**, 1–9.

Rot-Nikcevic, I., Downing, K. J., Hall, B. K. and Kablar, B. (2007). Development of the mouse mandibles and clavicles in the absence of skeletal myogenesis. *Histology and Histopathology*, **22**, 51–60.

Rowe, T. (1996). Coevolution of the mammalian middle ear and neocortex. *Science*, **273**, 651–4.

Sánchez-Villagra, M. R. and Smith, K. K. (1997). Diversity and evolution of the marsupial mandibular angular process. *Journal of Mammalian Evolution*, **4**, 119–44.

Sanford, L. P., Ormsby, I., Gittenberger-de Groot, A. C., *et al.* (1997). TGFbeta2 knockout mice have multiple developmental defects that are non-overlapping with other TGFbeta knockout phenotypes. *Development*, **124**, 2659–70.

Sears, K. E. (2011). Novel insights into the regulation of limb development from 'natural' mammalian mutants. *BioEssays*, **33**, 327–31.

Shibata, S., Suda, N., Fukada, K., *et al.* (2003). Mandibular coronoid process in parathyroid hormone-related protein-deficient mice shows ectopic cartilage formation accompanied by abnormal bone modeling. *Anatomy and Embryology (Berlin)*, **207**, 35–44.

Smith, K. K. (2001). Early development of the neural plate, neural crest and facial region of marsupials. *Journal of Anatomy*, **199**, 121–31.

Smith, K. K. (2006). Craniofacial development in marsupial mammals: developmental origins of evolutionary change. *Developmental Dynamics*, **235**, 1181–93.

Smith, K. K. and van Nievelt, A. F. (1997). Comparative rates of development in *Monodelphis* and *Didelphis*. *Science*, **275**, 683–4.

ten Berge, D., Brouwer, A., Korving, J., Martin, J. F. and Meijlink, F. (1998). Prx1 and Prx2 in skeletogenesis: roles in the craniofacial region, inner ear and limbs. *Development*, **125**, 3831–42.

Tomo, S., Ogita, M. and Tomo, I. (1997). Development of mandibular cartilages in the rat. *Anatomical Record*, **249**, 233–9.

Tucker, A. S., Watson, R. P., Lettice, L. A., Yamada, G. and Hill, R. E. (2004). Bapx1 regulates patterning in the middle ear: altered regulatory role in the transition from the proximal jaw during vertebrate evolution. *Development*, **131**, 1235–45.

Vinkka, H. (1982). Secondary cartilages in the facial skeleton of the rat. *Proceedings of the Finnish Dental Society*, **78**, Suppl 7, 1–137.

Vinkka-Puhakka, H. and Thesleff, I. (1993). Initiation of secondary cartilage in the mandible of the Syrian hamster in the absence of muscle function. *Archives of Oral Biology*, **38**, 49–54.

Wilson, J. and Tucker, A. S. (2004). Fgf and Bmp signals repress the expression of Bapx1 in the mandibular mesenchyme and control the position of the developing jaw joint. *Developmental Biology*, **266**, 138–50.

9

Flexibility and constraint: patterning the axial skeleton in mammals

EMILY A. BUCHHOLTZ

Introduction

Over the past 200 million years, the mammalian vertebral column has been adapted to the functional demands of animals as diverse as giraffes and bats, giant ground sloths and whales. Despite its impressive morphological and functional adaptability, the column also exhibits unmistakable signs of evolutionary constraint.

The co-existence of flexibility and constraint makes the mammalian vertebral column an intriguing subject for anatomical and evolutionary analysis. Vertebral counts vary from just over 30 to nearly 100 (Flower 1885; Narita and Kuratani 2005), and vertebral lengths from millimetres to tens of centimetres; shape is even more disparate. Yet, the morphological range of the column is limited by the patterning and sequence of its five component subunits (cervical, thoracic, lumbar, sacral, caudal) and by fixed or nearly fixed counts in the cervical and combined thoracolumbar series. Fortunately, the column's accessibility as a subject for study is enhanced by its composition of discrete units that are relatively simple anatomically and easily counted, by the diversity of its adaptations, and by its frequent preservation in the fossil record.

Evolution by natural selection is dependent on both the production of phenotypic variation by the developmental process and by the sorting of that variation by the environment (Raff 1996; Beldade *et al.* 2002; Brakefield 2006). In simplistic presentation, variation is viewed as the product of a random generating process, with non-random natural selection operating external to the organism. If variation were strictly random, however, evolution should be able to generate almost any morphology, given enough time. But evidence for the existence of bias in the generation of variation has been acknowledged since the early days of evolutionary study, although it was not addressed

From Clone to Bone: The Synergy of Morphological and Molecular Tools in Palaeobiology,
ed. Robert J. Asher and Johannes Müller. Published by Cambridge University Press.
© Cambridge University Press 2012.

systematically until William Bateson's landmark *Materials for the Study of Variation* (1894). The century of subsequent work has reinforced the important roles that genetic channelling and/or developmental bias can play in constraining the production of phenotypic variants and the effects that they can have on both the direction and the rate of evolution (Porto *et al.* 2009). Further, the biases can themselves evolve, generating variation in evolvability (Brakefield 2006).

Nineteenth-century anatomists wrestled with the modes of column evolution responsible for the observed patterns of mammalian column morphology. The variable segmental location of morphologically similar and apparently homologous vertebrae was of particular concern, and seemed to require repeated addition and/or subtraction of somites. In 1913, Goodrich laid out three competing hypotheses that could explain the observed diversity, which he identified as intercalation, redivision and transposition. He was able to eliminate all but transposition, concluding that 'the various regions of the vertebral column, in so far as possessed by a common ancestor, are homologous, but they are not necessarily composed of the same segments.'

The molecular processes that control segmental location and morphological identity, and thus govern transposition, have been detailed by molecular biologists working with model organisms over the last 20 years. The process of segmentation is now understood as the product of oscillating signals (the 'segmentation clock') that subdivide the presomitic mesoderm into a set number of subunits characteristic of each species (Pourquié 2003; Dubrulle and Pourquié 2004; Dequéant *et al.* 2006). Changing the periodicity of the clock changes the number of segments generated (Tam 1981). The morphological identity of a segment is the result of regional expression of the colinear *Hox* and other patterning genes, and morphological transitions from one vertebral series to the next are associated with the same *Hox* expression domain transitions across multiple taxa with different counts (e.g. Gaunt 1994; Burke *et al.* 1995; Carapuço *et al.* 2005). Somite generation and pattern assignment are apparently largely independent processes (Burke *et al.* 1995; Richardson *et al.* 1998; Müller *et al.* 2010; but see Zákány *et al.* 2001), allowing the transposition of the same morphology to different column locations.

Recent experimental work has expanded our understanding of how the structures that articulate with the column are patterned. Most critical is the explicit recognition and mapping of separate primaxial and abaxial patterning domains (Nowicki and Burke 2000; Burke and Nowicki 2003; Durland *et al.* 2008). Anteroposterior regions of the somitic column are classically recognized not only by their inherent characteristics, but also by their relationships to structures patterned independently (some rib units, sternum, ilium) in the abaxial domain. An alteration in either patterning domain can therefore affect series counts and generate unique morphologies. Some morphologically discrete structures are patterned complexly. For example, three different compartments have been identified in ribs (Aoyama *et al.* 2005), providing the possibility of partial or independently modified proximal, vertebral-distal, and sternodistal subunits. Finally, although *Hox* genes are involved in the patterning of lateral plate as well as axial structures, their abaxial expression domains are independently organized and not colinear (McIntyre *et al.* 2007).

The insights that comparative morphology and palaeontology bring to the study of the column are complementary to and synergistic with those of molecular biology. The most important is diversity itself, which documents the range of morphology that has been generated over evolutionary time. The limits of this diversity, especially under strong selection pressures and over long time periods, reveal apparently 'forbidden' morphologies (e.g. Raup 1961, 1966) that are signals of evolutionary constraints. Variants that occur repeatedly, both within and across taxa, reflect the shape and hierarchy of the developmental process, as do integrated (covarying) structures (Marroig *et al.* 2009). Stratigraphic sequence and phylogenetic reconstruction are key in establishing the time and sequence of trait origin as well as the number of times a given innovation has evolved.

The improving understanding of the evolution of the mammalian vertebral column draws heavily on both molecular and morphological data. Here a discussion of the definition of column series is followed by an argument for their interpretation as developmental modules. This is followed by a presentation of the morphological range and categories of developmental change observed in each vertebral series, and by an analysis of the exceptional cases where the limits to that change apparently have been broken. These patterns are then used to address the evolution of the modular hierarchy of the column over geologic time.

Abbreviations used in this chapter: FMNH, Field Museum of Natural History, Chicago, IL; MCZ, Museum of Comparative Zoology, Cambridge, MA; NMNZ, Museum of New Zealand Te Papa Tongarewa, Wellington, New Zealand; NUVC, Northeastern University Vertebrate Collection, Boston, MA; SAM, South Australia Museum, Adelaide, Australia.

Column series definition

Definition of the component anatomical series of the vertebral column is critical to a discussion of its evolution. A well-established set of characters has been used to identify series boundaries since the nineteenth century (e.g. Turner 1847; Owen 1859; Flower and Lydekker 1891; Le Double 1912):

Cervical: the atlas, axis and the immediately following vertebrae that lack movable rib articulations. Cervical vertebrae frequently bear fused proximal ribs that enclose the vertebrarterial foramen.

Thoracic: those vertebrae that articulate with movable ribs. The first thoracic vertebra is defined by its articulation with the sternum.

Lumbar: vertebrae that lie between the thorax and the sacrum. They lack movable ribs and typically bear transverse processes of diverse origin.

Sacral: synostosed vertebrae that articulate via transverse processes and/or modified ribs with the pelvic girdle. The first sacral vertebra articulates with the ilium.

Caudal: vertebrae posterior to the sacrum.

Despite nearly universal acceptance of these boundary markers, unambiguous application is not always possible. Goodrich's (1930) description of column series as 'arbitrarily defined for descriptive purposes' is probably overcritical, but the series remain as human constructs imposed upon biological reality. A trivial (but frequent) example of ambiguous series allocation is caused by bilateral asymmetry of a series-defining character such as costal/sternal articulation (Figure 9.1A). The different segmental occurrences of the multiple traits that define some boundaries are more problematical (Figure 9.1B). The lumbosacral transition provides an extreme example. Filler (2007) highlighted five traits that change at or near the thoracolumbar transition: rib reduction, zygopophysis orientation, laminapophysis subdivision, transverse processes elaboration, and horizontal septum. A sixth, neural spine orientation, could be added to this list. These rarely occur at exactly the same anteroposterior location (Figure 9.1B). A further set of problems can be traced to the fact that boundaries established for subdivision of the somitic vertebrae were historically established by the location and/or presence of lateral plate structures. Examples include the first articulation of a somitic rib to the lateral plate sternum to define the cervicothoracic junction (Figure 9.1C) and the articulation of vertebral processes with the ilium to define the first sacral vertebra. The current understanding that primaxial and abaxial domains are patterned independently (Burke and Nowicki 2003) helps to explain the origin, if not the resolution, of these problematic boundaries.

The important roles that lateral plate structures play in series definitions, as well as in column evolution, are reinforced by column morphology in taxa in which the lateral plate structures are secondarily reduced or lost. The lack of pelves, and thus of sacral vertebrae, in the Cetacea and some Sirenia provides extreme examples. An extensive literature addresses the search for the

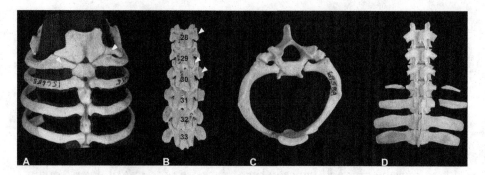

Figure 9.1 Recurrent problems in vertebral series assignment. (A) Bilateral asymmetry of the costosternal articulation (arrowheads) at the cervicothoracic boundary of *Choloepus hoffmanni* (FMNH 156655). The first rib with a sternal articulation is associated with vertebra 6 on the left and with vertebra 7 on the right. (B) The multiple traits used to identify a series transition may occur at different vertebral locations. Here, laminapophysis subdivision occurs at vertebra 28, zygopophysis reorientation occurs at vertebra 29, and transverse process elaboration occurs at vertebra 30 across the thoracolumbar boundary of *Choloepus hoffmanni* (FMNH 69576). (C) The first thoracic vertebra is historically defined by the location of the first rib that articulates with the sternum, a structure with abaxial patterning (*Bradypus variegatus*, FMNH 69588). (D) Identity of series boundaries between the posterior thorax and anterior tail are often ambiguous and asymmetrical in taxa with hind limb and sacrum loss (*Trichechus manatus*, USNM 551671).

current axial location of these 'lost' sacrals (e.g. Brandt 1862; Slijper 1936). Except in transitional fossils, where vestigial sacral morphology can be seen (e.g. *Georgiacetus*; Hurlbert 1998), they must now lie within the lumbar or caudal series. In *Trichechus* the problem is exacerbated by lumbar reduction, with the result that the complex transition from posterior thoracic to anterior caudal series occurs over a very small number of highly variable and often bilaterally asymmetrical vertebrae (Buchholtz *et al.* 2007; Figure 9.1D).

Vertebral series as morphological and evolutionary modules

The developmental integration of traits within the same individual has been a topic of biological debate since the eighteenth century (Rolian and Willmore 2009). One extreme possibility is that all traits are autonomous and open to independent evolutionary modification. Organisms with such particulate or 'parcellated' anatomy would be infinitely plastic when subject to selection, but would fail to respond to selection in a co-ordinated fashion (Lovejoy *et al.* 1997; Porto *et al.* 2009). At the other extreme, all traits are seen as interdependent or co-ordinated. Cuvier's principle of the correlation of parts emphasized the functional interdependence of each organ with every other, leading to his claim that the whole organism could be reconstructed from a single part (Mayr 1982). Organisms are now understood to exist between these extremes, and to consist of multiple internally integrated modules with variable levels of interaction. Schlosser and Wagner (2004) argue compellingly that organisms must have such independent or semi-independent subunits to evolve. Without modular autonomy, selection on one subunit would disrupt the functional organization of the whole. The patterns of modular interactions, like the developmental processes that generate them, are hierarchical in organization (Bolker 2000). Because modularity can affect the direction and rate of variation, it can have profound effects on evolution (Marroig *et al.* 2009; Porto *et al.* 2009).

The vertebral column is an overtly modular structure: each of its recurrent column subunits retains characteristic anatomy and internal consistency across taxa and evolutionary history while at the same time undergoing modification in response to selection independent of other series. Four categories of vertebral module modification are commonly recognized (Raff 1996; Polly *et al.* 2001; Figure 9.2):

1. Homologous variations (Figure 9.2A) alter the size and/or shape of particular column subunits that do not differ in count and series identity. Regionally expressed growth factors, acting after vertebral count and series boundaries are set, may have a role in homologous variation (Johnson and O'Higgins 1996; McPherron *et al.* 1999; Oostra *et al.* 2005).
2. Meristic variations (Figure 9.2B) alter the total count of somites generated from the presomitic mesoderm during somitogenesis, largely controlled by the oscillating somite clock. The rate of clock oscillation governs the number of somites generated (Hirsinger *et al.* 2000; Pourquié 2003; Sanger and Gibson-Brown 2004).
3. Homeotic variations (Figure 9.2C) alter the count of one series at the expense of an adjacent series, and reflect changes in the expression domains of *Hox* and other patterning genes (Burke *et al.* 1995).
4. Associational variations (Figure 9.2D) alter the architecture of the developmental process, with the result that the number or interrelationships of modules change (Polly *et al.* 2001; Buchholtz 2007).

Morphological range and modular change in mammalian vertebral series

The five column series differ with respect to the modes of modular change observed. All series provide multiple examples of the homologous adaptations of size or shape that are adaptive for column use in a given species (see, e.g. Slijper 1946; Filler 1986, 2007).

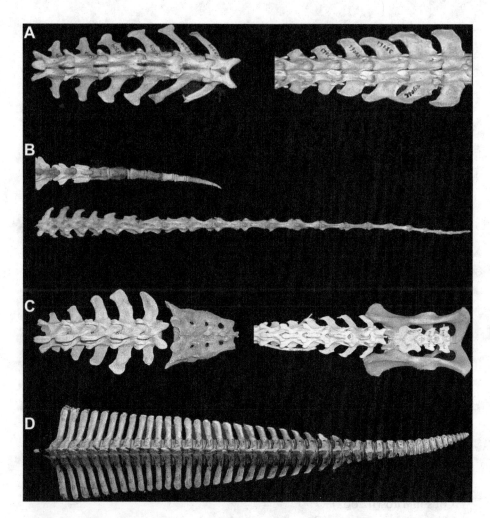

Figure 9.2 Categories of change in modular units of the column in mammals.
(A) Homologous change: the carnivores *Atilax paludinosus* (MCZ 38063) and *Canis mesomelas* (MCZ 58309) have identical lumbar counts of 6 but differ in centrum shape and transverse process structure. (B) Meristic change: the same modular unit (caudal series) in two members of the same Order (Carnivora) is represented by vertebral counts of 15 (*Phoca groenlandica*, MCZ 28682) and 23 (*Fossa fossa*, MCZ 29404). (C) Homeotic change: lumbar+sacral counts of 9 are allocated L5 S4 in *Crocuta crocuta* (MCZ 20968) and L6 S3 in *Procyon lotor* (MCZ 62164).
(D) Associational change: anterior caudal vertebrae of modern whales
(*Lagenorhynchus acutus*, NUVC 1968) closely resemble posterior lumbars,
but terminal caudal vertebrae that support the fluke form a distinct morphological subunit. Vertebrae shown are posterior lumbars (L16–22), anterior tail caudals (Cd1–20) and fluke caudals (Cd21–39). (See also colour plate.)

Cervical series

The mammalian cervical series is an extreme example of stasis in count over long geological periods and diverse selection regimes. Palaeontological evidence indicates that cervical count has been fixed at seven for at least 200 million years (Jenkins 1971; Crompton and Jenkins 1973). Eighteenth-century anatomists documented both the near fixity of mammalian cervical count and the rare apparent exceptions to it (Buffon 1769; Cuvier 1798).

Infrequent examples of the homeotic occurrence of movable ribs on cervical vertebrae are known in a variety of eutherian species, but have been most thoroughly studied in humans (Le Double 1912; Oostra et al. 2005; Galis et al. 2006). Similar morphologies can also be generated by manipulation of Hox gene expression domains in model organisms (e.g. Horan et al. 1994). When these abnormalities are restricted to ribs, they do not formally change the cervicothoracic boundary, which is defined by sternal articulation. Galis et al. (2006) documented the occurrence of supernumerary proximal ribs on posterior cervical vertebrae in a large database of human stillbirths. They found that many of these individuals also displayed a variety of other congenital abnormalities. They further observed elevated frequencies of paediatric cancers in individuals that survived birth. These data support their hypothesis that, in humans, the cervical Hox patterning genes are linked to genes with deleterious effects, and that the cervical constraint is the indirect pleiotropic product of that linkage (Galis 1999; Galis et al. 2006).

Aberrant anatomy that includes non-traditional location of the first costo-sternal articulation occurs almost universally in three genera: Bradypus (3-toed sloths, 8–10 cervical vertebrae), Choloepus hoffmanni (2-toed sloth, 5–6 cervical vertebrae) and Trichechus (manatees, 6 cervical vertebrae). A fourth genus, Caperea (pygmy right whale), lacks ribs on the first 8 vertebrae (Beddard 1901), and has been less widely and more questionably included as an exception to the cervical constraint. All four genera have unusual lifestyles, have exceptionally long or short necks, display marked disruption of the column posterior to the neck, and exhibit unusually high intraspecific variation in column morphology. These observations suggest that the fixed location of the first costosternal articulation is resistant to all but extreme selection pressures and may be linked to the patterning of other column series. The anatomy of all four genera is addressed below.

Cervical anatomy was examined in a large database of tree sloths by Buchholtz and Stepien (2009). Tree sloths are unusual in their suspensory locomotion and low metabolic rates. Their very long (Bradypus) or very short (Choloepus) necks are usually interpreted as different functional responses to head support while

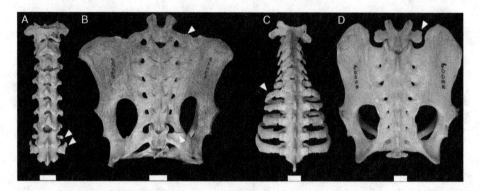

Figure 9.3 Vertebral anatomy in tree sloths is associated with disrupted sacral anatomy. All views are dorsal, with anterior up. (A, B) *Bradypus variegatus* (FMNH 69588) (A) The two most posterior of nine 'cervical' vertebrae articulate with small proximal riblets (arrowheads). (B) Transverse processes of the terminal lumbar vertebra (arrowhead) angle posteriorly to articulate with the ilium and the first caudal vertebra (arrowhead) is incompletely integrated into the sacrum. (C, D) *Choloepus hoffmanni* (FMNH 60058). (C) Only six vertebrae lie anterior to the first vertebra with full sternal ribs (arrowhead). (D) At the lumbosacral boundary, the terminal lumbar vertebra (arrowhead) lies between the ilia, but does not articulate with them. All scale bars represent 10 mm.

inverted (Miller 1935). Buchholtz and Stepien were able to rule out both simple homeotic transformations of the cervicothoracic boundary and meristic changes in precaudal counts as the source of non-traditional cervical counts. Instead, they documented recurrent examples of mediolaterally disjunct anatomy at all vertebral series boundaries. These include small partial riblets on 'extra' cervicals (vertebrae 8 and 9 in *Bradypus*), deformed anterior ribs with aberrant sternal connections on 'extra' thoracics (vertebrae 6 and 7 in *Choloepus*) and striking (and opposite) patterns of anteroposterior disruption at the lumbosacral boundary. They proposed a hypothesis of global frame shifts in abaxial patterning that brings segmentally 'mismatched' structures into conjunction (Figure 9.3). In the case of *Bradypus*, the posterior shift of abaxial patterning displaces the distal vertebral ribs, sternal ribs, sternum and pelvis relative to their traditional segmental locations. The resulting phenotype includes incomplete proximal riblets without sternal articulations on vertebrae with anterior thoracic patterning (Figure 9.3A), 'sacral' vertebrae with transverse processes that must be extended posteriorly to articulate with their ilial targets (Figure 9.3B), and anterior 'caudal' vertebrae that are incorporated incompletely into the posterior sacrum (Figure 9.3B). A similar, but opposite, shift in abaxial patterning is observed in *Choloepus* (Figure 9.3C, D).

This interpretation of tree sloth axial anatomy has at least two surprising implications. The first is that the observed shifts in abaxial patterning are not local, but rather global. Although selection pressures may have been for changes in neck length, the morphological disruptions impact the entire torso and pelvis as well as cervical structures. This coordinated alteration of multiple lateral plate structures supports molecular evidence that abaxial domain patterning is both independent of the primaxial domain and non-colinear (McIntyre *et al.* 2007; Wellik 2007). Secondly, it suggests that somitic tissues have retained traditional patterning: the vertebral cervicothoracic boundary still occurs between vertebrae 7 and 8, even though the (abaxial) sternal location has been shifted. This interpretation has subsequently been supported by analysis of vertebral ossification sequence (Hautier *et al.* 2010). If this interpretation is correct, the resistance of the (primaxial) 'rule of seven' to evolutionary change must be stronger than previously understood, and variation of lateral plate structures has played the key role in providing morphological flexibility to mammals with extreme selection pressures for neck length.

Cervical anatomy in the manatee *Trichechus manatus* is unusual both in its count and in its lack of variability. Despite highly variable lumbosacrocaudal patterning, there is essentially no variation in the cervical count of six (Buchholtz *et al.* 2007). A large nineteenth-century literature addresses possible mechanisms for non-traditional cervical counts in manatees (e.g. Brandt 1862; Murie 1872; Chapman 1875). These early works often focused on identification of 'the missing vertebra' in the cervical series (Murie 1872), reflecting the now discarded theory of intercalation. The presence of both unusual and highly variable sternal anatomy (E. A. Buchholtz, unpublished data) again raises the possibility of abaxial repatterning, although it is more difficult to test in manatees than in terrestrial taxa because of the lack of articulating pelvic structures.

The pygmy right whale *Caperea marginata* exhibits a greatly expanded thorax, an extremely short lumbar series, plate-like posterior ribs and highly unusual cervical patterning (Figure 9.4A). In his *Osteology of the Pigmy Whale*, Beddard (1901) reported that in the two mounted specimens he examined, 'the first dorsal [= thoracic] appears to have no rib.' Slijper (1936) disputed the interpretation of *Caperea* as a fourth exception to the rule of seven. He described another individual with normal cervical counts and suggested that both of Beddard's specimens had been incorrectly mounted.

This disagreement was resolved by an analysis of ontogenetic variation in *Caperea marginata*, which indicates that costal anatomy differs in juvenile and adult individuals (Buchholtz 2010). Each thoracic vertebra is associated with a single rib in very small juveniles but not in adults (Figure 9.4B–F). At small

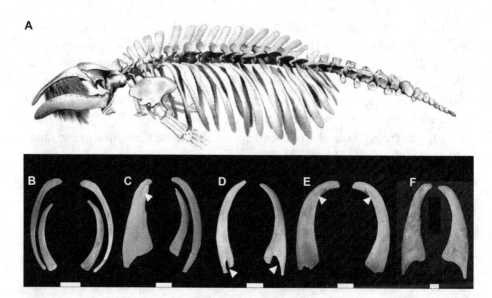

Figure 9.4 The pygmy right whale, *Caperea marginata*, displays unusual costal anatomy and vertebral counts. (A) Diagram of skeletal anatomy (modified from Beddard 1901). (B–F) Variation in anatomy of ribs 1 and 2 in five individuals of different ontogenetic stage. (B) Neonate (NMNZ Uncatalogued B) with separate ribs. (C) Dependent calf (SAM M17364) with left/right asymmetry of rib fusion. (D) Dependent calf (NMNZ 2049) with incomplete distal fusion (arrowheads). (E) Dependent calf (NMNZ 2119) with incomplete proximal fusion (arrowheads). (F) Adult (NMNZ Uncatalogued A) with complete rib fusion. All scale bars represent 5 cm.

body sizes interpreted as indicative of the dependent calf life stage (Kemper and Leppard 1999), the first two ribs fuse to form a composite triangular rib. The rib 1 component of the composite articulates with the sternum but has no vertebral articulation. The rib 2 component articulates with thoracic vertebra 2, but lacks a sternal articulation. The selection forces that may have favoured this unusual configuration are unknown, but the adult anatomy apparently allows the first thoracic vertebral centrum to be functionally incorporated into the neck. Analogous to the sloths, morphological innovation appears to have been possible in the pygmy right whale despite constraints on cervical counts by means of the adaptation of lateral plate structures patterned within the abaxial domain.

Thoracic and lumbar series

The synapsid ancestors of mammals had an undifferentiated series of vertebrae between the limbs, each with an articulating bicipital rib (Romer 1955;

Carroll 1988). In modern mammals, the separation of this dorsal series into thoracic and lumbar series is associated with the independence of respiratory and locomotor functions (Carrier 1987). Most thoracic vertebrae are either directly or indirectly connected to the sternum via ribs, although the most posterior ribs may float; rib homologues are reduced or absent in the lumbar series. Phylogenetic analysis of Mesozoic mammals indicates that the reduction and vertebral fusion of lumbar ribs evolved multiple times homoplastically (Luo *et al.* 2007). Filler (2007) further demonstrated that in some mammalian orders lumbar transverse processes are not rib homologues at all, but modified laminapophyses or diapophyses.

Mammalian thoracolumbar count is conserved but not fixed at 19–20 (Todd 1922; Narita and Kuratani 2005; Müller *et al.* 2010). Afrotherians exhibit high thoracolumbar counts, a trait that has been identified as a synapomorphy (Sanchez-Villagra *et al.* 2007). In the Order Carnivora, thoracolumbar count is nearly fixed at 20 despite adaptations to diverse terrestrial and marine lifestyles. Counts from both artiodactyls and carnivores (Figure 9.5, Appendix 9.1) indicate less variation in thoracolumbar count than in either thoracic or lumbar count, suggesting a reciprocal homeotic relationship between the two series. At least one of the rare exceptions (*Mephitis*) appears to be augmented by a homeotic exchange with the sacral series, although others with reduced counts (e.g. *Helarctos*, *Alcinonyx*) have normal sacral counts of three. In contrast, thoracolumbar count in Rodentia (Figure 9.5, 9.6A; Appendix 9.1) is more variable than either thoracic or lumbar count. This appears to reflect more frequent homeotic exchange of thoracolumbar with sacral vertebrae, as well as more frequent deviation of the precaudum from the typical mammalian count of 30. Importantly, these interspecific conclusions are based on counts gathered here and previously reported in the literature (e.g. Flower 1885; Narita and Kuratani 2005) from one or few individuals per species. An intraspecific analysis of presacral count based on a large number of individuals (9–71) from a much smaller number of species (Asher *et al.* 2011) indicates important differences of intraspecific variation across certain high-level mammalian clades. Among other conclusions, Asher *et al.* (2011) found unusually low levels of intraspecific variation in eight rodent species, suggesting that rodents exhibit fewer departures from median species counts than other mammalian clades. Low intraspecific variation need not necessarily conflict with relatively frequent departures of rodent species from conserved mammalian counts.

Large increases in dorsal (thoracolumbar) count are associated with reduction of the hind limbs and secondary adoption of axial locomotion in many groups of vertebrates. Of these, the most striking examples are limbless squamates, snakes and marine reptiles (Richardson *et al.* 1998; Müller *et al.* 2010). Augmentation of

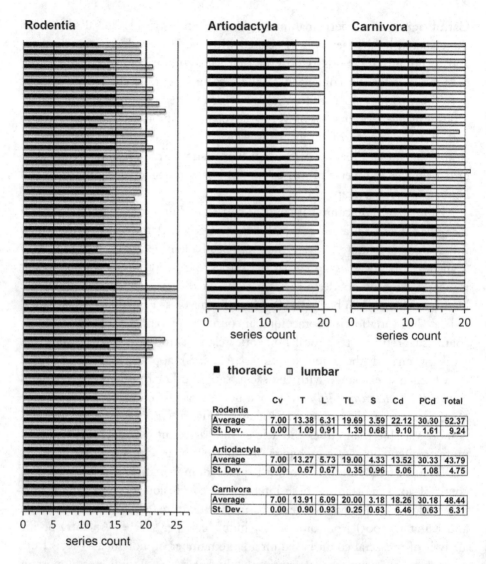

Figure 9.5 Precaudal counts in Rodentia, Artiodactyla and Carnivora. Thoracolumbar counts are conserved at 19 (Rodentia, Artiodactyla) or at 20 (Carnivora). Precaudal counts are conserved at 30 in all three orders. Carnivoran and rodent raw data in Appendix 9.1. Cv, cervical; T, thoracic; L, lumbar; TL, thoracolumbar; S, sacral; Cd, caudal. Specimens without caudal counts had incomplete tails, so full counts could not be determined.

lumbar count is also typical of delphinoid cetaceans, where thoracolumbar counts of over 40 are known (*Lissodelphis, Phocoenoides*). Despite extreme hind limb reduction or loss, living sirenians present a contrasting pattern. The posterior margin of the short thoracolumbar series in *Dugong* is recognized by both bony and soft tissue identifiers of the single sacral vertebra. *Trichechus* lacks any trace of a sacral series, and the thoracolumbar/caudal transition is identified by the first occurrence of haemal arch facets. In both species, thoracolumbar counts are at or very near traditional values (Buchholtz *et al.* 2007).

Sacral series

The anterior boundary of the sacral series is defined by the first vertebra that articulates with the ilium. Two or more additional vertebrae with ilial contacts and/or synostosed centra are typically incorporated into the sacrum. Many Cretaceous mammals have sacral counts of two or three (e.g. Ji *et al.* 2002; Li and Luo 2006; Luo *et al.* 2007), and counts as low as one (e.g. *Cercopithecus*) and as high as nine (e.g. *Dasypus*) are known in the modern fauna (Flower and Lydekker 1891; Narita and Kuratani 2005). Species with high sacral counts usually retain traditional precaudal counts, which are quite stable not only within but also among mammalian orders at 29–31 (Figure 9.6A). This suggests that elevated sacral counts are the product of incorporation of lumbar vertebrae into the sacrum, likely as a means of limiting lateral bending of the trunk and/or mechanical reinforcement of the sacrum (Carrier 1987).

Caudal series

The caudal series is unique among mammalian series, and similar to those of non-synapsid amniotes, in the flexibility of its count. Caudal counts as low as two (some monkeys; Narita and Kuratani 2005) and as high as the mid 40s (*Manis tetradactyla*; Flower and Lydekker 1891) are known. Graphs of total column counts against series counts of terrestrial mammalian orders (Figure 9.6A) make clear the unique meristic flexibility of the caudal series.

Although typically undifferentiated in terrestrial mammals, the cetacean caudal series offers two examples of the associational mode of change. The first to occur historically was the subdivision of the series into discrete anterior tail and fluke units. The fluke is first clearly known in basilosaurid archaeocetes, where it is recognizable on the basis of a characteristic pattern of reversal in centrum width and height dimensions (Uhen 2004). Modular independence of the fluke in living cetaceans is suggested by the marked morphological discontinuity at the anterior tail/fluke boundary and by limitation (but not fixation) of its meristic range across a broad taxonomic radiation (Buchholtz 2007).

The dissociation of the cetacean fluke was followed by a second associational change, the integration of the lumbar and prefluke caudal vertebrae into a single novel module that crosses the precaudal/caudal boundary (Buchholtz 2007). The developmental integration of lumbar vertebrae and prefluke caudal

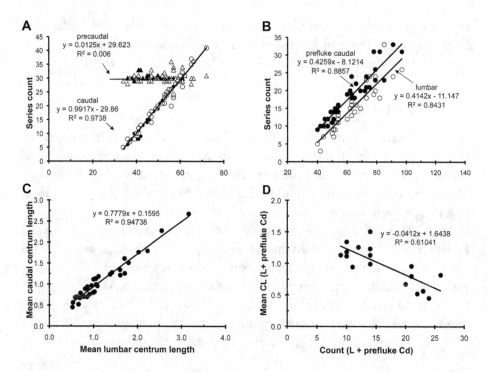

Figure 9.6 Patterns of mammalian caudal count variation. (A) Caudal counts vary much more widely than precaudal counts in Rodentia and Carnivora. (B) Counts of prefluke caudal and lumbar vertebrae increase in parallel in delphinoid cetaceans. (C) Normalized mean lumbar centrum length predicts mean caudal centrum length with great accuracy in delphinoid cetaceans. (D) The reciprocal relationship between count and centrum length in delphinoid cetaceans suggests localized changes in the 'segmentation clock'. (Data for A in Appendix 9.1; data for B, C and D from Buchholtz 2007.)

vertebrae is clear from their shared patterns of morphology, size and count in neocete cetaceans (Figure 9.6B, C, D). This integration is particularly clear in delphinoids, which vary widely in count and size. Delphinoid lumbars and prefluke caudals increase on a one-to-one basis (Figure 9.6B), in stark contrast to the pattern in terrestrial mammals (Figure 9.6A); they also mimic each other in normalized size (Figure 9.6B). Even more surprisingly, count and normalized centrum length vary inversely across this unit (Figure 9.6D). Added vertebrae do not therefore confer increased body length, and similar body proportions are maintained despite wide variation in count and taxonomic placement (Buchholtz 2007). This may reflect localized (i.e. lumbar + anterior tail) changes in the periodicity of the segmentation 'clock' (Tam 1981; Pourquié 2003).

Historical patterns of evolutionary change in the mammalian column

The key feature of vertebral column morphology is its progressive differentiation over geologic time. The nearly uniform sequence of vertebrae that articulate with movable ribs in the aquatic ancestors of tetrapods was replaced by a column with abrupt morphological transitions in terrestrial vertebrates (Romer 1955; Carroll 1988). These transitions in axial anatomy occur coincident with the anteroposterior sites at which the somitic column is integrated with structures of lateral plate mesoderm origin. Chief among these are the limbs, which are abaxial markers of the cervical, dorsal and caudal series boundaries. Historically, a single sacral vertebra marked the tetrapod dorsocaudal transition, but a multisegment sacrum has evolved convergently in mammals and birds, most likely by means of variable incorporation of posterior dorsal vertebrae. The resulting four series (cervical, dorsal, sacral, caudal) were typical of early tetrapods, and are retained by living non-synapsid tetrapods. The dorsal series was further differentiated in later synapsids (and inherited by crown group mammals) to yield separate thoracic and lumbar series in the Triassic. Mammals also exhibit internal subdivision of both the cervical and thoracic series. The atlas/axis complex was an amniote innovation (Goodrich 1930), and thus predates the synapsids, but cervical vertebrae 3–5 share distinctive morphology, and are now known to be patterned by a unique suite of genes (Johnson and O'Higgins 1996). Similarly, distinctive patterning genes are found in the subregions of the mammalian thorax that articulate with sternal and 'false' ribs (Wellik 2007).

The progressive regionalization of the column outlined above is not primarily linear, as might be predicted from the anterior to posterior sequence of somitogenesis. Instead, differentiation has had a hierarchical pattern. Romer (1955) explicitly inferred such a hierarchy of differentiation over time when he charted the stratigraphic sequence of vertebral series origins (Figure 9.7A). As typically presented today (Figure 9.7B), this hierarchy documents the early subdivision of the column into precaudal and caudal units, and the almost exclusive location of subsequent differentiation within the precaudal column. With a few exceptions, the caudal column has remained an undifferentiated unit, with vertebrae representing size variants of a single basic phenotype. The ancient segregation of precaudum and caudum is reflected in the separate developmental origins of the two regions from the primitive streak and the tail bud, respectively, and in the separate genetic controls of their axial extension (Tam and Tan 1992; Economides *et al.* 2003; Dubrulle and Pourquié 2004; Wilson *et al.* 2009; Aulehla and Pourquié 2009).

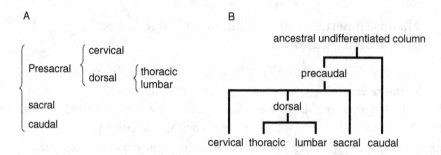

Figure 9.7 Differentiation of the vertebral series in time is hierarchical, not linear. (A) The differentiation of the column based on stratigraphic occurrence presented by Romer (1955). (B) Modular relationships of vertebral series as currently understood from morphological and developmental data. Note the inclusion of the sacral series within the precaudum.

Progressive regionalization of the anterior column over geologic time may reflect a parallel regionalization of *Hox* gene expression domains. Several authors (e.g. Pollock *et al.* 1995; Wellik 2007; Gehring *et al.* 2009) have suggested that most or all of the *Hox* genes were ancestrally coexpressed across broad regions of the column. The resulting morphology was also uniform and probably 'thoracic-like', with a pair of movably articulating ribs in each segment. Subsequent restriction of the expression domains of some of the genes to subunits of the column then provided the genetic basis for regional repression of ribs and morphological differentiation. Emerging experimental evidence supports this scenario, at least in part. Wellik and Capecchi (2003) demonstrated that *Hox10* genes act to repress rib formation posterior to the thorax; in the absence of all *Hox10* function, ribs appear on these segments. They also demonstrated that *Hox11* genes act to promote sacral morphology; no sacral vertebrae form in the absence of *Hox11* function. Wellik (2007) interpreted this as partial suppression of *Hox10* function. However, if this suppression is interpreted as historically subsequent to lumbar patterning, it is difficult to reconcile with the chronological occurrence of sacral vertebrae many millions of years before the origin of the lumbar series in the stratigraphic record. The recent generation (Mallo *et al.* 2009) and suppression (Carapuço *et al.* 2005) of ribs throughout large regions of the column by ectopic expression of *Hox* genes in the presomitic (as opposed to somitic), mesoderm indicates that the molecular basis of vertebral identity assignment is even more complex than currently understood.

A second key observation is that greater column differentiation is linked to restricted column variability. The greatest differentiation, and the most limited meristic variation, occurs in the anterior part of the column. Homologous variation is unrestricted in all vertebral series, but meristic variation occurs freely and frequently only in the caudal series. The tail retains both the lack of differentiation and the meristic flexibility seen in non-synapsid amniotes. As a whole, the precaudal unit has only limited meristic flexibility, and the cervical series has none at all.

Goodrich (1913) noted both the posterior increase in meristic variation and its relationship to morphological differentiation almost a century ago:

Generally it [segmental correspondence of structure and count] is more definite and invariable in the anterior than in the posterior region, and in regions of animals composed of few than in those composed of many segments. It is just as if Nature got tired of counting towards the tail end of a developing animal, and as if her arithmetic became uncertain when dealing with large numbers. (Goodrich 1913)

An 'evolutionary trade-off' between flexibility in count and morphological differentiation is a pattern repeated in other segmented taxa, and has been addressed most forcefully in arthropods (Cisne 1974). The recent work of Hughes (Hughes et al. 1999; Hughes 2003a, 2003b) is of particular interest in that it traces the transformation of the flexible pattern of multiple, identical segments seen in basal euarthropods to several different highly differentiated and less flexible patterns in late Palaeozoic forms across a period of several hundred million years. This arthropod trend, the exchange of 'flexible segment numbers for greater regional autonomy' (Hughes 2003a), is analogous to that seen in the mammalian vertebral column. The dramatic parallel raises questions about the advantages and costs of differentiation.

As discussed above, current modular theory suggests that there is a relationship between regional differentiation (modularity) and evolutionary flexibility, the capacity of a population or morphologically complex structure to respond in the direction that selection is pushing (Marroig et al. 2009). A highly differentiated morphology reflects a highly modularized develop-mental architecture, allowing response to selection pressures acting on one module without generating possibly non-adaptive morphologies in another. This flexibility may be one of the foundations that underlies the marked morphological and ecological diversity of the mammalian radiation. In the vertebral column, the accompanying 'price' of reduced meristic variability is absolute only in the cervical series. Despite this constraint, homologous variation of the mammalian cervical series has been sufficient to allow evolu-tion of morphologies that support extremely different functions and lifestyles. While the cervical constraint has limited the meristic evolution of the column, adaptation has still been possible via change in integrated lateral structures that are patterned independently.

Summary

The vertebral column of mammals is regionally differentiated into series, each composed of vertebrae with characteristic morphology. Transitions in column morphology occur at anteroposterior locations at which the somitic column is integrated with structures of lateral plate mesoderm origin. The regional differentiation of the column is now understood as the product of the modular organization of the developmental process that generated it. Column modularity allows each series to respond to selection pressures independently of the others, and may have been a key factor in the great morphological and ecological diversity of the mammalian radiation. High levels of modularity are also associated with restricted variability of count. In the mammalian column, numerical constraint is most rigid in the cervical series, which also exhibits the most marked internal morphological differentiation.

Acknowledgements

I thank Robert Asher and Johannes Müller for their invitations to participate in the Molecules and Morphology Symposium at the 2009 Annual Meeting of the Society of Vertebrate Paleontology and to contribute to this volume. I owe special debts to Farish A. Jenkins for his inspiration and support over many years and to Ann C. Burke for her generosity in sharing her insights on vertebral patterning. The projects on which this review is based could not have been completed without the contributions of my students Kate Webbink, Rachel Schneider, Amy Booth, Anthea Maslin, Courtney Stepien, Jennifer Yang, and Grady Bailin. My special thanks to Grady Bailin, who collected much of the data in Appendix 9.1. I also thank Linda Gordon, Charley Potter, Jim Mead, Darren Lunde, Elaine Westwig and Judy Chupasko, who have repeatedly and cheerfully made collections under their care available for study. Funding for this work was supplied in part by the National Science Foundation Grant IOS-0842507 and by Wellesley College.

REFERENCES

Aoyama, H., Mizutani-Koseki, Y. and Koseki, H. (2005). Three developmental compartments involved in rib formation. *International Journal of Developmental Biology*, **49**, 325–33.

Asher, R. J., Lin, K. H., Kardijilov, N. and Hautier, L. J. (2011). Variability and constraint in the mammalian vertebral column. *Journal of Evolutionary Biology*, **24** (5), 1080–90.

Aulehla, A. and Pourquié, O. (2009). More than patterning – *Hox* genes and the control of posterior axial elongation. *Developmental Cell*, **17**, 439–40.

Bateson, W. (1894). *Materials for the Study of Variation*. London: MacMillan and Company.

Beddard, F. E. (1901). Contribution toward knowledge of the osteology of the pigmy whale (*Neobalaena marginata*). *Transactions of the Zoological Society of London*, **16**, 87–108.

Beldade, P., Koops, K. and Brakefield, P. M. (2002). Developmental constraints versus flexibility in morphological evolution. *Nature*, **416**, 844–7.

Bolker, J. (2000). Modularity in development and why it matters to evo-devo. *American Zoologist*, **40**, 770–6.

Brakefield, P. M. (2006). Evo-devo and constraints on selection. *Trends in Ecology and Evolution*, **21**, 362–8.

Brandt, J. F. (1862). Bemerkungen über die Zahl der Halswirbel bei den Sirenien. *Mélanges biologiques tirés du Bulletin de l'Académie impériale des Sciences de St. Pétersbourg*, **5**, 7–10.

Buchholtz, E. A. (2007). Modular evolution of the cetacean vertebral column. *Evolution and Development*, **9**, 278–89.

Buchholtz, E. A. (2010). Vertebral and rib anatomy in *Caperea marginata*: implications for evolutionary patterning of the mammalian vertebral column. *Marine Mammal Science*, doi: 10.1111/j.1748–7692.2010.00411.x

Buchholtz, E. A. and Stepien, C. C. (2009). Anatomical transformation in mammals: developmental origin of aberrant cervical anatomy in tree sloths. *Evolution and Development*, **11**, 69–79.

Buchholtz, E. A., Booth, A. C. and Webbink, K. (2007). Vertebral anatomy in the Florida manatee, *Trichechus manatus latirostris*: a developmental and evolutionary analysis. *Anatomical Record*, **290**, 624–37.

Buffon, G.-L. L. Comte de. (1769). *Histoire Naturelle*. Paris: L'Imprimerie Royale.

Burke, A. C. and Nowicki, J. L. (2003). A new view of patterning domains in the vertebrate mesoderm. *Developmental Cell*, **4**, 159–65.

Burke, A. C., Nelson, C. E., Morgan, B. A. and Tabin, C. (1995). *Hox* genes and the evolution of vertebrate axial morphology. *Development*, **121**, 333–46.

Carapuço, M., Nóvoa, A., Bobola, N. and Mallo, M. (2005). *Hox* genes specify vertebral types in the presomitic mesoderm. *Genes and Development*, **19**, 2116–21.

Carrier, D. (1987). The evolution of locomotor stamina in tetrapods: circumventing a mechanical constraint. *Paleobiology*, **13**, 326–41.

Carroll, R. L. (1988). *Vertebrate Paleontology and Evolution*. New York: Freeman.

Chapman, H. C. (1875). Observations on the structure of the manatee. *Proceedings of the Academy of Natural Sciences of Philadelphia*, **27**, 452–62.

Cisne, J. L. (1974). Evolution of the world fauna of aquatic free-living arthropods. *Evolution*, **28**, 337–66.

Crompton, A. W. and Jenkins, F. A., Jr. (1973). Mammals from reptiles: a review of mammalian origins. *Annual Review of Earth and Planetary Science*, **1**, 131–55.

Cuvier, G. (1798). Extrait d'un Memoire sur les Ossemens fossils de quadrupeds. *Bulletin des Sciences, par la Societe Philomathique, Paris*, **1**, 138.

Dequéant, M.-L., Glynn, E., Gaudenz, K., *et al.* (2006). A complex oscillating network of signaling genes underlies the mouse segmentation clock. *Science*, **314**, 1595–8.

Dubrulle, J. and Pourquié, O. (2004). Coupling segmentation to axis formation. *Development*, **131**, 5783–93.

Durland, J. L., Sferlazzo, M., Logan, M. and Burke, A. C. (2008). Visualizing the lateral somitic frontier in the *Prx1Cre* transgenic mouse. *Journal of Anatomy*, **212**, 590–602.

Economides, K., Zeltser, L. and Capecchi, M. (2003). *Hoxb13* mutations cause overgrowth of caudal spinal cord and tail vertebrae. *Developmental Biology*, **256**, 317–30.

Filler, A. G. (1986). Axial character seriation in mammals: An historical and morphological exploration of the origin, development, use and current collapse of the homology paradigm. Ph.D. thesis, Harvard University, Cambridge, MA.

Filler, A. G. (2007). Homeotic evolution in the Mammalia: diversification of therian axial seriation and the morphogenetic basis of human origins. *One*, **2**(10), e1019.

Flower, W. H. (1885). *Osteology of the Mammalia*. London: Macmillan.

Flower, W. H. and Lydekker, R. (1891). *An Introduction to the Study of Mammals, Living and Extinct*. London: Adam and Charles Black.

Galis, F. (1999). Why do almost all mammals have seven cervical vertebrae? Developmental constraints, *hox* genes, and cancer. *Journal of Experimental Zoology–Molecular and Developmental Evolution*, **285**, 19–26.

Galis, F., Van Dooren, T. J. M., Feuth, J. D., *et al.* (2006). Extreme selection in humans against homeotic transformations of cervical vertebrae. *Evolution*, **60**, 2643–54.

Gaunt, S. J. (1994). Conservation in the *Hox* code during morphological evolution. *International Journal of Developmental Biology*, **38**, 549–52.

Gehring, W. J., Kloter, U. and Suga, H. (2009). Evolution of the *Hox* gene complex from an evolutionary ground state. *Current Topics in Developmental Biology*, **88**, 35–61.

Goodrich, E. S. (1913). Metameric segmentation and homology. *Quarterly Journal of Microscopical Science*, **59**, 227–50.

Goodrich, E. S. (1930). *Studies on the Structure and Development of Vertebrates*. London: MacMillan.

Hautier, L., Weisbecker, V., Sánchez-Villagra, M. R., Goswami, A. and Asher, R. J. (2010). Skeletal development in sloths and the evolution of mammalian vertebral patterning. *Proceedings of the National Academy of Sciences of the United States of America*, **44**, 18903–8.

Hirsinger, E., Jouve, C., Dubrulle, J. and Pourquié, O. (2000). Somite formation and patterning. *International Review of Cytology*, **198**, 1–65.

Horan, G. S. B., Wu, K., Wolgemuth, D. J. and Behringer, R. R. (1994). Homeotic transformation of cervical vertebrae in *Hoxa*-4 mutant mice. *Proceedings of the National Academy of Sciences of the United States of America*, **91**, 12 644–8.

Hughes, N. (2003a). Trilobite body patterning and the evolution of arthropod tagmosis. *BioEssays*, **25**, 386–95.

Hughes, N. (2003b). Trilobite tagmosis and body patterning from morphological and developmental perspectives. *Integrative and Comparative Biology*, **43**, 185–206.

Hughes, N., Chapman, R. and Adrain, J. (1999). The stability of thoracic segmentation in trilobites: a case study in developmental and ecological constraints. *Evolution and Development*, **1**, 24–35.

Hurlbert, R. C. (1998). Postcranial osteology of the North American Middle Eocene protocetid *Georgiacetus*. In *The Emergence of Whales*, ed. J. G. M. Thewissen. New York: Plenum Press, pp. 235–67.

Jenkins, F. A., Jr. (1971). The postcranial skeleton of African cynodonts. *Bulletin of the Peabody Museum of Natural History*, **36**, 1–216.

Ji, Q., Luo, Z.-X., Yuan, C.-X., *et al.* (2002). The earliest known eutherian mammal. *Nature*, **416**, 816–22.

Johnson, D. R. and O'Higgins, P. (1996). Is there a link between changes in the vertebral 'hox code' and the shape of vertebrae? A quantitative study of shape change in the cervical vertebral column of mice. *Journal of Theoretical Biology*, **183**, 89–93.

Kemper, C. M. and Leppard, P. (1999). Estimating body length of pygmy right whales (*Caperea marginata*) from measurements of the skeleton and baleen. *Marine Mammal Science*, **15**, 683–700.

Le Double, A. F. (1912). *Traité des Variations des Os de la Colonne Vertébrale*. Paris: Vigot Frères.

Li, G. and Luo, Z.-X. (2006). A Cretaceous symmetrodont therian with some monotreme-like postcranial features. *Nature*, **439**, 195–200.

Lovejoy, O., Cohn, M. J. and White, T. D. (1997). Morphological analysis of the mammalian postcranium: a developmental perspective. *Proceedings of the National Academy of Sciences of the United States of America*, **96**, 13 247–52.

Luo, Z.-X., Chen, P., Li, G. and Chen, M. (2007). A new eutriconodont mammal and evolutionary development in early mammals. *Nature*, **446**, 288–93.

Mallo, M., Vinagre, T. and Carapuço, M. (2009). The road to the vertebral formula. *International Journal of Developmental Biology*, **53**, 1469–81.

Marroig, G., Shirai, L. T., Porto, A., de Oliveira, F. B. and De Conto, V. (2009). The evolution of modularity in the mammalian skull. II. evolutionary consequences. *Evolutionary Biology*, **36**, 136–48.

Mayr, E. (1982). *The Growth of Biological Thought*. Cambridge, MA: Harvard University Press.

McIntyre, D. C., Rakshit, S., Yallowitz, A. R., *et al.* (2007). *Hox* patterning of the vertebrate rib cage. *Development*, **134**, 2981–9.

McPherron, A. C., Lawler, A. M. and Lee, S.-J. (1999). Regulation of anterior/posterior patterning of the axial skeleton by growth/differentiation factor 11. *Nature Genetics*, **22**, 260–4.

Miller, R. A. (1935). Functional adaptations in the forelimbs of sloths. *Journal of Mammalogy*, **16**, 38–51.

Müller, J., Scheyer, T. M., Head, J. J., *et al.* (2010). Homeotic effects, somitogenesis and the evolution of vertebral numbers in recent and fossil amniotes. *Proceedings of the National Academy of Sciences of the United States of America*, **107**, 2118–23.

Murie, J. (1872). On the form and structure of the manatee (*Manatus americanus*). *Transactions of the Zoological Society of London*, **8**, 127–202.

Narita, Y. and Kuratani, S. (2005). Evolution of the vertebral formulae in mammals: a perspective on developmental constraints. *Journal of Experimental Zoology–Molecular and Developmental Evolution*, **304B**, 91–106.

Nowicki, J. L. and Burke, A. C. (2000). *Hox* genes and morphological identity: axial versus lateral patterning in the vertebrate mesoderm. *Development*, **127**, 4265–75.

Oostra, R.-J., Hennekam, R. C. M., de Rooij, L. and Moorman, A. (2005). Malformations of the axial skeleton in Museum Vrolik. I. homeotic transformations and numerical anomalies. *American Journal of Medical Genetics*, **134A**, 268–81.

Owen, R. (1859). *The Principle Forms of the Skeleton and the Teeth*. London: Houlston and Wright.

Pollock, R. A., Sreenath, T., Ngo, L. and Bieberich, C. J. (1995). Gain of function mutations for paralogous *Hox* genes: implications for the evolution of *Hox* gene function. *Proceedings of the National Academy of Sciences of the United States of America*, **92**, 4492–6.

Polly, P. D., Head, J. J. and Cohn, M. J. (2001). Testing modularity and dissociation: the evolution of regional proportions in snakes. In *Beyond Heterochrony: The Evolution of Development*, ed. M. L. Zelditch. New York: Wiley-Liss, pp. 305–35.

Porto, P., de Oliveira, F. B., Shirai, L. T., De Conta, V. and Marroig, G. (2009). The evolution of modularity in the mammalian skull. I. Morphological integration patterns and magnitudes. *Evolutionary Biology*, **36**, 118–35.

Pourquié, O. (2003). The segmentation clock: converting embryonic time into spatial pattern. *Nature*, **301**, 328–30.

Raff, R. A. (1996). *The Shape of Life*. Chicago, IL: University of Chicago Press.

Raup, D. (1961). The geometry of coiling in gastropods. *Proceedings of the National Academy of Sciences of the United States of America*, **24**, 602–9.

Raup, D. (1966). Geometric analysis of shell coiling: general problems. *Journal of Paleontology*, **40**, 1178–90.

Richardson, M., Allen, S., Wright, G., Raynaud, A. and Hanken, J. (1998). Somite number rand vertebrate evolution. *Development*, **125**, 151–60.

Rolian, C. and Willmore, K. E. (2009). Morphological integration at 50: patterns of integration in biological anthropology. *Evolutionary Biology*, **36**, 1–4.

Romer, A. S. (1955). *The Vertebrate Body*, 2nd edn. Philadelphia, PA: W. B. Saunders.

Sánchez-Villagra, M. R., Narita, Y. and Kuratani, S. (2007). Thoracolumbar vertebral number: the first skeletal synapomorphy for afrotherian mammals. *Systematics and Biodiversity*, **5**, 1–17.

Sanger, T. J. and Gibson-Brown, J. J. (2004). The developmental bases of limb reduction and elongation in squamates. *Evolution*, **58**, 2103–6.

Schlosser, G. and Wagner, G. P. (2004). Introduction: the modularity concept in developmental and evolutionary biology. In *Modularity in Development and Evolution*, ed. G. Schlosser and G. P. Wagner. Chicago, IL: University of Chicago Press, pp. 1–11.

Slijper, E. J. (1936). *Die Cetaceen, vergleichend-anatomisch und systematisch*. Amsterdam, Netherlands: M. Nijhoff.

Slijper, E. J. (1946). Comparative biologic-anatomical investigations on the vertebral column and spinal musculature of mammals. *Koninklijke Nederlandse Akademie van Wetenschappen, Verhandelingen (Tweede Sectie)*, **42**, 1–128.

Tam, P. (1981). The control of somitogenesis in mouse embryos. *Journal of Embryology and Experimental Zoology*, **65**(Supplement), 103–28.

Tam, P. and Tan, S.-S. (1992). The somitogenic potential of cells in the primitive streak and the tail bud of the organogenesis-stage mouse embryo. *Development*, **115**, 703–15.

Todd, T. W. (1922). Numerical significance in the thoracicolumbar vertebrae of the Mammalia. *Anatomical Record*, **24**, 260–86.

Turner, H. N. (1847). Observations on the distinction between the cervical and dorsal vertebrae in the Class Mammalia. *Proceedings of the Zoological Society of London*, **15**, 110–14.

Uhen, M. (2004). Form, function and anatomy of *Dorudon atrox* (Mammalia, Cetacea): an archaeocete from the Middle to Late Eocene of Egypt. *University of Michigan Papers in Paleontology*, **34**, 1–222.

Wellik, D. M. (2007). *Hox* patterning of the vertebrate axial skeleton. *Developmental Dynamics*, **236**, 2454–63.

Wellik, D. M. and Capecchi, M. R. (2003). *Hox10* and *Hox11* genes are required to globally pattern the mammalian skeleton. *Science*, **301**, 363–7.

Wilson, V., Olivera-Martinez, I. and Storey, K. G. (2009). Stem cells, signals and vertebrate body axis extension. *Development*, **136**, 1591–604.

Zákány, J., Kmita, M., Alardon, P., de la Pompa, J. L. and Duboule, D. (2001). Localized and transient transcription of *Hox* genes suggests a link between patterning and the segmentation clock. *Cell*, **106**, 207–17.

Appendix 9.1

	Museum No.	C	T	L	S	Cd
Carnivora						
Canis latrans	MCZ 62197	7	13	7	3	20
Cuon alpinus	MCZ 19566	7	13	7	3	22
Urocyon cineroargenteus	MCZ 57137	7	13	7	3	21
Vulpes vulpes	MCZ 61731	7	13	7	3	19
Nyctereutes procyonoides	MCZ 24860	7	13	7	3	19
Crocuta crocuta	MCZ 20968	7	15	5	4	15
Hyaena brunnea	MCZ 57136	7	15	5	3	19
Cystophora cristata	MCZ 1084	7	15	5	3	11
Arctocephalus ursinus	MCZ 1785	7	15	5	3	6
Phoca groenlandica	MCZ 28682	7	15	5	4	14
Ommatophoca rossi	MCZ 51852	7	15	5	3	11
Callorhinus ursinus	MCZ 128	7	15	5	3	8
Eumetopias jubatus	MCZ 129	7	15	5	3	10
Odobenus rosmarus	MCZ 1720	7	14	6	4	8
Salanoia concolor	MCZ 27827	7	13	7	3	26
Fossa fossa	MCZ 29404	7	14	6	3	23
Viverricula indica	MCZ 5138	7	13	7	3	25
Mephitis mephitis	MCZ 27881	7	15	6	2	24
Melogale everetti	MCZ 36114	7	13	7	3	18
Lutra longicaudis	MCZ 37845	7	15	5	3	25
Martes pennanti	MCZ 59280	7	14	6	3	13
Enhydra lutris	MCZ 61578	7	14	6	4	19
Helarctos malayanus	MCZ 34152	7	15	4	3	10
Ursus americanus	MCZ 59938	7	14	6	6	8
Bassiricyon alleni	MCZ 37923	7	13	7	3	29
Bassiriscus astutus	MCZ 42161	7	13	7	3	26
Procyon lotor	MCZ 62164	7	14	6	3	18
Atilax paludinosus	MCZ 38063	7	14	6	3	27
Suricata suricata	MCZ 5115	7	15	5	3	26
Herpestes auropunctatus	MCZ 5123	7	13	7	3	19
Alcionyx jubatus	MCZ 58142	7	13	7	3	23
Caracal caracal	MCZ 58305	7	13	7	3	15
Rodentia						
Rhizomys sumatrensis	USNM 270291	7	14	6	4	16
Myospalax fontanieri	USNM 240750	7	13	6	4	
Dendromus insignis	USNM 589857	7	13	6	4	36

	Museum No.	C	T	L	S	Cd
Cricetomys gambianus	USNM 395836	7	13	6	4	
Brachytarsomys albicauda	USNM 449215	7	13	7	3	35
Neofiber alleni	MCZ 59695	7	13	6	4	23
Ondatra zibethicus	MCZ 59182	7	13	6	3	25
Phenacomys longicaudus	USNM 271153	7	13	7	3	
Lemmiscus pauperrimus	USNM 266584	7	13	6	3	13
Microtus longicaudus	USNM 67607	7	13	6	3	18
Microtus pennsylvanicus	MCZ 59228	7	12	7	3	17
Icthyomys hydrobates	USNM 172941	7	13	6	4	
Rheomys raptor	USNM 179026	7	13	6	3	24
Sigmon arizonae	USNM 510045	7	12	7	4	
Oryzomys capito	MCZ 37956	7	13	6	3	
Oecomys bicolor	MCZ 37899	7	12	7	2	37
Oligoryzomys fulvescens	MCZ 61800	7	12	7	3	31
Neotoma albigula	MCZ 59477	7	13	6	3	26
Neotoma fuscipes	MCZ 41056	7	13	6	3	32
Mesocricetus auratus	MCZ 56965	7	12	7	3	12
Rattus rattus	MCZ 36520	7	13	6	3	32
Hydromys fuligenosus	USNM 237315	7	14	7	4	
Hydromys chryogaster	USNM 221203	7	14	7	3	
Lophiomys imhausi	USNM 395134	7	16	7	3	24
Allactaga elater	USNM 341595	7	13	6	4	29
Jaculus jaculus	USNM 267371	7	13	6	4	26
Jaculus jaculus	USNM 308387	7	13	6	4	28
Zapus hudsonius	USNM 551925	7	13	6	4	36
Pedetes caffer	USNM 221381	7	12	7	3	
Anomalurus pelii	USNM 429832	7	15	10	4	
Anomalurus dirbianus	MCZ 43043	7	15	10	4	20
Crategeomys castanops	MCZ 59727	7	12	7	5	16
Thomomys umbrinus	MCZ 6304	7	13	7	4	
Castor canadensis	MCZ 60881	7	14	5	4	24
Aplodontia rufa	MCZ 49103	7	13	6	5	10
Marmota monax	MCZ 60874	7	12	7	4	19
Sciurus aberti	MCZ 46439	7	12	7	3	
Tamiasciurus hudsonicus	MCZ 2990	7	14	6	3	
Glis italicus	USNM 105291	7	13	6	3	
Graphiurus murinus	USNM 548526	7	13	6	3	13
Eliomys quercinus	USNM 38653	7	13	6	3	26
Ctenodactylus gundi	USNM 325851	7	13	6	4	10
Heliophobius argenteocinereus	USNM 376266	7	13	5	3	8

	Museum No.	C	T	L	S	Cd
Dasyprocta punctata	USNM 49946	7	13	6	5	
Myoprocta pratti	USNM 491289	7	13	6	5	
Hydrochoerus hydrochaeris	MCZ 1014	7	14	5	4	8
Cavia porcellus	MCZ 6945	7	13	6	3	5
Cavia tschudii	MCZ 46439	7	13	6	4	7
Coendou prehensilis	USNM 362242	7	15	6	3	31
Erethizon dorsatum	MCZ 61102	7	15	5	3	
Abrocoma cinerea	USNM 583254	7	16	5	3	21
Octodontomys gliroides	USNM 397002	7	12	7	4	28
Ctenomys sp.	USNM 132279	7	13	6	4	
Capromys melanurus	MCZ 25840	7	16	7	4	27
Capromys pilorides	MCZ13111	7	16	6	5	23
Geocapromys thoracatus	MCZ 12818	7	15	6	5	14
Plagiodontia hylaem	MCZ 49611	7	15	6	4	21
Hoplomys gymnurus	USNM 578395	7	13	6	4	26
Dactylomys dactylinus	USNM 549596	7	15	6	3	41
Echimys chrysurus	USNM 549549	7	14	7	3	
Myocastor coypus	USNM 549549	7	14	5	4	
Lagidium peruanium	MCZ 7289	7	13	6	3	
Lagostomus tridachtylus	MCZ 7285	7	12	7	4	20

Abbreviations: C, cervical; T, thoracic; L, lumbar; S, sacral; Cd, caudal.

10

Molecular determinants of marsupial limb integration and constraint

KAREN E. SEARS, CAROLYN K. DOROBA, XIAOYI CAO, DAN XIE AND SHENG ZHONG

Introduction

What are the intrinsic factors responsible for shaping mammalian biodiversity? This question is highly relevant to discussion of the two groups of therian mammals alive today, marsupials (kangaroos, possums, etc.) and placentals (humans, bats, whales, etc.). Despite arising at the same time, marsupials have never achieved the taxonomic or morphologic diversity of their sister group, the placentals (Lillegraven 1975; Kirsch 1977; Sears 2004; Cooper and Steppan 2010). To explain this phenomenon, scientists hypothesized that the evolution of marsupials had been constrained relative to that of placentals as a result of marsupials' unique mode of reproduction (Lillegraven 1975; Klima 1987; Sanchez-Villagra and Maier 2003).

Subsequent research confirmed that marsupials are less morphologically diverse and specialized than placentals (Sears 2004; Cooper and Steppan 2010; Kelly and Sears 2011a). Sears (2004) found that the shoulder girdles of living and extinct adult marsupials are less diverse than those of adult placentals, and adult marsupial scapulae are less morphologically diverse than adult marsupial pelves, as predicted by the marsupial constraint. Cooper and Steppan (2010) and Kelly and Sears (2011a) found that this pattern extended to forelimb versus hind limb comparisons in living mammals. Sears (2004) also linked this reduction in morphological diversity to a reduction in morphologic variation during development as a result of the functional requirements on the marsupial newborn.

Marsupials give birth after very short gestations to highly immature newborns that must immediately make a life-or-death crawl from the birth canal to the teat (Tyndale-Biscoe 1973; Lyne 1974; Hutson 1976; Parker 1977; Renfree *et al.* 1989; Gemmell *et al.* 1999, 2002; Veitch *et al.* 2000; Shaw and Renfree 2006).

From Clone to Bone: The Synergy of Morphological and Molecular Tools in Palaeobiology, ed. Robert J. Asher and Johannes Müller. Published by Cambridge University Press.
© Cambridge University Press 2012.

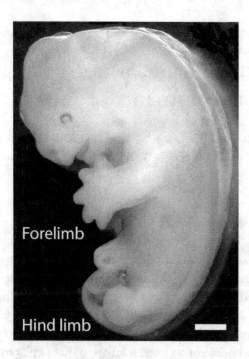

Figure 10.1 *Monodelphis domestica* embryo shortly before birth (embryonic stage 33). Note the relative developmental state of the limbs. The forelimb is relatively enlarged and advanced in its development, while the hind limb is at a much earlier developmental state. Scale bar represents 10 mm.

Subsequent to the crawl, they attach to the teat and complete their development. In general, marsupial newborns are under strong selective pressure to independently complete the crawl as their mothers do not assist them in their journey, and if they do not reach the teat they die (Reynolds 1952; Lyne *et al.* 1959; Sharman *et al.* 1965; Lyne 1974; Poole 1975; Hutson 1976; Renfree *et al.* 1989; Gemmell *et al.* 1999, 2002; Veitch *et al.* 2000; Nelson and Gemmell 2003). Furthermore, several marsupials give birth to more neonates than there are teats available (Nelson and Gemmell 2003). As a result, the young often have to compete for teat access and therefore survival.

The newborn's crawl is entirely powered by its massively developed forelimbs and highly modified shoulder girdles (Sharman 1973; Lyne and Hollis 1977; Hughes and Hall 1988), while its less developed hind limbs hang passively from the body (Figure 10.1). The constraint therefore is hypothesized to function as follows: by having to form a specific morphology (advanced forelimbs and modified shoulder girdle) at a specific time in their ontogeny (birth), the ability of marsupial forelimb morphology to vary through development is reduced relative to placentals (Lillegraven 1975; Sears 2004, 2009). As briefly mentioned previously, Sears (2004) confirmed that the forelimb complex (forelimb and shoulder girdle) of marsupials whose newborns complete an extensive crawl to the teat is restricted to a common pattern of ontogenetic shape change, strongly supporting the hypothesis that the morphological development of the marsupial forelimb complex has been limited evolutionarily by its obligate role in the crawl to the teat. As natural selection can only act on existing variation, this, in turn, is hypothesized to have reduced, or constrained, the ability of the group to evolve new morphologies of the forelimb complex. It is important to emphasize that, as defined in this chapter, constraints shape evolutionary history by making the evolution of certain morphologies more difficult, not by concretely preventing specific evolutionary changes (Alberch 1982; Maynard Smith *et al.* 1985; Schluter 1996; Yampolsky and Stoltzfus 2001).

Given the differential success of marsupial and placental mammals, it is possible that the marsupial constraint has had a profound effect on mammalian evolutionary history. Therefore, it is critical that we understand the mechanisms underlying the constraint. As a first step toward this understanding, we (Kelly and Sears 2011b), along with other authors (Goswami *et al.* 2009; Bennett and Goswami 2011), recently investigated limb integration in marsupial and placental mammals. We found that marsupial fore- and hind limbs, in general, exhibit significantly less phenotypic covariation than do the fore- and hind limbs of placentals and monotremes (the most immediate, extant therian outgroup). Furthermore, we reconstructed this difference in phenotypic covariation as having arisen via a reduction early in marsupial evolution along the branch leading to modern marsupials (Kelly and Sears 2011b). Our results are, in general, consistent with those of other recent investigations. Patterns of phenotypic covariation can be the result of many factors, including function, development or genetics (Hallgrímmson *et al.* 2002; Young and Hallgrímsson 2005; Wilmore *et al.* 2006; Schmidt and Fischer 2009). By combining these findings with previous research on the marsupial constraint and our knowledge of the marsupial mode of reproduction, we can speculate that the differential selective pressures exerted on marsupial fore- and hind limbs as a result of the functional requirements of the crawl led to a breakdown of the developmentally and functionally driven phenotypic covariation of the limbs early in marsupial evolution (Sears 2009; Kelly and Sears 2011b). As marsupial limbs became more modular, the forelimb was free to specialize for the crawl to the teat (Wagner 1996; Young and Hallgrímsson 2005; Goswami and Polly 2010). A side-effect of this specialization would have been the reduced developmental variation that we previously documented (Sears 2004), and the resulting constraint on the morphological evolution of marsupial limbs.

Although our examination of adult mammals has yielded many interesting results, to fully understand the mechanisms operating to shape marsupial limb modularity and generate evolutionary constraints, we need to go beyond traditional morphologic approaches, and incorporate additional data from embryologic, molecular and genetic sources into our research. In doing so, we can begin to investigate the molecular basis of the reduction of phenotypic covariation in adult marsupial limbs, and how the development of the marsupial limbs has been modified to generate their unique morphology at birth. Here we detail the preliminary results of our forays into these diverse areas, and how our findings contribute to our understanding of the modifications in marsupial development that have resulted in the reduced phenotypic covariation between adult marsupial limbs and the unique limb morphology of newborn marsupials.

The unique limb phenotype of marsupial newborns arises through independent developmental changes in both the fore- and hind limbs

Although the developmentally advanced forelimb of the marsupial newborn has generally aroused more scientific curiosity than its less-developed hind limb (e.g. Cheng 1955; Lillegraven 1975; Hughes and Hall 1988; Sears 2004), our research suggests that independent evolutionary changes in the development of both the fore- and hind limbs contributed to the establishment of the marsupial newborn's unique morphology (Sears 2009; Doroba and Sears 2010; Hübler *et al.* 2010). This is consistent with recent research by other labs that has highlighted the importance of evolutionary changes in hind limb development in the establishment of the marsupial newborn's phenotype (e.g. Weisbecker *et al.* 2008; Harrison and Larsson 2008; Keyte and Smith 2010; see also chapters in this volume by Kuratani and Nagashima, Mitgutsch *et al.*, and Richardson for data on limb development in non-marsupial vertebrates). In subsequent paragraphs we detail the evolutionary changes in the morphological development of both the fore- and hind limbs of marsupials that underlie the newborn's unique phenotype.

Marsupial forelimb

It has long been assumed that evolutionary changes in the timing of forelimb development have contributed to the advanced phenotype of the marsupial newborn's forelimb (Lillegraven 1975; Tyndale-Biscoe and Renfree 1987; Sanchez-Villagra 2002; Bininda-Emonds *et al.* 2003b). Separate analyses of a limited subset of prechondrogenic and chondrogenic events (e.g. limb bud formation, AER formation and digital plate crenation) in both American (Bininda-Emonds *et al.* 2003a) and Australasian (Bininda-Emonds *et al.* 2007) marsupials supported this hypothesis. In our own research into this area, we have taken two different approaches to study the timing of marsupial limb development. In the first, we analysed the evolution of the sequence of several early (i.e. prechondrogenic) limb developmental events among amniotes using a phylogenetic approach and the PGi program (Harrison and Larsson 2008). PGi uses sequences of developmental events coupled with phylogenetic data to infer the ancestral temporal sequences, and quantify sequence heterochronies (i.e. relative accelerations and decelerations). In the second, we compared the relative rates of prechondrogenic limb development in marsupial and placental mammals. In both of these analyses, we included representatives of American and Australasian marsupials, several placentals and a selection of outgroup taxa. We found that the earliest outgrowth of the marsupial forelimb was accelerated relative to that in placentals and outgroup taxa, and that the pre-birth rate of marsupial forelimb development is significantly greater than that of the marsupial hind limb and the limbs of placentals (Sears 2009), supporting the hypotheses and observations of earlier researchers (e.g. Hughes and Hall 1988; Lillegraven 1975; McCrady 1938). These data suggest that the marsupial forelimb achieves its highly developed state at birth by beginning its outgrowth earlier, and accelerating its subsequent rate of development, relative to marsupial hind limbs and the limbs of placentals.

However, the marsupial forelimb's story doesn't end with changes in developmental timing. Study of the gross structure of the developing marsupial limbs revealed that the physical manifestation of one of the major signalling centres of the limb, the apical ectodermal ridge (AER), is greatly altered in the forelimbs of at least one marsupial – the grey short-tailed opossum *Monodelphis domestica* (Doroba and Sears 2010). The AER manifests as a physical ridge along the dorsoventral boundary of the developing limb, formed through a thickening of the distal epithelium (Capdevila and Belmonte 2001; Niswander 2003). All amniotes with functional limbs that have been examined possess a physical AER (Cooper *et al.* 2011), consistent with the AER playing an important role in limb development. As a signalling centre, the AER plays a large role in distal outgrowth and patterning of the limb (Niswander 2003; Fernandez-Teran and Ros 2008; Towers and Tickle 2009a). For example, surgical removal of the physical manifestation of the AER from the developing limb bud results in truncation of the limb and the loss of distal elements (Saunders 1948; Summerbell 1974). Additionally, multiple instances of tetrapod limb reduction (e.g. in whales and snakes) are thought to have occurred, at least in part, through reductions in the physical AER (Cohn and Tickle 1999; Thewissen *et al.* 2006).

However, it is important to note that although the molecules produced by the AER (e.g. *Fgf4*, *Fgf8*) are required for limb outgrowth (Niswander *et al.* 1993; Fallon *et al.* 1994; Sun *et al.* 2002), experimental manipulation suggests that the physical ridge itself is not (Errick and Saunders 1974; Errick and Saunders 1976). Despite this, the fact that all amniotes with functional limbs retain a physical AER suggests that either the ridge itself does have some role in limb development (see Fernandez-Teran and Ros 2008 for some possibilities), or that reduction of the ridge negatively affects limb development to a degree that the ridge is always maintained.

In contrast to its apparent conservation across amniotes, we found the physical manifestation of the forelimb AER to be greatly reduced in the only marsupial in which the AER has been examined, *Monodelphis* (Doroba and Sears 2010). This conclusion was reached through examination of gross limb morphology at several developmental stages using scanning electron microscopy and histological sections. Specifically, we found that where the *Monodelphis* forelimb should have an AER, it has instead only a few isolated clumps of cells (Figure 10.2). In contrast, the AER of the *Monodelphis* hind limb, while less protuberant than that of some other amniotes (e.g. chick and mouse) is perfectly formed (Figure 10.2). These results are consistent with two theories. First, that the development of the marsupial forelimb is divergent from that of all other amniotes that form a typical physical AER. Second, that the development of the fore- and hind limbs of marsupials is highly divergent, consistent with the reduced phenotypic covariation observed in adult marsupial limbs (Kelly and Sears 2011b). This second theory leads us to speculate that a developmental and/or genetic divergence between marsupial fore- and hind limbs led to their reduced phenotypic covariation. This speculation will be further addressed below during discussion of the genes expressed during marsupial limb development. We acknowledge that these theories are based on examination of the AER of only one marsupial, *Monodelphis*. It is therefore possible that the AERs of the forelimbs of other marsupials resemble those of other amniotes, and that *Monodelphis* is unique. However, it seems likely that the reduction of the marsupial forelimb's AER was a byproduct of the divergence of the forelimb's development as a result of the crawl to the teat. As all marsupial newborns possess enlarged and, therefore, presumably developmentally divergent forelimbs (Hughes and Hall 1988; Gemmell *et al.* 2002), it is likely that this developmental divergence, and the reduction of the AER, occurred very early in marsupial evolution. These events possibly occurred synchronously with the reduction in adult phenotypic covariation we reconstructed as occurring along the basal branch leading to crown-group marsupials. As such, we predict that the forelimb AER of at least most and likely all extant marsupials will be found to be reduced.

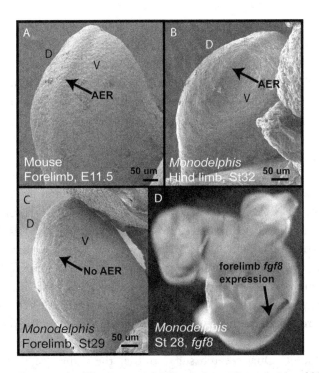

Figure 10.2 The *Monodelphis'* apical ectodermal ridge (AER). The limbs of
mouse (A; embryonic day 11.5) and other amniotes that have been
examined possess a well-formed and protuberant physical manifestation of
the AER. The *Monodelphis* hind limb (B; embryonic stage 32) possesses a
similar AER. However, in stark contrast, the physical AER of the *Monodelphis*
forelimb (C; embryonic stage 29) from a comparable limb developmental
stage is greatly reduced. (D) Despite its lack of a physical AER, the
Monodelphis forelimb possesses the molecular manifestation of the AER,
a typical, solid band of *fgf8* expression along the limb's dorsal–ventral
boundary. (Adapted from Doroba and Sears 2010.) D, dorsal, V, ventral.

In contrast to the physical AER, we found that the molecular AER, at least
as indicated by *Fgf8* expression, appears normal in *Monodelphis* forelimbs
(Figure 10.2). This explains how the marsupial forelimb successfully completes
its outgrowth, despite lacking a physical AER. However, our recent microarray
research suggests that the level of *Fgf8* is twofold less ($p < 0.001$) in marsupial
forelimbs than hind limbs at comparable stages of limb development (Sears
et al., unpublished data; see below for a description of the microarray analyses).
In similar microarrays performed in mouse, *Fgf8* was not found to be
differentially expressed in fore- and hind limbs (Shou *et al.* 2005). This
difference in expression is consistent with the fore- and hind limbs of

marsupials having experienced a developmental and genetic divergence, and an associated reduction in their phenotypic covariation.

We are currently investigating the molecular and genetic basis of the differences in limb development between *Monodelphis* and mouse forelimbs, and *Monodelphis* fore- and hind limbs. This research is discussed in greater detail in the 'Expression of limb developmental genes' section below.

Marsupial hind limb

Initial analysis of the timing of early limb developmental events (e.g. limb bud formation, AER formation, etc.) in both American (Bininda-Emonds *et al.* 2003a) and Australasian (Bininda-Emonds *et al.* 2007) marsupials found no evidence for evolutionary changes in the timing of marsupial hind limb development. However, other recent analyses of developmental timing focusing on later developmental events (e.g. ossification) found evidence that hind limb development is delayed in marsupials (Harrison and Larsson 2008; Weisbecker *et al.* 2008). Our research on the timing of early developmental events in mammals (see the 'Marsupial forelimb' section above for details) has helped to resolve these studies. Our research suggests that the timing of the initial outgrowth of the marsupial hind limb remains unchanged in marsupials relative to placentals and outgroup taxa, consistent with the results of Bininda-Emonds *et al.* (2003a, 2007) (Sears 2009). This finding is consistent with the earliest outgrowth of tetrapod limbs being constrained (for a discussion of this idea see Galis *et al.* 2001; Galis and Metz 2001; Hamrick 2002; Bininda-Emonds *et al.* 2003a, 2007). However, our results also indicate that marsupial hind limb development subsequently slows down and almost pauses during the limb bud stage (when the semi-circular limb is as long as it is wide), resulting in the timing of subsequent stages (e.g. formation of the foot plate, digits, etc.) being relatively delayed (Sears 2009). These findings are consistent with those of Harrison and Larsson (2008) and Weisbecker *et al.* (2008). Therefore, the developmental timing of the marsupial hind limb, like that of the forelimb, has experienced evolutionary changes that contribute to the unique limb phenotype of the marsupial newborn, and likely the reduction in phenotypic covariation between adult marsupial limbs.

We have begun to investigate the cellular and molecular basis of the slow-down in marsupial hind limb development at the bud stage. To date we have assayed the cellular processes of proliferation and death (i.e. apoptosis) in the developing fore- and hind limbs of the marsupial, *Monodelphis*, and the placental, *Mus musculus* (mouse) (K. E. Sears, unpublished data). To do this, we cryosectioned the limbs at a constant thickness (10 μm), and performed standard immunohistological techniques using antibodies known to label proliferating (phosphohistone H3) and apoptotic (TUNEL) cells (Hübler *et al.* 2010). While patterns of cell death were consistent in all limbs (data not shown), we found that the *Monodelphis* hind limb exhibits a significantly decreased rate of cellular proliferation during the bud stage ($p < 0.001$) (Figure 10.3). Therefore, it is likely that the slowing of *Monodelphis* hind limb development during the bud stage is at least in part due to a reduction in cellular proliferation. We are currently investigating the molecular basis of these evolutionary changes in proliferation rate.

Expression of limb developmental genes is highly divergent in marsupial fore- and hind limbs

Microarray assay

An increasingly common approach to investigating gene expression during development is to examine gene expression levels using high-throughput methods such as microarray or RNA sequencing (Mardis 2008; Shendure 2008; Lu *et al.* 2009). The advantage of these methods is that they quickly assay the mRNA expression levels of many genes ($> 40\,000$), and represent a relatively unbiased method of identifying the genes that differ in their expression levels in two samples. Because of these advantages, we recently used microarray analysis to investigate the differences in gene expression levels between *Monodelphis* fore- and hind limbs from limb stage 3 (i.e. the bud stage) (Wanek *et al.* 1989; K. E. Sears *et al.*, unpublished data).

To do this, we dissected and pooled limbs separately for eight litters: four for the forelimb (embryonic stage 28), and four for the hind limb (embryonic stage 30) (K. Smith, personal communication; McCrady 1938; Mate *et al.* 1994). Because marsupial limbs develop at different rates, different embryonic developmental stages were used to obtain *Monodelphis* fore- and hind limbs at comparable stages of limb development. We then extracted the total RNA from the limbs, using standard techniques (Kalinka *et al.* 2010). This allowed four replicate arrays. Arrays were run on a custom Agilent chip we designed for *Monodelphis*, following the Agilent Two-Color Microarray-Based Gene Expression Analysis protocol version 5.7 (Agilent Technologies, www.agilent.com). Results across assays were remarkably consistent, and many genes ($n = 92$) exhibited significant differential expression ($p < 0.01$) on the order of twofold or greater between *Monodelphis* fore- and hind limbs. Of these genes, 27 are known to either be expressed in the limb or have a known role in limb development. These include genes known to be differentially expressed in mouse limbs, such

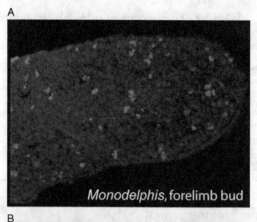

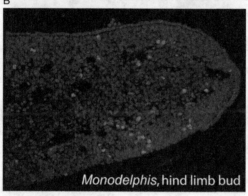

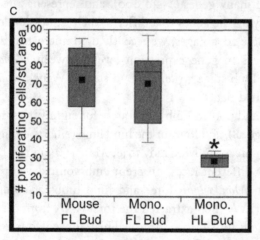

Figure 10.3 Cell proliferation in *Monodelphis* limbs. Red (To-pro-3) indicates non-proliferating cells; green (Phospho-histone H3) indicates proliferating cells. (A) *Monodelphis* forelimb at bud stage (embryonic stage 29). (B) *Monodelphis* hind limb at bud stage (embryonic stage 32). (C) The *Monodelphis* (Mono.) and mouse forelimb (FL, Embryonic Day 10.5) buds have significantly more proliferating cells in a given standard area than the *Monodelphis* hind limb (HL) (p = 0.0004). (A) and (B) are at 20×. (See also colour plate.)

as *Tbx4*, *Tbx5*, and *Pitx1*. In mouse, *Tbx5* is expressed at higher levels in the forelimbs, while *Tbx4* and *Pitx1* are more highly expressed in the hind limbs (Lanctot *et al.* 1999; Rodriguez-Esteban *et al.* 1999; Margulies *et al.* 2001; Takeuchi *et al.* 2003). Our microarray results indicate that the relative expression levels of these genes in *Monodelphis* limbs are consistent with those previously documented in mouse: *Tbx4* levels are 2× higher in *Monodelphis* hind limbs than forelimbs ($p < 0.001$), *Tbx5* levels are 4.6× higher in *Monodelphis* forelimbs than hind limbs ($p < 0.001$), and *Pitx1* levels are 7× higher in *Monodelphis* hind limbs than forelimbs ($p < 0.001$). This validates the results of our microarray.

A similar microarray experiment comparing gene expression levels in the fore- and hind limbs of embryonic day 12.5 mice found 10 genes (13 probe sets) that were differentially expressed at > twofold and at a significance level of $p < 0.01$ (Shou *et al.* 2005). Of these genes, three are known to have a role in limb development (*Tbx4*, *Tbx5*, and *HoxC10*). When we compare these results with those of our *Monodelphis* microarray, it appears that an order of magnitude more genes (92 total; 27 limb) are differentially expressed at the same significance level in *Monodelphis* than in mouse fore- and hind limbs. Even at a significance level of $p < 0.05$, mouse limbs only differ in the expression levels of 37 total genes (44 probe sets), five of which are limb genes (Shou *et al.* 2005). However, given differences in chips, tissues used, etc., caution should be taken when comparing the results of microarray assays across species. Furthermore, caution should also be taken when comparing these results as many of the genes identified by the *Monodelphis* array are likely differentially expressed as a result of the *Monodelphis* fore- and hind limbs being from different overall stages of embryonic development. As a result, although the fore- and hind limbs themselves are at a comparable developmental stage, other organs such as the skin that covers the limbs are likely not. This would tend to inflate the number of total genes that are differentially expressed in *Monodelphis* limbs relative to those of mouse. However, this should not inflate the number of limb patterning genes that are differentially expressed, and an order of magnitude more limb patterning genes are differentially expressed in *Monodelphis* than mouse limbs. Another possible confounding factor is that the mouse microarray was performed on limbs from a slightly later stage of development than the *Monodelphis* microarray, approximately limb stage 6 (Wanek *et al.* 1989; Shou *et al.* 2005). Nevertheless, we might expect that the molecular and developmental divergence of the fore- and hind limbs would tend to increase through ontogeny in at least placentals, as limb-specific morphologies begin to emerge (Wanek *et al.* 1989; Galis and Metz 2001; Galis *et al.* 2001, 2002). Consistent with this, placental limb development appears fairly conserved during initial outgrowth and patterning, with many species-specific morphologies appearing later (Sears 2011). This is even generally true for the hind limbs of whales, which form and then are lost at later developmental stages (Thewissen *et al.* 2006; Cooper *et al.* 2011), and the forelimbs of bats, which initially resemble those of more generalized mammals (Chen *et al.* 2005; Sears *et al.* 2006; Hockman *et al.* 2008). However, evidence supporting this hypothesis remains limited, and further investigation is needed to clarify when limb specialization occurs during development.

In summary, even with many possibly confounding factors taken into account, the comparison of the microarray results remains intriguing, and supports the hypothesis that the molecular controls of *Monodelphis*, and potentially marsupial, fore- and hind limb development are more divergent than those of mouse, and potentially placental, limbs. Furthermore, these findings suggest that the reduced phenotypic covariation we have documented in marsupial limbs may have its origin in a divergence of the genetic and molecular controls of limb development.

The genes identified by the microarray as being differentially expressed in *Monodelphis* limbs include some interesting candidates that are possibly associated with the reduced AER of the *Monodelphis* forelimb (Niswander 2003; Fernandez-Teran and Ros 2008). Additionally, the microarray pulled out several genes that are known to regulate cellular proliferation in mammalian tissues (Lee *et al.* 1999; Lai *et al.* 2007; Motomura *et al.* 2007), and therefore may have been evolutionarily modified to control the relatively rapid outgrowth of the *Monodelphis* forelimb or the delay in *Monodelphis* hind limb development at the bud stage. Also intriguing are the several genes ($n = 65$) pulled out by the microarray that have no known role in limb development. Although it is likely that many of these genes are differentially expressed in *Monodelphis* limbs as a result of the embryos being at different overall developmental stages, it is possible that at least some of these genes may have a previously undiscovered role in limb development. We are currently investigating the expression patterns, levels and functions in limb development of the genes identified by the microarray. The goal of this research is to identify the specific differences in gene regulation and function that underlie the divergent development of *Monodelphis* fore- and hind limbs, and *Monodelphis* and placental forelimbs.

To provide one specific example of the findings that are emerging from the microarray, several of the genes identified by the microarray as being differentially expressed in *Monodelphis* fore- and hind limbs are involved in the regulation of AP patterning (*Hand2*, *Hoxd9*, *Fgf8*, *Gli3*, etc.; Figure 10.4). As patterning and outgrowth along the limb's major axes (e.g. AP and proximal-distal [PD]) are molecularly linked, it is possible that these differences in

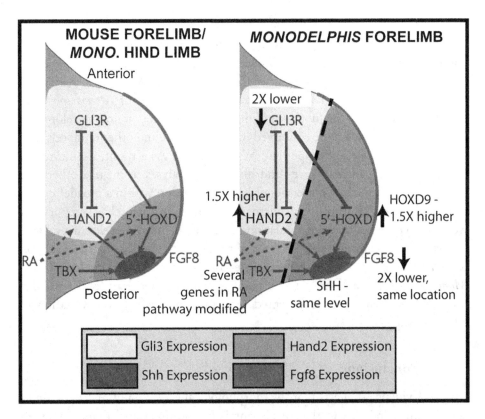

Figure 10.4 Predicted expression differences in AP patterning genes between *Monodelphis* forelimbs (FL), and mouse forelimbs and *Monodelphis* hind limbs. Mouse FL/*Monodelphis* hind limb (HL) is on the left, *Monodelphis* forelimbs on the right. In mouse, interplay between *gli3*, *hand2*, retinoic acid (RA), 5′ *HoxD* genes, *fgf8*, *tbx3* and *shh* sets up anterior–posterior (AP) polarization. The same genes are present in *Monodelphis* limbs, but several are expressed in different levels and/or locations in the *Monodelphis* FL relative to the *Monodelphis* HL and mouse FL. For example, in the *Monodelphis* FL, *hand2* (orange) is anteriorly expanded relative to mouse (extent of *hand2* expression indicated by dashed line). Moreover, differences in gene expression levels between the *Monodelphis* FL and HL were documented by the *Monodelphis* microarray (vertical up or down arrows in the figure represent the fold differences from the microarray in *Monodelphis* FL and HL). As *Monodelphis* HL development is in many ways more similar to that of the mouse than *Monodelphis* FL, we predict that these genes will also exhibit differential expression between *Monodelphis* and mouse FLs. Additional research is under way to test these predictions. However, even just given results so far, it is likely that even a process as fundamental as AP patterning varies between the *Monodelphis* FL, and the mouse FL and *Monodelphis* hind limb. (See also colour plate.)

AP patterning are related to the differences in rate of limb growth along the PD axis documented in marsupial fore- and hind limbs (Bénazet *et al.* 2009; Sears 2009; Towers and Tickle 2009b; Zeller *et al.* 2009). We are currently investigating the extent of the difference in AP pattern regulation in *Monodelphis* fore- and hind limbs, and the forelimbs of *Monodelphis* and mouse. Furthermore, the functions of many of these genes in limb development (e.g. *Fgf8*) are known to be conserved in mouse, chick and other tetrapods (e.g. anurans and urodeles) (Stopper and Wagner 2005). As a result, we ultimately would like to expand our fore- and hind limb analyses to include other taxa (e.g. chick), and thereby place the marsupial condition within a broader evolutionary context. However, the microarray results by themselves support our working hypothesis that the regulation of one of the major developmental axes of the limb, specifically the AP axis, varies extensively between *Monodelphis* fore- and hind limbs. These results are consistent with the development of *Monodelphis* fore- and hind limbs being incredibly divergent, and with this possibly underlying the documented reduction in phenotypic covariation of adult marsupials.

Conclusions

Evidence suggests that the morphological evolution of marsupial limbs is constrained relative to that of placentals by the functional requirements of the marsupial newborn's forelimb-driven crawl to the teat (Sears 2004; Cooper and Steppan 2010). We have begun to investigate the mechanisms underlying this constraint to better understand how it might have arisen, and how it might have shaped mammalian evolutionary history, as captured by the fossil record.

By taking a traditional approach incorporating study of adult morphometric data, we determined that the phenotypes of marsupial fore- and hind limbs covary significantly less than those of placentals, and that this decrease in covariation occurred early in marsupial evolution (Kelly and Sears 2011b). From study of gross embryology and the timing of developmental events, we determined that this breakdown in adult phenotypic covariation, and the unique morphology of the marsupial newborn (i.e. developmentally advanced forelimb and delayed hind limb), likely arose through evolutionary changes in the development of both the marsupial fore- and hind limb. Within the forelimb, the initial outgrowth and rate of development is relatively accelerated, while the hind limb temporarily slows its development during the bud stage, likely through a decrease in cellular proliferation, such that its subsequent stages of development are delayed (Sears 2009). Additionally, our studies indicate that the formation of the physical manifestation of one of the major signalling centres in limb development, the apical ectodermal ridge (AER), is disrupted in the forelimbs of at least one marsupial, *Monodelphis domestica*, and likely others (Doroba and Sears 2010).

By broadening our research to include molecular data, we have begun to address the molecular mechanisms behind the generation of the unique pheno-type of the limbs of marsupial newborns, and the reduced covariation among the limbs of marsupial adults. We have found that the regulation of AP patterning is substantially different in the forelimbs of at least one marsupial, *Monodelphis domestica*, relative to *Monodelphis* hind limbs and the limbs of at least one placental, *Mus musculus*. Moreover, our results suggest that the overall suite of genes that are expressed during early limb development is much more divergent in *Monodelphis*, and probably marsupial, limbs, than in mouse, and probably placental, limbs. Taken together, these molecular findings suggest that the reduced phenotypic covariation between the fore- and hind limbs of adult marsupials may be a manifestation of a breakdown in the genetic links between the limbs early in their development. Additionally, these findings coupled with those obtained from analysis of gross embryological morphology suggest that multiple evolutionary changes in key signalling pathways (e.g. AER, AP patterning) may underlie the unique and highly divergent morphology of marsupial newborn limbs.

Although the research presented here is still in progress, our investigations to date suggest a possible hypothesis for how the constraint arose and subse-quently played a role in shaping marsupial evolution. Specifically, we hypothe-size that a breakdown in the genetic connections between the fore- and hind limbs occurred early in marsupial evolution, possibly as a result of differential selective pressures on the marsupial newborn's fore- and hind limbs as a result of the forelimb-driven crawl to the teat (Figure 10.5). We speculate that this breakdown increased the independent variation of the limbs during develop-ment (i.e. their modularity), thereby freeing up the forelimb to (continue to) specialize for the functional requirements of the newborn's crawl. With the forelimb's specialization would have come increased divergence of the fore- and hind limb morphology in marsupial newborns, and a reduction in the pheno-typic covariation of the limbs in adults. Most importantly, the marsupial forelimb's specialization for the crawl would have led to a reduction in its developmental variation as a result of the forelimb's requirement to make one particular morphology at one particular point in ontogeny. In this way, the specialization of the forelimb could have led to the previously identified constraint on marsupial limb evolution (Sears 2004; Cooper and Steppan 2010), and the differential taxonomic and morphologic diversification of marsupial and placental mammals. Obviously much of this hypothesis remains speculative, and

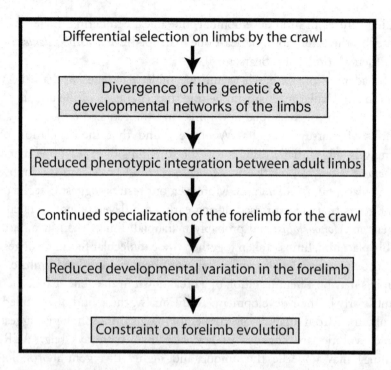

Figure 10.5 Flowchart describing our working hypothesis for the mechanisms by which the constraint on marsupial limb evolution arose and has influenced mammalian evolution. Hypotheses in grey boxes have been supported with preliminary data.

we are currently testing its expectations. Nevertheless, we believe that our research highlights the novel insights that can be gained by combining molecular and morphology approaches to address long-standing mechanistic questions in evolutionary biology.

Summary

The functional requirements of the marsupial newborn's crawl to the teat have likely constrained the evolution of marsupial limbs, and thereby marsupials in general, relative to placentals. We combine study of morphology and molecules to investigate the mechanisms by which this constraint arose and subsequently shaped mammalian evolution. Our results suggest that the unique, crawl-facilitating limb morphology of the marsupial newborn and the associated reduction in the phenotypic covariation between adult marsupial limbs arose early in marsupial evolution through evolutionary changes in the developmental timing of both limbs, and through fundamental changes in major signalling pathways regulating limb development. In line with this, additional results suggest that the transcriptomes of the forelimbs and hind limbs of *Monodelphis domestica*, a marsupial, are much more divergent than those of *Mus musculus*, a placental. Taken together, our results are consistent with the hypothesis that differential selective pressures imposed by the crawl broke the genetic connections between the forelimbs and hind limbs early in marsupial evolution, thereby freeing up the forelimb to specialize for the crawl. This, in turn, reduced the developmental variation in marsupial forelimbs and resulted in the constraint on their evolution. Although we are still testing these ideas, our results illustrate the synergistic insights that can be gained by combining morphological and molecular approaches to investigate major questions in evolutionary biology.

REFERENCES

Alberch, P. (1982). Developmental constraints in evolutionary processes. In *Evolution and Development*, ed. J. T. Bonner. New York: Springer-Verlag, pp. 313–32.

Bénazet, J. D., Bischofberger, M., Tiecke, E., *et al.* (2009). A self-regulatory system of interlinked signaling feedback loops controls mouse limb patterning. *Science*, **323**, 1050–3.

Bennett, V. and Goswami, A. (2011). Does developmental strategy drive limb integration in marsupials and monotremes? *Mammalian Biology*, **76**, 79–83.

Bininda-Emonds, O. R. P., Jeffery, J. E. and Richardson, M. K. (2003a). Inverting the hourglass: quantitative evidence against the phylotypic stage in vertebrate development. *Proceedings of the Royal Society B–Biological Sciences*, **270**, 341–6.

Bininda-Emonds, O. R. P., Jeffrey, J. E. and Richardson, M. K. (2003b). Is sequence heterochrony an important evolutionary mechanism in mammals. *Journal of Mammalian Evolution*, **10**(4), 335–61.

Bininda-Emonds, O. R. P., Jeffery, J. E., Sánchez-Villagra, M. R., *et al.* (2007). Forelimb-hindlimb developmental timing changes across tetrapod phylogeny. *BMC Evolutionary Biology*, **7**, 182–9.

Capdevila, J. and Belmonte, J. (2001). Patterning mechanisms controlling vertebrate limb development. *Annual Review of Cellular and Developmental Biology*, **17**, 87–132.

Chen, C. H., Cretekos, C. J., Rasweiler, J. J. and Behringer, R. R. (2005). Hoxd13 expression in the developing limbs of the short-tailed fruit bat, *Carollia perspicillata*. *Evolution and Development*, **7**(2), 130–41.

Cheng, C. (1955). The development of the shoulder region of the opossum, *Didelpihis virginiana*, with special reference to the musculature. *Journal of Morphology*, **97**, 415–71.

Cohn, M. J. and Tickle, C. (1999). Developmental basis of limblessness and axial patterning in snakes. *Nature*, **399**, 474–9.

Cooper, J. and Steppan, S. J. (2010). Developmental constraint on the evolution of marsupial forelimb morphology. *Australian Journal of Zoology*, **58**, 1–15.

Cooper, L. N., Armfield, B. A. and Thewissen, J. G. M. (2011). Phenotypic variation and conserved FGF gene expression in the vertebrate limb ectoderm. In *Epigenetics: Linking Genotype and Phenotype in Development and Evolution*, ed. B. Hallgrímmson and B. K. Hall, pp. 238–55.

Doroba, C. K. and Sears, K. E. (2010). The divergent developmental of the apical ectodermal ridge in the marsupial *Monodelphis domestica*. *Anatomical Record*, **293**(8), 1325–32.

Errick, J. E. and Saunders, J. W. (1974). Effects of an 'inside-out' limb-bud ectoderm on development of the avian limb. *Developmental Biology*, **41**, 338–51.

Errick, J. E. and Saunders, J. W. (1976). Limb outgrowth in the chick embryo induced by dissociation and reaggregated cells of the apical ectodermal ridge. *Developmental Biology*, **50**, 26–34.

Fallon, J. F., Lopez, A., Ros, M. A., *et al.* (1994). Fgf-2: apical ectodermal ridge growth signal for chick limb development. *Science*, **264**, 104–7.

Fernandez-Teran, M. and Ros, M. A. (2008). The apical ectodermal ridge: morphological aspects and signaling pathways. *International Journal of Developmental Biology*, **52**, 857–71.

Galis, F. and Metz, J. A. J. (2001). Testing the vulnerability of the phylotypic stage: on modularity and evolutionary conservation. *Journal of Experimental Zoology–Molecular and Developmental Evolution*, **291**, 195–204.

Galis, F., van Alphen, J. J. M. and Metz, J. A. J. (2001). Why five fingers? Evolutionary constraints on digit numbers. *Trends in Ecology and Evolution*, **16**(11), 637–46.

Galis, F., van Alphen, J. J. M and Metz, J. A. J. (2002). Digit reduction: via repatterning or developmental arrest. *Evolution and Development*, **4**(4), 249–51.

Gemmell, R. T., Veitch, C. and Nelson, J. (1999). Birth in the northern brown bandicoot, *Isoodon macrourus* (Marsupialia: Peramelidae). *Australian Journal of Zoology*, **47**, 517–28.

Gemmell, R. T., Veitch, C. and Nelson, J. (2002). Birth in marsupials. *Comparative Biochemistry and Physiology Part B*, **131**, 621–30.

Goswami, A. and Polly, P. D. (2010). The influence of modularity on cranial morphological disparity in Carnivora and Primates (Mammalia). *PLOS One*, **5**(3), e9517.

Goswami, A., Weisbecker, V. and Sánchez-Villagra, M. R. (2009). Developmental modularity and the marsupial-placental dichotomy. *Journal of Experimental Zoology B*, **312B**, 186–95.

Hallgrímmson, B., Willmore, K. E. and Hall, B. K. (2002). Canalization, developmental stability, and morphological integration in primate limbs. *Yearbook of Physical Anthropology*, **45**, 131–58.

Hamrick, M. W. (2002). Developmental mechanisms of digit reduction. *Evolution and Development*, **4**(4), 247–8.

Harrison, L. B. and Larsson, H. C. E. (2008). Estimating evolution of temporal sequence changes: a practical approach inferring ancestral developmental sequences and sequences heterochrony. *Systematic Biology*, **57**(3), 378–87.

Hockman, D., Cretekos, C. J., Mason, M. K., *et al.* (2008). A second wave of sonic hedgehog expression during the development of the bat limb. *Proceedings of the National Academy of Sciences of the United States of America*, **105**(44), 16 982–7.

Hübler, M., Niswander, L., Peters, J. and Sears, K. E. (2010). The developmental reduction of the marsupial coracoid: a case study in *Monodelphis domestica*. *Journal of Morphology*, **271**(7), 769–76.

Hughes, R. L. and Hall, L. S. (1988). Structural adaptations of the newborn marsupial. In *The Developing Marsupial: Models for Biomedical Research*, ed. C. H. Tyndale-Biscoe and P. A. Janssens. Berlin: Springer, pp. 8–27.

Hutson, G. D. (1976). Grooming behaviour and birth in the dasyurid marsupial *Dasyuroides byrnei*. *Australian Journal of Zoology*, **24**, 277–82.

Kalinka, A. T., Varga, K. M., Gerrard, D. T., *et al.* (2010). Gene expression divergence recapitulates the developmental hourglass model. *Nature*, **468**(7325), 811–14.

Kelly, E. M. and Sears, K. E. (2011a). Limb specialization in living marsupial and eutherian mammals: an investigation of constraints on mammalian limb evolution. *Journal of Mammalogy*, **92**(5), 1038–49.

Kelly, M. and Sears, K. E. (2011b). Limb integration in New World marsupials. *Biological Journal of the Linnean Society*, **102**, 22–36.

Keyte, A. L. and Smith, K. K. (2010). Developmental origins of precocial forelimbs in marsupial neonates. *Development*, **137**, 4283–94.

Kirsch, J. A. W. (1977). The six-percent solution: second thoughts on the adaptedness of the marsupialia. *American Scientist*, **65**, 276–88.

Klima, M. (1987). Early development of the shoulder girdle and sternum in marsupials (Mammalia: Metatheria). *Advances in Anatomy Embryology and Cell Biology*, **47**(2), 1–80.

Lai, D. M., Tu, Y. K., Hsieh, Y. H., *et al.* (2007). Angiopoietin-like protein 1 expression is related to intermuscular connective tissue and cartilage development. *Developmental Dynamics*, **236**, 2643–52.

Lanctot, C., Moreau, A., Chamberland, M., *et al.* (1999). Hindlimb patterning and mandible development require the *Ptx1* gene. *Development*, **126**, 1805–10.

Lee, K. K. H., Tang, M. K., Yew, D. T. W., *et al.* (1999). gas2 is a multifunctional gene involved in the regulation of apoptosis and chondrogenesis in the developing mouse limb. *Developmental Biology*, **207**, 14–25.

Lillegraven, J. A. (1975). Biological considerations of the marsupial-placental dichotomy. *Evolution*, **29**, 707–22.

Lu, Y., Huggins, P. and Bar-Joseph, Z. (2009). Cross species analysis of microarray expression data. *Bioinformatics*, **25**(12), 1476–83.

Lyne, A. G. (1974). Gestation period and birth in the marsupial *Isoodon macrourus*. *Australian Journal of Zoology*, **22**, 303–9.

Lyne, A. G. and Hollis, D. E. (1977). The early development of marsupials, with special references to bandicoots. In *Reproduction and Evolution*, ed. J. H. Calaby and C. H. Tyndale-Biscoe. Canberra, Australia: Australian Academy of Science, pp. 293–302.

Lyne, A. G., Pilton, P. E. and Sharman, G. B. (1959). Oestrous cycle, gestation period and parturition in the marsupial *Trichosurusvulpecula*. *Nature*, **183**(4661), 622–3.

Mardis, E. R. (2008). Next-generation DNA sequencing methods. *Annual Review of Genomics and Human Genetics*, **9**, 328–402.

Margulies, E. H., Kardia, S. L. R. and Innia, J. W. (2001). A comparative molecular analysis of developing mouse forelimbs and hindlimbs using serial analysis of gene expression (SAGE). *Genome Research*, **11**, 1686–98.

Mate, K. E., Robinson, E. S., VandeBerg, J. L. and Pederson, R. A. (1994). Timetable of in vivo embryonic development in the gray short-tailed opossum (*Monodelphis domestica*). *Molecular Reproduction and Development*, **39**, 365–74.

Maynard Smith, J., Burian, R., Kauffman, S., *et al.* (1985). Developmental constraints and evolution: a perspective from the Mountain Lake Conference on Development and Evolution. *Quarterly Review of Biology*, **60**(3), 265–87.

McCrady, E. (1938). *The Embryology of the Opossum*. Philadelphia, PA: Wistar Institute of Anatomy and Biology, 233 pp.

Motomura, H., Niimi, H., Sugimori, K., Ohtsuka, T., Kimura, T. and Kitajima, I. (2007). Gas6, a new regulator of chondrogenic differentiation from mesenchymal cells. *Biochemical and Biophysical Research Communications*, **15**(4), 997–1003.

Nelson, J. E. and Gemmell, R T. (2003). Birth in the northern quoll, *Dasyurus hallucatus* (Marsupialia: Dasyuridae). *Australian Journal of Zoology*, **51**(2), 187–98.

Niswander, L. (2003). Pattern formation: old models out on a limb. *Nature Reviews Genetics*, **4**(2), 133–43.

Niswander, L., Tickle, C., Vogel, A., Booth, I. and Martin, G. R. (1993). FGF4 replaces the apical ectodermal ridge and directs outgrowth and patterning of the limb. *Cell*, **75**, 579–87.

Parker, P. (1977). The evolutionary comparison of placental and marsupial patterns of reproduction. In *The Biology of Marsupials*, ed. B. Stonehouse and D. Gilmore. London: Macmillan Press, pp. 273–86.

Poole, W. E. (1975). Reproduction in two species of grey kangaroos, *Macropus giganteus* Shaw and *Macropus fuliginosus* (Desmarest). 2. Gestation, parturition and pouch life. *Australian Journal of Zoology*, **23**(3), 333–53.

Renfree, M., Fletcher, T. P. and Lewis, D. R. (1989). Physiological and behavioral events around the time of birth in macropodid marsupials. In *Kangaroos, Wallabies and Rat-Kangaroos*, ed. P. Jarman, I. Hume and G. Grigg. Sydney, Australia: Surrey Beatty, pp. 323–7.

Reynolds, H. C. (1952). Studies on reproduction in the opossum (*Didelphis virginiana*). *University of California Publications in Zoological Sciences*, **52**, 223–84.

Rodriguez-Esteban, C., Tsukui, T., Yonei, S., *et al.* (1999). The T-box genes Tbx4 and Tbx5 regulate limb outgrowth and identity. *Nature*, **398**, 814–18.

Sánchez-Villagra, M. R. (2002). Comparative patterns of postcranial ontogeny in therian mammals: an analysis of relative timing of ossification events. *Journal of Experimental Zoology*, **294**, 264–73.

Sánchez-Villagra, M. R. and Maier, W. (2003). Ontogenesis of the scapula in marsupial mammals, with special emphasis on perinatal stages of *Didelphis* and remarks on the origin of the therian scapula. *Journal of Morphology*, **258**, 115–29.

Saunders, J. W. J. (1948). The proximo-distal sequence of origin of the parts of the chick wing and the role of the ectoderm. *Journal of Experimental Zoology*, **108**, 363–403.

Schluter, D. (1996). Adaptive radiation along genetic lines of least resistance. *Evolution*, **50**(5), 1766–74.

Schmidt, M. and Fischer, M. S. (2009). Morphological integration in mammalian limb proportions: dissociation between function and development. *Evolution*, **63**, 749–66.

Sears, K. E. (2004). Constraints on the morphological evolution of marsupial shoulder girdles. *Evolution*, **58**(10), 2353–70.

Sears, K. E. (2009). Differences in the timing of prechondrogenic limb development in mammals: the marsupial-placental dichotomy resolved. *Evolution*, **63**(8), 2193–200.

Sears, K. E. (2011). Novel insights into the regulation of limb development from 'natural' mammalian mutants. *BioEssays*, **33**(5), 327–31.

Sears, K. E., Behringer, R. R., Rasweiler, J. J. and Niswander, L. A. (2006). Development of bat flight: morphologic and molecular evolution of bat wing digits. *Proceedings of the National Academy of Sciences of the United States of America*, **103**(17), 6581–6.

Sharman, G. B. (1973). Adaptations of marsupial pouch young for extra-uterine existence. In *The Mammalian Fetus in Vitro*, ed. C. R. Austin. London: Chapman and Hall, pp. 67–90.

Sharman, G. B., Calaby, J. H. and Poole, W. E. (1965). Patterns of reproduction in female diprotodont marsupials. *Journal of Reproduction and Fertility*, **9**(3), 375–6.

Shaw, G. and Renfree, M. B. (2006). Parturition and perfect prematurity: birth in marsupials. *Australian Journal of Zoology*, **54**, 139–49.

Shendure, J. (2008). The beginning of the end for microarrays? *Nature Methods*, **5**(7), 585–87.

Shou, S., Scott, V., Reed, C., Hitzemann, R. and Stadler, H. S. (2005). Transcriptome analysis of the murine forelimb and hindlimb autopod. *Developmental Dynamics*, **234**, 74–89.

Stopper, G. F. and Wagner, G. P. (2005). Of chicken wings and frog legs: a smorgasbord of evolutionary variation in mechanisms of tetrapod limb development. *Developmental Biology*, **288**, 21–39.

Summerbell, D. A. (1974). A quantitative analysis of the effect of excision of the AER from the chick limb-bud. *Journal of Embryology and Experimental Morphology*, **32**, 651–60.

Sun, X., Mariani, F. V. and Martin, G. R. (2002). Functions of FGF signalling from the apical ectodermal ridge in limb development. *Nature*, **418**, 501–8.

Takeuchi, J., Koshiba-Takeuchi, K., Suzuki, T., *et al.* (2003). Tbx5 and Tbx4 trigger limb initiation through activation of the Wnt/Fgf signaling cascade. *Development*, **130**(12), 2729–39.

Thewissen, J. G. M., Cohn, M. J., Stevens, M. E., *et al.* (2006). Developmental basis for hind-limb loss in dolphins and origin of the cetacean bodyplan. *Proceedings of the National Academy of Sciences of the United States of America*, **103**(22), 8414–18.

Towers, M. and Tickle, C. (2009a). Generation of pattern and form in the developing limb. *International Journal of Developmental Biology*, **53**, 805–12.

Towers, M. and Tickle, C. (2009b). Growing models of vertebrate limb development. *Development*, **136**, 179–90.

Tyndale-Biscoe, C. H. (1973). *Life of Marsupials*. London: Edward Arnold, 254 pp.

Tyndale-Biscoe, C. H. and Renfree, M. (1987). *Reproductive Physiology of Marsupials*. Cambridge, UK: Cambridge University Press.

Veitch, C., Nelson, C. E. and Gemmell, R. T. (2000). Birth in the brushtail possum, *Trichosurus vulpecula* (Marsupialia: Phalangeridae). *Australian Journal of Zoology*, **48**, 691–700.

Wagner, G. P. (1996). Homologues, natural kinds and the evolution of modularity. *American Zoologist*, **36**(1), 36–43.

Wanek, N., Muneoka, K., Holler-Dinsmore, G., Burton, R. and Bryant, S. V. (1989). A staging system for mouse limb development. *Journal of Experimental Zoology*, **249**, 41–9.

Weisbecker, V., Goswami, A., Wroe, S. and Sánchez-Villagra, M. R. (2008). Ossification heterochrony in the therian postcranial skeleton and the marsupial-placental dichotomy. *Evolution*, **62**(8), 2027–41.

Wilmore, K. E., Leamy, L. J. and Hallgrímsson, B. (2006). Effects of develpmental and functional interactions on mouse cranial variability through late ontogeny. *Evolution and Development*, **8**(6), 550–67.

Yampolsky, L. Y. and Stoltzfus, A. (2001). Bias in the introduction of variation as an orienting factor in evolution. *Evolution and Development*, **3**(2), 73–83.

Young, N. M. and Hallgrímsson, B. (2005). Serial homology and the evolution of mammalian limb covariance structure. *Evolution*, **59**(12), 2691–704.

Zeller, R., Lopez-Rios, J. and Zuniga, A. (2009). Vertebrate limb bud development: moving towards integrative analysis of organogenesis. *Nature Reviews Genetics*, **10**, 845–58.

11

A developmental basis for innovative evolution of the turtle shell

SHIGERU KURATANI AND HIROSHI NAGASHIMA

Introduction

Turtles are characterized by the possession of shells. For acquisition of this structure, this animal group appears to have undergone various types of anatomical changes in their body plan, not only in their skeletal system, but also in the muscular, nervous and respiratory systems (Bojanus 1819; Thomson 1932). These features often lead to confusion in determining homology, especially of amniote ribs, as well as in establishing the phylogenetic position of this animal group (Goodrich 1930; Remane 1936). To understand the origins of the morphology of turtles, a number of embryologists and morphologists have studied their embryonic developmental patterns (e.g. Rathke 1848; Agassiz 1857; Mitsukuri and Ishikawa 1887; Mitsukuri 1894, 1896; Ogushi 1911, 1913; Ruckes 1929; Walker 1947; Burke 1989, 1991; Gilbert *et al.* 2001, 2008; Nagashima *et al.* 2005, 2007; 2009; Sánchez-Villagra *et al.* 2009; Werneburg *et al.* 2009; reviewed by Gilbert *et al.* 2008; Kuratani *et al.* 2011).

The phylogenetic position of turtles remains controversial, but considerable progress has been made. Although recent molecular phylogenetics and genomic analyses have placed this taxon close to or even within the archosaurians, including birds and crocodiles (Caspers *et al.* 1996; Zardoya and Meyer 1998, 2001; Hedges and Poling 1999; Kumazawa and Nishida 1999; Mannen and Li 1999; Mindell *et al.* 1999; Cao *et al.* 2000; Iwabe *et al.* 2005; Matsuda *et al.* 2005; Kuraku *et al.* 2006; Hugall *et al.* 2007; Chapus and Edwards 2009), morphological analyses do not always agree with this conclusion (reviewed by Kuratani *et al.* 2011). However, some early embryologists supported an affinity to archosaurians (Haeckel 1891; de Beer 1937; see also Figure 1.1 in Asher and Müller, this volume). Typically, the turtles tend to be placed at a basal position among amniotes, as a representative of the Parareptilia, the most

From Clone to Bone: The Synergy of Morphological and Molecular Tools in Palaeobiology, ed. Robert J. Asher and Johannes Müller. Published by Cambridge University Press.

basal lineage of amniotes (see Tsuji and Müller 2009; also see Romer 1966; Gaffney 1980; Carroll 1988; Reisz and Laurin 1991; Laurin and Reisz 1995; Lee 1993, 1996, 1997, 2001; Reisz 1997). Of the post-cranial skeletal features, possessing dermal armour and an acromion process on the scapula are counted as two major synapomorphies uniting the turtles and pareiasaurs (Lee 1993; see Scheyer and Sander 2009), both of which would be linked to each other, not only functionally, but also developmentally at least in the turtles (see below).

The present chapter is intended to focus on the evolutionary acquisition of the turtle shell from an evolutionary developmental perspective. The aim is to understand how a novel structure can be obtained by a shifted pattern of ontogenetic developmental process. In the acquisition of the turtle shell, especially the carapace, we will show how this process involved repatterning of not only the primordium of the shell per se (ribs), but also some adjacent skeletal elements and certain groups of muscles connecting them.

Even in turtles, some of these structures show conserved relative positions seen in the amniote body plan. This conservation has partly originated from developmental inevitability, such as the fixed topographical relationships of primordial tissues and tissue interactions required for normal development. Some of these factors are regarded as a manifestation of developmental constraints, which have yielded morphological homologies of some organs (Nagashima *et al.* 2009). However, evolutionary novelties – especially those originating by a heterotopic shift of developmental patterns – tend to disturb the topographical relationships, leading to the loss of morphological homologies. Such topographically shifted tissue interactions often lead to ectopic gene regulation bringing about a chain reaction in further tissue interactions, which is the nature of organogenesis (Müller and Wagner 1991; Shigetani *et al.* 2002). Although this strategy is apt to assume the saltatory establishment of a new pattern by a few abrupt changes in developmental programmes, it is also possible that a novelty can be achieved through a stepwise manner that could be traced in the fossil record.

Below, we will show how small changes could have been introduced into different stages of a hypothetical ancestral lineage of turtles and how these sequential changes could – possibly – result in the topographically conspicuous morphology of the turtles. We will also emphasize how important it is to understand the developmental nature of evolutionary novelty in evaluating morphological characters. The key innovative change that enabled turtle shell development is not yet fully understood. However, it is clear that the developmental innovations for carapacial formation take place surprisingly early in development.

Novelty of the turtle: the shell

The skeletal shell of the turtle is composed of dorsal and ventral moieties: the carapace and plastron, respectively. The whole shell is covered by keratinized epidermal structures, the scales (Figure 11.1). The carapace is based developmentally upon the ribs, which have grown laterally, and upon the vertebral column. These ribs are associated with vertebral segments developmentally specified as 'thoracic' in terms of *Hox* gene family expression patterns, which determine the vertebral formula (Burke *et al.* 1995; Ohya *et al.* 2005; reviewed by Burke 2000). The skeletal element found in the carapace consists of neural plates on the midline, bilateral costal plates growing from the ribs, a peripheral series of bones along the margin and a few accessorial elements called the nuchal, pygial and suprapygial plates (Figure 11.1).

It is widely accepted that the peripheral elements and some other structures such as the pygal and nuchal plates are dermal in nature (Figure 11.1; reviewed by Rieppel and Reiz 1999; also see Gilbert *et al.* 2001). However, a question remains as to the developmental and evolutionary nature of the costal plates, because the initial anlagen of the costals arise as endochondral rib primordia,

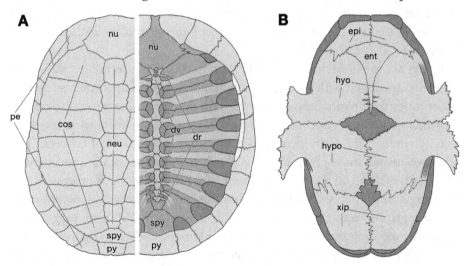

Figure 11.1 The turtle shell (redrawn from Bojanus 1819). (A) Dorsal (left) and ventral (right) views of the carapacial skeleton of *Emys orbicularis*. The carapace is composed of fused ribs and vertebrae. The ventral view is based on a juvenile specimen, and unossified fontanelles (dark grey) remain between distal portions of the ribs. (B) Dorsal view of the plastron. Abbreviations: cos, costal plate; dr, dorsal ribs; dv, dorsal vertebrae; ent, entoplastron; epi, epiplastron; hyo, hyoplastron; hypo, hypoplastron; neu, neural plate; nu, nuchal plate; pe, peripheral plate; py, pygal plate; spy, suprapygal plate; xip, xiphiplastron.

which later grow appositional bony tissues in the dermis with a histological configuration similar to other dermal skeletal features (Suzuki 1963; reviewed by Gilbert *et al.* 2008). The membranously ossified plate-like ridges, or the primordia of 'costals', grow from the ribs both anteriorly and posteriorly until each plate meets with its neighbours to form a suture in between (Figure 11.1; see Scheyer *et al.* 2007).

With the atrophy of myotome-derived muscular tissue, the completed bony plate of the carapace comes to lie just beneath the epidermis with a slight amount of dermis: it appears as if the whole carapace represents a 'dermal' structure. Thus, the costal plates first arise as part of the endoskeleton as cartilaginous primordia, like the ribcage of other amniotes, and are completed with a histological configuration associated with typical dermal skeletons. In terms of comparative embryology, the major part of the turtle carapace is to be regarded as an endochondral, as well as an endoskeletal structure, covered by expanded scales as keratinized epidermal structures. No independent dermal ossification centres appear for the costals during this developmental process. The 'dermal bony tissues' associated with the costals arise through continuous growth of skeletal tissue that initially develops as endoskeletal elements: the ribs. In this connection, a recently discovered fossil species, *Odontochelys*, had independent neural plates, which, however, are thought to be displaced during fossilization (Li *et al.* 2008).

In the above connection, it should be noted that the term 'dermal' is not primarily associated with so-called 'membranous bone'. The former refers to a structure evolutionarily derived from certain types of mineralized exoskeletal elements in ancestral vertebrates and not to any modes of histogenetic ossification. Therefore, the term 'dermal' is primarily a morphological concept. The adjective 'membranous', by contrast, is clearly a type of ossification as a process of histogenesis and all the cartilage-based bones go through this process during the late phase of their development. Thus, the term 'membranous' never carries any connotations associated with evolution or anatomy (Patterson 1977; Starck 1979; also see Rieppel and Reiz 1999). In addition, endochondral ossification can be eliminated from the entire developmental process for certain endoskeletal elements (Bellairs and Gans 1983). The morphology of the turtle carapace is peculiar because it arises (both evolutionarily and developmentally) as a modified endoskeleton, but morphologically and functionally it appears as an exoskeletal structure (Li *et al.* 2008; Rieppel 2009; also see Joyce *et al.* 2009; Lyson *et al.* 2010).

Gilbert *et al.* (2008) has put forth a heterotopy hypothesis to explain this developmental and evolutionary shift in ossification ('heterotopy' refers to a shift in position in development during evolution, just as 'heterochrony' refers to a shift in the developmental timetable; reviewed by Hall 1998). As in other amniote embryos, the turtle ribs also appear to arise from somites and first grow medial to the myotomes (Emelianov 1936; Yntema 1970; and unpublished observations by H. Nagashima and S. Kuratani). Later in development, they grow more laterally than those in other amniotes, adhering to the lateral aspect of the muscle plate, which is the direct derivative of the embryonic myotome, if not entirely penetrating it. This is unlike the process seen in other amniotes where the rib primordia are embedded in the muscle plate (Figure 11.2). According to Gilbert *et al.* (2008), this heterotopic shift leads to a new interaction between the mesenchyme in the dermis and the rib cartilage. Although there has not been an evaluation of how similar the environments of the cranial dermal bone and trunk dermis are, it is true that various vertebrates are capable of dermal ossification in the trunk – as seen in crocodiles and some fossil mammals. It has also been suggested that the mesenchyme surrounding this newly growing rib cartilage is responding to bone morpho-genetic protein (BMP) signals apparently emanating from the cartilage, leading to ectopic ossification in the superficial mesenchyme (Cebra-Thomas *et al.* 2005; reviewed by Gilbert *et al.* 2008). Thus, the membranous bony tissue growing from the turtle ribs could be called 'dermal' in an anatomical and morphological sense, as it is obtained through a heterotopic shift of rib primordia to dermal positions. However, because its primordium arises as an endochondral skeleton, we do not have a proper term to describe the nature of this unique skeletal element. If heterotopy is the essence of the novelty associ-ated with the turtle carapace, one would not have to assume that ancient reptiles were the ancestors of the turtles – such as pareiasaurs, whose body was covered by 'osteoderms' (Lee 1993, 1996, 1997). Anatomically, the latter dermal armour type was closer to that found in *Glyptodon*, a fossil mammalian species. Rather, the membranously ossified part of the costals in the turtle carapace is truly an autoapomorphy of this group (reviewed by Rieppel and Reiz 1999; and by Rieppel 2009) and no equivalent skeletal elements would be found in other vertebrates if ectopic membranous ossification could reflect an ancestrally obtained developmental potential for dermal bone formation.

If the above scenario is correct, it will raise another question: from which cell lineage does the costal plate originate (reviewed by Kuratani *et al.* 2011)? In addition, it will require cell lineage labelling studies to determine whether the costal tissue truly originates from the cells belonging to the dermatome and not by self-differentiation of the perichondrium/periosteum of the ribs themselves. If the latter is the case, the costal plates would represent endochondral skeletal elements whose anatomical position was secondarily shifted to a dermal layer. This might justify the novel nature of the carapace, as has been assumed by

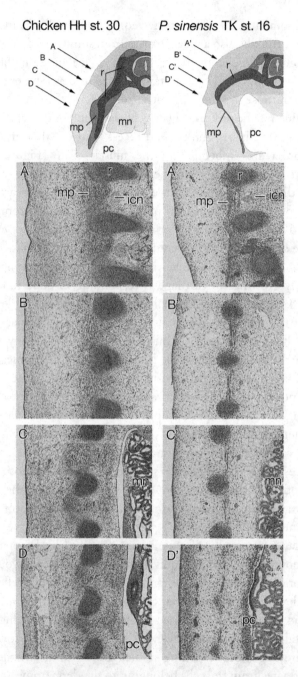

Figure 11.2 Embryonic comparison of rib primordia between chick and *Pelodiscus sinensis* embryos. (Top) Transverse sections of an HH stage 30 (Hamburger and Hamilton 1951) chicken embryo and a TK stage 16 (Tokita and Kuratani 2001) Chinese soft-shelled turtle (*P. sinensis*) embryo are shown diagrammatically to indicate the topographical relationships between rib (r)

some morphologists (reviewed by Rieppel and Reiz 1999; Rieppel 2009). To show that the costal plate is dermatome in origin would require long-term cell labelling studies, preferably with genetic markers – except that these are not yet available in the turtle system. Although green fluorescent protein (GFP) labelling has been applied to embryos of the Chinese soft-shelled turtle (*Pelodiscus sinensis*) (Nagashima *et al.* 2007), the success rate is much lower than for studying avian systems (Momose *et al.* 1999; Sato *et al.* 2002, 2007) and it cannot be employed as a stable labelling method for developmental studies.

The above difficulty also applies to the study of the cell lineage origin of the costal plates (even when the tissue is ectopically induced), because there are two or more potential candidates for the origin of the membranously ossified part of the costals. First, it might come from the periosteum of the translocated ribs themselves, as noted above. In this case, the costals are derived very probably from the sclerotome of the somites, which are not formed genuinely of dermis, but represent peculiar endoskeletal elements that have shifted secondarily to the surface. Second, they might represent dermatome-derived tissues; or third, they might even be derived from the trunk neural crest cells (Clark *et al.* 2001; Cebra-Thomas *et al.* 2007; Gilbert *et al.* 2007; reviewed by Gilbert *et al.* 2008).

Because most of the dermal cranial roof is made of crest-derived ectomesenchyme in avians (Couly *et al.* 1993), there has been an idiosyncratic view that *all* the exoskeletal elements of vertebrates should be derived from the neural crest, which has not been proven (see Hall 1998). Although the cephalic crest usually shows overt skeletogenic capability (Le Douarin 1982; Noden 1988), the crest-derived cells are potentially osteogenic if they are cephalic crest- or trunk crest-derived (reviewed by Le Douarin 1982; Nakamura and Ayer-le Lievre 1982; Noden 1988; McGonnell and Graham 2002; Abzhanov *et al.* 2003; Ido and Ito 2006; also see Kuratani 2005). Some of the membranous bones arise from the mesoderm (Noden 1986; Jiang *et al.* 2002; Matsuoka *et al.* 2005) and division between the mesoderm- and crest-derived cranial roof might differ in each animal lineage (Le Lièvre 1974, 1978; Noden 1982, 1984; Couly *et al.* 1993; Le Douarin and Kalcheim 1999; Chai *et al.* 2000; Jiang *et al.* 2002; Matsuoka *et al.* 2005). Furthermore, the morphological nature of the skeleton may not

Caption for figure 11.2 (*cont.*) and muscle plate (mp) as well as the levels of the sections (A–D, A′–D′) shown below. (Below) Histological sections of the chicken (A–D) and *P. sinensis* (A′–D′) embryos. Note that the chicken rib primordia grow into the body wall (at the level of the peritoneal cavity, pc) embedded in the muscle plate, whereas those in *P. sinensis* shift laterally from the muscle plate and never grow into the body wall. Other abbreviations: icn, intercostal nerve; mn, mesonephros. (See also colour plate.)

even be associated strictly with any cell lineages as suggested by Schneider (1999); for example, the mesodermal tissue destined to form the neurocranium can generate visceral cranial elements when inserted laterally in the place normally occupied by crest cells. Thus, the distinction between the exo- and endoskeleton also reflects differences in the signalling mechanisms involved in different tissue interactions for skeletogenesis in different places in embryos. For further discussion on the possibility of crest involvement in the turtle shell, see Kuratani *et al.* (2011).

The scapula problem and the folding theory for the origin of the turtle shell

An even more conspicuous trait specific to the turtle is in the topographical relationships between some skeletal elements. Unlike in other amniotes, in which the scapula is found outside the ribcage, the turtle scapula is situated on the inside (Figure 11.3). Thus, the topographical relationship between these skeletal elements is reversed in the turtle relative to other amniotes. Because the morphological homology of organs and structures usually depends on their relative topographical positions, the situation observed in the turtle skeleton raises the question of the morphological identities of these elements. This question is pertinent not only to skeletal elements, but also to

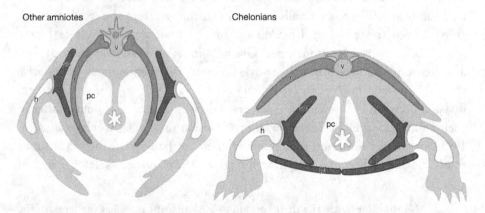

Figure 11.3 Comparison of the positions of ribs and scapulae between turtles and other amniotes. Schematic drawings of avian (left) and turtle (right) skeletons showing topographical relationships between the ribs (r) and scapulae (sc). In the typical body plan of amniotes, the scapula is found outside the ribcage, whereas in the turtle the scapula is located inside it. Other abbreviations: h, humerus; pc, pleural cavity; pl, plastron; v, vertebrae.

the muscular elements that apparently show evolutionarily non-conserved positions relative to those skeletal elements.

There are two different categories among the forelimb-associated muscles connecting the main body and shoulder. One comprises the trunk muscle derivatives such as the serratus anterior (SA) and the rhomboid/levator scapulae (RLS) muscle complex, both of which connect the blade part of the scapula with the ribcage or the vertebral column. During development, these muscles appear to arise directly from the myotomes and their connections are already established with appropriate skeletal elements at early stages of development. Curiously, it appears that the somite-derived part of the scapula (or the blade; Huang *et al.* 2000) is attached selectively by myotome-derived muscles (Valasek *et al.* 2010), whereas the neck muscles derived from migrating muscle precursor (MMP) cells (Dietrich 1999; Dietrich *et al.* 1998, 1999; see Matsuoka *et al.* 2005; reviewed by Kuratani 2008) would attach to the lateral plate-derived part of the scapula. Although the latter idea has not been shown experimentally, it is highly plausible as far as the anatomical pattern implies. The other group is a subset of limb muscles such as the pectoralis and latissimus dorsi muscles (Romer and Parsons 1977), connecting the humerus and the trunk. As shown by Patel and his colleagues (Valasek *et al.* 2005; reviewed by Evans *et al.* 2006), these muscles first arise as *Pax3/Lbx-1*-positive MMP cells that migrate once into the limb bud and then grow secondarily out of the bud to establish their connections onto the trunk. Thus, for the late development, they are located more superficially to the trunk muscle components (the SA and RLS muscles). During turtle evolution, both of the above-noted groups of muscles appear to have undergone shifts of morphological patterning, in association with the establishment of the shell.

If we compare the trunk and limb bud muscles in turtle anatomy, it is clear that the trunk muscles retain morphologically homologous (ancestral) connections to similar skeletal elements seen in turtle development, whereas the limb muscles have changed their attachments (non-homologous connections; Nagashima *et al.* 2009). Although the mechanical background for this difference remains unknown, it is true that the trunk muscles establish connections to skeletal elements quite early in development and show no further detachment or shifting, except for possible minor secondary expansion. Thus, as the ribs grow laterally as well as rostrally over the scapular anlage, the serratus anterior follows this developmental movement and rotates dorsoventrally to finish beneath the carapace, when the scapula becomes covered by the growing ribs (Nagashima *et al.* 2009). This muscular rotation is created by the inward folding of the lateral body wall characteristic of the turtles (Figure 11.4). Thus, although the apparent shape differs conspicuously, compared with the

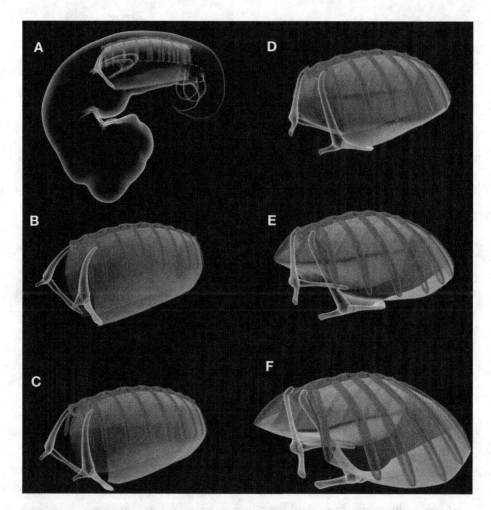

Figure 11.4 Turtle shell formation by folding of the body wall. Three-dimensional diagrams to show the developmental sequence of carapacial formation. The vertebral column and ribs are coloured red, scapula and coracoid are yellow and myotome-derived muscle plates are green. Serratus anterior (coloured blue in E and F) is derived from the muscle plate. Note that the earliest stage (A, B) resembles the late pharyngula embryo of the chicken, with the scapula anlage located laterally and rostrally to the ribcage. At stage C, the ribs start to expand laterally, not invading the body wall. At stage D, the boundary between the axial part and the body wall is clearly seen as the laterally protruding 'folding line', which corresponds to the turtle-specific embryonic structure, the carapacial ridge (CR, broken line in A). At stages E and F, the covering of the scapula is completed with a conspicuous medial folding of the lateral body wall. (See also colour plate.)

morphology of turtle limb bud muscles, the SA muscle retains ancestral connectivity through this folding.

Along the line of this folding is found a longitudinal ridge called the carapacial ridge (CR). This consists of a condensed undifferentiated mesenchyme covered by the thickened epithelium. Because the distal tip of the growing rib of the turtle embryo points toward this structure, it has been assumed by Burke (1989, 1991; reviewed by Burke 2009) that this structure might function as an inductive centre to determine the turtle-specific growth pattern of the ribs. This hypothesis also assumed that the histological features of the CR resemble those of the apical ectodermal ridge of the limb bud functioning in the growth and developmental patterning of the vertebrate limb (reviewed by Tabin and Wolpert 2007). At any rate, it is true that the CR is specific to the turtle embryo, and no other vertebrate embryos possess similar structures at the equivalent position of the body. Thus, the CR can be called an evolutionary novelty of the turtle that arises in association with the folding of the embryonic body to form the carapace.

Using labelling experiments, it has been shown that the CR mesenchyme is exclusively derived from the somites as an axial structure and does not arise from the lateral body wall containing the lateral plate-derived mesenchyme. In non-turtle amniotes, a longitudinal ridge also arises in the flank of the pharyngula stage embryo. However, the latter ridge – called the Wolffian ridge – is purely of lateral wall origin and does not have anything to do with the CR (Burke 1989; Nagashima et al. 2007; reviewed by Kuratani et al. 2011).

The turtle-specific development of the CR appears to be based on turtle-specific gene regulation patterns. Gene expression analyses have so far identified a number of regulatory genes expressed in the CR (Loredo et al. 2001; Vincent et al. 2003; Kuraku et al. 2005; Moustakas 2008; reviewed by Kuratani et al. 2011). In particular, the expression patterns of Lef-1, APCDD1 and Sp5 suggest the involvement of a canonical Wnt signalling pathway and the nuclear localization of β-catenin is consistently associated with CR development (Kuraku et al. 2005). However, no Wnt genes have been identified as being expressed specifically in the turtle embryo in association with the appearance of the CR.

The question arises: what is the function of the CR in carapacial patterning during development? Introduction of the dominant-negative form of Lef-1 results in partial reduction of the carapacial anlage, implying that Lef-1 would be functional as an upstream factor in the gene regulatory cascade that maintains carapacial development (Nagashima et al. 2007). Similarly, surgical removal of the CR results in partial loss of flabellate expansion of the turtle-specific rib growth pattern, implying that the CR is responsible for the marginal growth of the carapacial anlage, which is also consistent with the active incorporation of BrdU labelling into the CR mesenchyme. Importantly, the axial arrest of the rib, or the superficial/dorsal position of the rib primordia, did not change after the CR removal (Burke 1991; Nagashima *et al.* 2007). Thus, the axial arrest of the turtle ribs per se might not be caused by the CR.

Scenario of turtle shell evolution – recapitulation?

For understanding turtle shell evolution, the most curious finding is the ancestral fossil of the turtle recently found from China. This Triassic animal, named *Odontochelys*, did not possess a complete carapace, but did exhibit a well-developed plastron (Li *et al.* 2008). Such an anatomical pattern itself is reminiscent of the late-embryo stage of modern turtles (Li *et al.* 2008). *Odontochelys* thus appears to have possessed laterally growing ribs as in modern turtles, but their distal ends did not expand to form a flabellate pattern: a characteristic of carapacial ribs in modern turtles (see Nagashima *et al.* 2009). Such a peculiar morphology of *Odontochelys* can be explained reasonably from the above developmental considerations. First, the lateral growth of the ribs in *Odontochelys* is suggestive of the axial arrest of the ribs that inhibited the rib primordia to penetrate into the lateral body wall. Second, the lack of a flabellate pattern is highly suggestive of the absence of the late function of the CR that is specifically responsible for marginal growth of the carapacial primordium in the turtle embryo (Nagashima *et al.* 2007). In other words, *Odontochelys* appears to have developed up to the halfway point in the developmental process of modern turtles.

It is plausible that CR was also present in *Odontochelys* embryos because in modern turtles the CR and carapacial arrest of the ribs appear to be coupled developmentally. However, the adult anatomy of *Odontochelys* is highly suggestive of the lack of any late function for the CR. As a result, the second thoracic rib of this animal pointed caudally, just like the first rib in modern turtles. Moreover, the scapula of this animal was not encapsulated by the ribs, but was located rostrally to the carapacial precursor (Figure 11.5). Curiously, this pattern is very reminiscent of the *Pelodiscus sinensis* embryo at the stage when the second rib has not yet covered the scapula (Figure 11.5). If the SA muscle of *Odontochelys* retained evolutionarily conserved connections, as inferred from the muscle's developmental nature, it should have connected to the distal tips of one or two rostral ribs of the thoracic region and to the blade of the scapula. The reconstructed anatomy of *Odontochelys* strikingly resembles the late embryo of *P. sinensis* (Figure 11.5). Thus, the modern turtle appears to develop into the adult by way of a stage that has the pattern of its putative ancestor.

Can a phenomenon such as this be called recapitulation? Does an ontogenetic sequence of developmental events simply recapitulate that of evolutionary

Modern turtles + *Proganochelys*   ***Odontochelys***

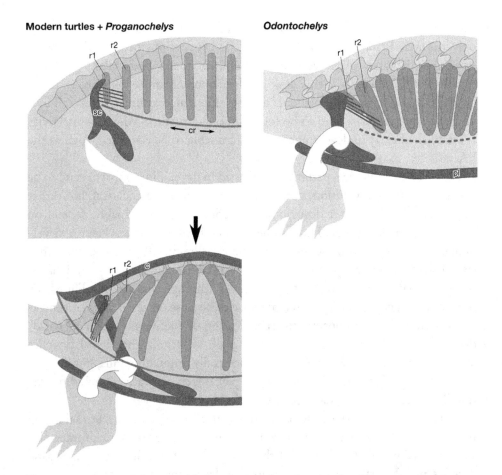

Figure 11.5 *Odontochelys* resembled embryos of modern turtles. Diagrams showing developmental sequences at the trunk level in *Pelodiscus sinensis* (left) and the reconstructed morphology of an adult *Odontochelys* (right). Both in the turtle embryo (left top) and in *Odontochelys*, the scapula (sc) is located rostral to the ribs, and serratus anterior (red) spans rostrocaudally between the scapula and rostral ribs (r1 and r2). The CR (carapacial ridge) might also have developed in *Odontochelys*, but without a late function of the CR to expand the margin of the carapace (c). Note that *Odontochelys* resembled the embryo of modern turtles more than it resembled the adults. Because of the enhanced function of the CR, the ribs in the adult modern turtle become fanned during late development to cover the scapula dorsally (left bottom). Serratus anterior is folded secondarily under the carapace. Abbreviations: c, carapace; pl, plastron; r, ribs. (See also colour plate.)

events? If we cannot postulate simple recapitulation as a general rule of development, in the case of the turtle embryo it appears likely that the turtle's ancestor used adjustable parts of the developmental programme to attain the turtle-specific body plan. For example, encapsulation of the scapula was only possible once ribs did not grow into the lateral body wall. Moreover, only the MMP muscles could change their connections because their proximal growth takes place very late when the turtle body plan is almost established. If there is a rule linking development and evolution, it could be that changes can be introduced only upon an existing pattern. Thus, unchangeable patterns tend to be found among structures whose connectivity is established early in development and are unable to be cancelled, such as the SA muscle.

Certain morphological patterns established in developing embryos tend to be conserved through evolution, not simply because they are ancestral, but rather because some of the later developmental programmes inescapably depend on early-established patterns. Thus, some early embryonic patterns carry a 'developmental burden' as a prerequisite for the developmental processes thereafter (Riedl 1978), and these patterns have often established a structural and causal network that also tends to be conserved through evolution. Thus, the notochord appears in every vertebrate species because, as the source of ventralizing signals, it is responsible for specification of the neural tube and mesoderm and has thereby been favoured through selection. Pharyngeal pouches also appear in amniote embryos, not as ancestrally originating gills, but because they are all required for various pharyngeal pouch derivatives including the middle ear cavity, thymus and parathyroid gland.

During turtle evolution, the basic embryonic anatomy of this animal group has been conserved as a member of the amniotes. The trunk myotomes also form normally at first, but most of the tissues do not remain as the intercostal muscles in the turtle. Possibly, the normal developmental programme might have been favoured evolutionarily to maintain the initial differentiation of ribs, even in the absence of their direct derivatives in the adult. In fact, in other amniotes the myotome is reported to function as the source of inductive, trophic and/or patterning interactions required for proper formation of the ribs (Braun *et al.* 1994; Grass *et al.* 1996; Olson *et al.* 1996; Yoon *et al.* 1997; Brent *et al.* 2003, 2005; Vinagre *et al.* 2010). It was apparently only possible to introduce the change in the later phase of the developmental programme of the turtle ancestor, and this change can be recognized as the axial arrest of rib primordia (and associated acquisition of the CR). The latter apparently facilitated the inward folding of the lateral body wall to encapsulate the scapula. However, this did not alter the basic architecture of the vertebrate body. Namely, the traditional skeletal muscles retain their ancestral connections. In the sense of Geoffroy's *principe des connexions* (Geoffroy Saint-Hilaire, 1818), both the ribs and scapula are morphologically homologous with their counterparts in other amniotes, even when their topographical relationships are reversed. These structures retain ancestral connections including the muscles, and their basic configuration is set up in an early phase of development.

The evolutionary completion of the carapace can be understood as the enhancement of the late function of the CR, which is likely to have been added in the developmental programme in the animal lineage leading towards the modern turtles, after the divergence from *Odontochelys*. Therefore, acquisition of the carapace does not seem to have been a saltatory event, but instead consisted of a sequential series of changes in developmental programmes. Of these changes, the mechanical cause of the axial arrest of the ribs, which is likely to have been acquired by *Odontochelys*, remains unknown. Because formation of the CR does not appear to be upstream of this event, the ancestor before the *Odontochelys* would possibly have been an animal with extremely short ribs and a plastron (or its precursor). Further evolutionary developmental studies among amniotes with the concept of 'developmental burden' – especially for the early rib primordia – are thus expected to succeed in reconstructing the entire evolution of the turtle.

Summary

The dorsal moiety of the turtle shell, or the carapace, is made of modified ribs and the vertebral column. The turtle shell is regarded as an example of an evolutionary novelty because of the apparent topographical inconsistency of the position of the ribs with respect to the scapula, which is found – unusually – inside the ribcage. Its evolution was assumed to have been accomplished by developmental repatterning that resulted in an abrupt change in inside–outside relationships between the ribcage and the scapula. In contrast, our folding theory assumes that the turtle-specific body plan was achieved through a sequential series of developmental changes that proceeded in a stepwise manner. This involved the axial arrest of the rib primordia, peripheral growth of the carapacial anlage along the embryonic ridge called the carapacial ridge (CR), and inward folding of the ventral body wall leading to the reversed positions of the scapula and ribs. A recently discovered fossil animal, *Odontochelys*, assumed to have been in an ancestral lineage of the modern turtle, exhibited an anatomical pattern resembling the embryo of modern turtles before the folding. This animal possessed the ventral part of the shell, or plastron, but not the carapace. Such an anatomy is consistent with our proposed scenario that enhancement of the late CR function has led to the acquisition of a round carapace, resulting in covering of the scapula as an evolutionary novelty of the turtle lineage.

REFERENCES

Abzhanov, A., Tzahor, E., Lassar, A. B. and Tabin, C. J. (2003). Dissimilar regulation of cell differentiation in mesencephalic (cranial) and sacral (trunk) neural crest cells in vitro. *Development*, **130**, 4567–79.

Agassiz, L. (1857). Part III. Embryology of the turtle. *Contributions to the Natural History of the United States of America*, **2**, 451–643.

Bellairs, A. D'A. and Gans, C. (1983). A reinterpretation of the amphisbaenian orbitosphenoid. *Nature*, **302**, 243–4.

Bojanus, L. H. (1819). *Anatome Testudinis Europaeae*. Vilnae, Lithuania: Impensis Auctoris. Typis Josephi Zwadzki, Typographi Universitatis.

Braun, T., Bober, E., Rudnicki, M. A., Jaenisch, R. and Arnold, H. H. (1994). MyoD expression marks the onset of skeletal myogenesis in *Myf-5* mutant mice. *Development*, **120**, 3083–92.

Brent, A. E., Schweitzer, R. and Tabin C. J. (2003). A somitic compartment of tendon progenitors. *Cell*, **18**, 235–48.

Brent, A. E., Braun, T. and Tabin, C. J. (2005). Genetic analysis of interactions between the somitic muscle, cartilage and tendon cell lineages during mouse development. *Development*, **132**, 515–28.

Burke, A. C. (1989). Development of the turtle carapace: implications for the evolution of a novel bauplan. *Journal of Morphology*, **199**, 363–78.

Burke, A. C. (1991). The development and evolution of the turtle body plan: inferring intrinsic aspects of the evolutionary process from experimental embryology. *American Zoologist*, **31**, 616–27.

Burke, A. C. (2000). *Hox* genes and the global patterning of the somitic mesoderm. *Current Topics in Developmental Biology*, **47**, 155–81.

Burke, A. C. (2009). Turtles ... again. *Evolution and Development*, **11**, 622–4.

Burke, A. C., Nelson, C. E., Morgan, B. A., Tabin, C. (1995). *Hox* genes and the evolution of vertebrate axial morphology. *Development*, **121**, 333–46.

Cao, Y., Sorenson, M. D., Kumazawa, Y., Mindell, D. P. and Hasegawa, M. (2000). Phylogenetic position of turtles among amniotes: evidence from mitochondrial and nuclear genes. *Gene*, **259**, 139–48.

Carroll, R. L. (1988). *Vertebrate Paleontology and Evolution*. New York: Freeman.

Caspers, G.-J., Reinders, G.-J., Leunissen, J. A. M., Wattel, J. and de Jong, W. W. (1996). Protein sequences indicate that turtles branched off from the amniote tree after mammals. *Journal of Molecular Evolution*, **42**, 580–6.

Cebra-Thomas, J., Tan, F., Sistla, S., *et al.* (2005). How the turtle forms its shell: a paracrine hypothesis of carapace formation. *Journal of Experimental Zoology*, **304B**, 558–69.

Cebra-Thomas, J. A., Betters, E., Yin, M., Plafkin, C., McDow, K. and Gilbert, S. F. (2007). Evidence that a late-emerging population of trunk neural crest cells forms the plastron bones in the turtle *Trachemys scripta*. *Evolution and Development*, **9**, 267–77.

Chai, Y., Jiang, X., Ito, Y., *et al.* (2000). Fate of the mammalian cranial neural crest during tooth and mandibular morphogenesis. *Development*, **127**, 1671–9.

Chapus, C. and Edwards S. V. (2009). Genome evolution in Reptilia: *in silico* chicken mapping of 12,000 BAC-end sequences from two reptiles and a basal bird. *BMC Genomics*, **10** (Suppl. 2), S8.

Clark, K., Bender, G., Murray, B. P., *et al.* (2001). Evidence for the neural crest origin of turtle plastron bones. *Genesis*, **31**, 111–17.

Couly, G. F., Coltey, P. M. and Le Douarin, N. M. (1993). The triple origin of skull in higher vertebrates: a study in quail-chick chimeras. *Development*, **117**, 409–29.

de Beer, G. R. (1937). *The Development of the Vertebrate Skull*. London: Oxford University Press.

Dietrich, S. (1999). Regulation of hypaxial muscle development. *Cell and Tissue Research*, **296**, 175–82.

Dietrich, S., Schubert, F. R., Healy, C., Sharpe, P. T. and Lumsden, A. (1998). Specification of the hypaxial musculature. *Development*, **125**, 2235–49.

Dietrich, S., Abou-Rebyeh, F., Brohmann, H., *et al.* (1999). The role of SF/HGF and c-Met in the development of skeletal muscle. *Development*, **126**, 1621–9.

Emelianov, S. W. (1936). Die Morphologie der Tetrapodenrippen. *Zoologische Jahrbucher Abteilung für Anatomie und Ontogenie der Tiere*, **62**, 173–274.

Evans, D. J., Valasek, P., Schmidt, C. and Patel, K. (2006). Skeletal muscle translocation in vertebrates. *Anatomy and Embryology*, **211**, S43–S50.

Gaffney, E. S. (1980). Phylogenetic relationships of the major groups of amniotes. In *The Terrestrial Environment and the Origin of Land Vertebrates*, ed. A. L. Panchen. London: Academic Press, pp. 593–610.

Geoffroy Saint-Hilaire, E. (1818). *Philosophie Anatomique, tome premiere*. Paris: J.B. Baillière.

Gilbert, S. F., Loredo, G. A., Brukman, A. and Burke, A. C. (2001). Morphogenesis of the turtle shell: the development of a novel structure in tetrapod evolution. *Evolution and Development*, **3**, 47–58.

Gilbert, S. F., Bender, G., Betters, E., Yin, M. and Cebra-Thomas, J. A. (2007). The contribution of neural crest cells to the nuchal bone and plastron of the turtle shell. *Integrative and Comparative Biology*, **47**, 401–8.

Gilbert, S. F., Cebra-Thomas, J. A. and Burke, A. C. (2008). How the turtle gets its shell. In *Biology of Turtles*, ed. J. Wyneken, M. H. Godfrey and V. Bels. Boca Raton, FL: CRC Press, pp. 1–16.

Goodrich, E. S. (1930). *Studies on the Structure and Development of Vertebrates*. London: Macmillan.

Grass, S., Arnold, H. H. and Braun, T. (1996). Alterations in somite patterning of *Myf-5*-deficient mice: a possible role for FGF-4 and FGF-6. *Development*, **122**, 141–50.

Haeckel, E. (1891). *Anthropogenie oder Entwickelungsgeschichte des Menschen. Keimes-und Stammesgeschichte*, 4th edn. Leipzig, Germany: Wilhelm Engelmann.

Hall, B. K. (1998). *Evolutionary Developmental Biology*, 2nd edn. London: Chapman & Hall.

Hamburger, V. and Hamilton, H. L. (1951). A series of normal stages in the development of the chick embryo. *Journal of Morphology*, **88**, 49–92.

Hedges, S. B. and Poling, L. L. (1999). A molecular phylogeny of reptiles. *Science*, **283**, 998–1001.

Huang, R., Zhi, Q., Patel, K., Wilting, J. and Christ, B. (2000). Dual origin and segmental organisation of the avian scapula. *Development*, **127**, 3789–94.

Hugall, A. F., Foster, R. and Lee, M. S. Y. (2007). Calibration choice, rate smoothing, and the pattern of tetrapod diversification according to the long nuclear gene RAG-1. *Systematic Biology*, **56**, 543–63.

Ido, A. and Ito, K. (2006). Expression of chondrogenic potential of mouse trunk neural crest cells by FGF2 treatment. *Developmental Dynamics*, **235**, 361–7.

Iwabe, N., Hara, Y., Kumazawa, Y., *et al.* (2005). Sister group relationship of turtles to the bird-crocodilian clade revealed by nuclear DNA-coded proteins. *Molecular Biology and Evolution*, **22**, 810–13.

Jiang, X., Iseki, S., Maxson, R. E., Sucov, H. M. and Morriss-Kay, G. M. (2002). Tissue origins and interactions in the mammalian skull vault. *Developmental Biology*, **241**, 106–16.

Joyce, W. G., Lucas, S. G., Scheyer, T. M., Heckert, A. B. and Hunt, A. P. (2009). A thin-shelled reptile from the Late Triassic of North America and the origin of the turtle shell. *Proceedings of the Royal Society B*, **276**, 507–13.

Kumazawa, Y. and Nishida, M. (1999). Complete mitochondrial DNA sequences of the green turtle and blue-tailed mole skink: statistical evidence for archosaurian affinity of turtles. *Molecular Biology and Evolution*, **16**, 784–92.

Kuraku, S., Usuda, R. and Kuratani, S. (2005). Comprehensive survey of carapacial ridge-specific genes in turtle implies co-option of some regulatory genes in carapace evolution. *Evolution and Development*, **7**, 3–17.

Kuraku, S., Ishijima, J., Nishida-Umehara, C., *et al.* (2006). cDNA-based gene mapping and GC3 profiling in the soft-shelled turtle suggest a chromosomal size-dependent GC bias shared by sauropsid. *Chromosome Research*, **14**, 187–202.

Kuratani, S. (2005). Craniofacial development and evolution in vertebrates: the old problems on a new background. *Zoological Science*, **22**, 1–19.

Kuratani, S. (2008). Evolutionary developmental studies of cyclostomes and origin of the vertebrate neck. *Development, Growth and Differentiation*, **50** (Suppl. 1), S189–94.

Kuratani, S., Kuraku, S. and Nagashima, H. (2011). Evolutionary developmental perspective for the origin of the turtles: the folding theory for the shell based on the developmental nature of the carapacial ridge. *Evolution and Development*, **13**, 1–14.

Laurin, M. and Reisz, R. R. (1995). A reevaluation of early amniote phylogeny. *Zoological Journal of the Linnean Society*, **113**, 165–223.

Le Douarin, N. M. (1982). *The Neural Crest*. Cambridge, UK: Cambridge University Press.

Le Douarin, N. M. and Kalcheim, C. (1999). *The Neural Crest*, 2nd edn. Cambridge, UK: Cambridge University Press.

Lee, M. S. Y. (1993). The origin of the turtle body plan: bridging a famous morphological gap. *Science*, **261**, 1716–20.

Lee, M. S. (1996). Correlated progression and the origin of turtles. *Nature*, **379**, 812–15.

Lee, M. S. Y. (1997). Pareiasaur phylogeny and the origin of turtles. *Zoological Journal of the Linnean Society*, **120**, 197–280.

Lee, M. S. Y. (2001). Molecules, morphology, and the monophyly of diapsid reptiles. *Contributions to Zoology*, **70**, 1–22.

Le Lièvre, C. S. (1974). Rôle des cellules mesectodermiques issues des crêtes neurales céphaliques dans la formation des arcs branchiaux et du skelette viscéral. *Journal of Embryology and Experimental Morphology*, **31**, 453–77.

Le Lièvre, C. S. (1978). Participation of neural crest-derived cells in the genesis of the skull in birds. *Journal of Embryology and Experimental Morphology*, **47**, 17–37.

Li, C., Wu, X., Rieppel, O., Wang, L. and Zhao, L. (2008). An ancestral turtle from the Late Triassic of southwestern China. *Nature*, **45**, 497–501.

Loredo, G. A., Brukman, A., Harris, M. P., *et al.* (2001). Development of an evolutionarily novel structure: fibroblast growth factor expression in the carapacial ridge of turtle embryos. *Journal of Experimental Zoology*, **291B**, 274–81.

Lyson, T. R., Bever, G. S., Bhullar, B. A., Joyce, W. G. and Gauthier, J. A. (2010). Transitional fossils and the origin of turtles. *Biology Letters*, Published online [10.1098/rsbl.2010.0371].

Mannen, H. and Li, S. S. (1999). Molecular evidence for a clade of turtles. *Molecular Phylogenetics and Evolution*, **13**, 144–8.

Matsuda, Y., Nishida-Umehara, C., Tarui, H., *et al.* (2005). Highly conserved linkage homology between birds and turtles: bird and turtle chromosomes are precise counterparts of each other. *Chromosome Research*, **13**, 601–15.

Matsuoka, T., Ahlberg, P. E., Kessaris, N., *et al.* (2005). Neural crest origins of the neck and shoulder. *Nature*, **436**, 347–55.

McGonnell, I. M. and Graham, A. (2002). Trunk neural crest has skeletogenic potential. *Current Biology*, **12**, 767–71.

Mindell, D. P., Sorenson, M. D., Dimcheff, D. E., *et al.* (1999). Interordinal relationships of birds and other reptiles based on whole mitochondrial genomes. *Systematic Biology*, **48**, 138–52.

Mitsukuri, K. (1894). On the process of gastrulation in Chelonia (Contributions to the embryology of reptilia, IV). *The Journal of the College of Science, Imperial University, Japan*, **6**, 227–75.

Mitsukuri, K. (1896). On the fate of the blastopore, the relations of the primitive streak, and the formation of the posterior end of the embryo in Chelonia, together with remarks on the nature of meroblastic ova in vertebrates (Contributions to the embryology of reptilia, V). *The Journal of the College of Science, Imperial University, Japan*, **10**, 1–118.

Mitsukuri, K. and Ishikawa, C. (1887). On the formation of the germinal layers in Chelonia. *The Journal of the College of Science, Imperial University, Japan*, **1**, 211–46.

Momose, T., Tonegawa, A., Takeuchi, J., *et al.* (1999). Efficient targeting of gene expression in chick embryos by microelectroporation. *Development, Growth and Differentiation*, **41**, 335–44.

Moustakas, J. E. (2008). Development of the carapacial ridge: implications for the evolution of genetic networks in turtle shell development. *Evolution and Development*, **10**, 29–36.

Müller, G. B. and Wagner, G. P. (1991). Novelty in evolution: restructuring the concept. *Annual Review of Ecology, Evolution, and Systematics*, **22**, 229–56.

Nagashima, H., Uchida, K., Yamamoto, K., *et al.* (2005). Turtle–chicken chimera: an experimental approach to understanding evolutionary innovation in the turtle. *Developmental Dynamics*, **232**, 149–61.

Nagashima, H., Kuraku, S., Uchida, K., *et al.* (2007). On the carapacial ridge in turtle embryos: its developmental origin, function, and the chelonian body plan. *Development*, **134**, 2219–26.

Nagashima, H., Sugahara, F., Takechi, M., *et al.* (2009). Evolution of the turtle body plan by the folding and creation of new muscle connections. *Science*, **325**, 193–6.

Nakamura, H. and Ayer-le Lievre, C. S. (1982). Mesectodermal capabilities of the trunk neural crest of birds. *Journal of Embryology and Experimental Morphology*, **70**, 1–18.

Noden, D. M. (1982). Patterns and organization of cranial skeletogenic and myogenic mesenchyme: a perspective. In *Progress in Clinical and Biological Research: Factors and Mechanism Influencing Bone Growth*, ed. B. G. Sarnat. New York: Alan R. Liss, pp. 167–203.

Noden, D. M. (1984). The use of chimeras in analyses of craniofacial development. In *Chimeras in Developmental Biology*, ed. N. M. Le Douarin and A. McLaren. London: Academic Press, pp. 241–80.

Noden, D. M. (1986). Patterning of avian craniofacial muscles. *Developmental Biology*, **116**, 347–56.

Noden, D. M. (1988). Interactions and fates of avian craniofacial mesenchyme. *Development*, **103** (Suppl.), S121–40.

Ogushi, K. (1911). Anatomische Studien an der japanischen dreikralligen Lippenschildkröte (*Trionyx japonicus*). *Morphologisches Jahrbuch*, **43**, 1–106.

Ogushi, K. (1913). Anatomische Studien an der japanischen dreikralligen Lippenschildkröte (*Trionyx japonicus*). *Morphologisches Jahrbuch*, **46**, 299–562.

Ohya, Y. K., Kuraku, S. and Kuratani, S. (2005). *Hox* code in embryos of Chinese soft-shelled turtle *Pelodiscus sinensis* correlates with the evolutionary innovation in the turtle. *Journal of Experimental Zoology*, **304B**, 107–18.

Olson, E. N., Arnold, H.-H., Rigby, P. W. J. and Wold, B. J. (1996). Know your neighbors: three phenotypes in null mutants of the myogenic bHLH gene *MRF4*. *Cell*, **85**, 1–4.

Patterson, C. (1977). Cartilage bones, dermal bones and membrane bones, or the exoskeleton versus the endoskeleton. In *Problems in Vertebrate Evolution*, ed. S. M. Andrews, R. S. Miles and A. D. Walker. New York: Academic Press, pp. 77–121.

Rathke, H. (1848). *Ueber die Entwickelung der Schildkröten*. Braunschweig, Germany: Friedrich Vieweg.

Reisz, R. R. (1997). The origin and early evolutionary history of amniotes. *Trends in Ecology and Evolution*, **12**, 218–22.

Reisz, R. R. and Laurin, M. (1991). *Owenetta* and the origin of turtles. *Nature*, **349**, 324–6.

Remane, A. (1936). Wirbelsäule und ihre Abkömmlinge. In *Handbuch der vergleichenden Anatomie der Wirbeltiere*, Bd. 4, ed. L. Bolk, E. Göppert, E. Kallius and W. Lubosch. Berlin: Urban & Schwarzenberg, pp. 1–206.

Riedl, R. (1978). *Order in Living Organisms*. New York: Wiley Press.

Rieppel, O. (2009). How did the turtle get its shell? *Science*, **325**, 154–5.

Rieppel, O. and Reisz, R. R. (1999). The origin and early evolution of turtles. *Annual Review of Ecology, Evolution, and Systematics*, **30**, 1–22.

Romer, A. S. (1966). *Vertebrate Paleontology*, 3rd edn. Chicago, IL: University of Chicago Press.

Romer, A. S. and Persons, T. S. (1977). *The Vertebrate Body*. Philadelphia, PA: Saunders.

Ruckes, H. (1929). Studies in chelonian osteology. II. The morphological relationships between the girdles, ribs and carapace. *Annals of the New York Academy of Sciences*, **31**, 81–120.

Sánchez-Villagra, M. R., Müller, H., Sheil, C. A., *et al.* (2009). Skeletal development in the Chinese soft-shelled turtle *Pelodiscus sinensis* (Testudines: Trionychidae). *Journal of Morphology*, **270**, 1381–99.

Sato, Y., Yasuda, K. and Takahashi, Y. (2002). Morphological boundary forms by a novel inductive event mediated by Lunatic fringe and Notch during somitic segmentation. *Development*, **129**, 3633–44.

Sato, Y., Kasai, T., Nakagawa, S., *et al.* (2007). Stable integration and conditional expression of electroporated transgenes in chicken embryos. *Developmental Biology*, **305**, 616–24.

Scheyer, T. M. and Sander, P. M. (2009). Bone microstructures and mode of skeletogenesis in osteoderms of three pareiasaur taxa from the Permian of South Africa. *Journal of Evolutionary Biology*, **22**, 1153–62.

Scheyer, T. M., Sander, P. M., Joyce, W. G., Böhme, W. and Witzel, U. (2007). A plywood structure in the shell of fossil and living soft-shelled turtles (Trionychidae) and its evolutionary implications. *Organisms, Diversity and Evolution*, **7**, 136–44.

Schneider, R. A. (1999). Neural crest can form cartilages normally derived from mesoderm during development of the avian head skeleton. *Developmental Biology*, **208**, 441–55.

Shigetani, Y., Sugahara, F., Kawakami, Y., *et al.* (2002). Heterotopic shift of epithelial-mesenchymal interactions in vertebrate jaw evolution. *Science*, **296**, 1316–19.

Starck, D. (1979). *Vergleichende Anatomie der Wirbeltiere auf evolutionsbiologischer Grundlage*, Vol. 2, Berlin: Springer Verlag.

Suzuki, H. K. (1963). Studies on the osseus system of the slider turtle. *Annals of the New York Academy of Sciences*, **109**, 351–410.

Tabin, C. and Wolpert, L. (2007). Rethinking the proximodistal axis of the vertebrate limb in the molecular era. *Genes and Development*, **21**, 1433–42.

Thomson, J. S. (1932). The anatomy of the tortoise. *Scientific Proceedings of the Royal Dublin Society*, **20**, 359–462.

Tokita, M. and Kuratani, S. (2001). Normal embryonic stages of the Chinese softshelled turtle *Pelodiscus sinensis* (Trionychidae). *Zoological Sciences*, **18**, 705–15.

Tsuji, L. A. and Müller, J. (2009). Assembling the history of the Parareptilia: phylogeny, diversification, and a new definition of the clade. *Fossil Record*, **12**, 71–81.

Valasek, P., Evans, D. J., Maina, F., Grim, M. and Patel, K. (2005). A dual fate of the hindlimb muscle mass: cloacal/perineal musculature develops from leg muscle cells. *Development*, **132**, 447–58.

Valasek, P., Theis, S., Krejci, E., *et al.* (2010). Somitic origin of the medial border of the mammalian scapula and its homology to the avian scapula blade. *Journal of Anatomy*, **216**, 482–8.

Vinagre, T., Moncaut, N., Carapuço, M., *et al.* (2010). Evidence for a myotomal Hox/Myf cascade governing nonautonomous control of rib specification within global vertebral domains. *Developmental Cell*, **20**, 655–61.

Vincent, C., Bontoux, M., Le Douarin, N. M., Pieau, C. and Monsoro-Burq, A. H. (2003). *Msx* genes are expressed in the carapacial ridge of turtle shell: a study of the European pond turtle, *Emys orbicularis*. *Development Genes and Evolution*, **213**, 464–9.

Walker, W. F., Jr. (1947). The development of the shoulder region of the turtle, *Chrysemys picta marginata*, with special reference to the primary musculature. *Journal of Morphology*, **80**, 195–249.

Werneburg, I., Hugi, J., Müller, J. and Sánchez-Villagra, M. R. (2009). Embryogenesis and ossification of *Emydura subglobosa* (Testudines, Pleurodira, Chelidae) and patterns of turtle development. *Developmental Dynamics*, **238**, 2770–86.

Yntema, C. L. (1970). Extirpation experiments on the embryonic rudiments of the carapace of *Chelydra serpentina*. *Journal of Morphology*, **132**, 235–44.

Yoon, J. K., Olson, E. N., Arnold, H. H. and Wold, B. J. (1997). Different *MRF4* knockout alleles differentially disrupt Myf-5 expression: *cis*-regulatory interactions at the *MRF4/Myf-5* locus. *Developmental Biology*, **188**, 349–62.

Zardoya, R. and Meyer, A. (1998). Complete mitochondrial genome suggests diapsid affinities of turtles. *Proceedings of the National Academy of Sciences of the United States of America*, **95**, 14 226–31.

Zardoya, R. and Meyer, A. (2001). The evolutionary position of turtles revised. *Naturwissenschaften*, **88**, 193–200.

12

A molecular–morphological study of a peculiar limb morphology: the development and evolution of the mole's 'thumb'

CHRISTIAN MITGUTSCH, MICHAEL K. RICHARDSON, MERIJN A. G. DE BAKKER, RAFAEL JIMÉNEZ, JOSÉ EZEQUIEL MARTÍN, PETER KONDRASHOV AND MARCELO R. SÁNCHEZ-VILLAGRA

Introduction

Few areas of evolutionary investigation have benefited so much from the integration of molecular and morphological studies as that of the evolution of limbs (e.g. Shubin *et al.* 2009). Much effort has been concentrated on understanding the transition from 'fins to limbs' (Westoll 1943; Coates 1995; Shubin 1995; Shubin *et al.* 1997, 2004, 2006; Coates *et al.* 2002; Johanson *et al.* 2007; Larsson 2007) – for a recent summary of different aspects see Hall (2007) – but also major advances have been made concerning the origin and development of particularly specialized limbs such as those of bats (Sears *et al.* 2006; Sears 2008), whales (Thewissen *et al.* 2006), or their absence in snakes (Cohn and Tickle 1999). In fact, tetrapod limbs exhibit a stunning morphological diversity, reflecting their recruitment for a wide variety of functions ranging from walking to watch-making, often deployed as multipurpose structures serving in locomotion and support, personal hygiene, fighting, mating, manipulation and communication.

The reconstructed ground pattern of extant tetrapods may be viewed as a limb with a proximal stylopod (humeral/femoral region), a zeugopod (radial and ulnar/tibial and fibular region) and a distal, pentadactylous autopod (carpal/tarsal region), which contains the proximal mesopodium (wrist and ankle bones) and the terminal acropodium (the digits). With the notable exception of

From Clone to Bone: The Synergy of Morphological and Molecular Tools in Palaeobiology, ed. Robert J. Asher and Johannes Müller. Published by Cambridge University Press.

limbless tetrapods, most tetrapod limbs conservatively retain this pattern (Goodwin and Trainor 1983; Holder 1983). Based on these components, the diversity of skeletal morphology in tetrapod limbs is generated largely by differential growth and proportional changes of the common elements. Regarding the acropodia, hypodactyly, hyperphalangy and modifications of branching patterns involve various developmental mechanisms (Motani 1999; Hamrick 2001b, 2002; Richardson and Oelschläger 2002; Fedak and Hall 2004; Sears *et al.* 2007; Rolian 2008; Weisbecker and Nilsson 2008; Cooper and Dawson 2009). Phenotypic differences in limb proportions reflecting locomotory specializations have been considered to be the result of multiple cellular processes contributing to endochondral bone growth involving timing and rate of cellular events and separate developmental processes acting in proximal and distal limb bones (Rolian 2008). For example, the hyperphalangy of whales and dolphins was suggested to result from changes in developmental timing, heterochronies (Richardson and Oelschläger 2002; Fedak and Hall 2004). Seemingly additional digital rays, as observed in ichthyosaurs, have been shown to be the result of developmental branching in the phalangeal region (Motani 1999). Also, the rare condition of syndactyly seen in marsupials has been hypothesized to be due to a change in early developmental pattern based on the observation of ossification heterochronies between syndactylous and non-syndactylous species (Weisbecker and Nilsson 2008). Otherwise, digit reduction has been connected to developmental arrest and degeneration or initial failure to form their anlagen (Galis *et al.* 2001, 2002; Hamrick 2002).

Particular examples for morphological diversity are provided by the tetrapod mesopodium which has been evolutionarily diversified into various shapes to serve a variety of functions, a diversity that might even be expressed through sexual dimorphisms as shown for some marsupials (Lunde and Schutt 1999) and much effort has been put into understanding tarsal and carpal elements using a variety of approaches in various tetrapod taxa (Holmgren 1952; Lewis 1964; Burke and Alberch 1985; Rieppel 1993a; Buscalioni *et al.* 1997; Stafford and Thorington 1998; Fabrezi and Barg 2001; Thorington and Stafford 2001; Prochel and Sánchez-Villagra 2003; Prochel *et al.* 2004; Leal *et al.* 2010). Differences in the pattern of diversification seen in the mesopodium compared with the acropodium may be explained by both the evolutionary history of these structures (e.g. Wagner and Chiu 2001; Boisvert *et al.* 2008) as well as their distinct developmental programmes (e.g. Woltering and Duboule 2010). A developmental separation had previously been suggested to be based on specifications of domains in early limb buds (Richardson *et al.* 2004), supported by observations in hyperphalangous turtles (Delfino *et al.* 2010). Furthermore, limb specialization and diversification involves additions and extensive modifications of integumentary structures, such as hooves and nails, and yields considerable diversity even among closely related taxa (e.g. Hamrick 2001a, 2003).

Many limb specializations have generated questions about the identity and homology of digits and carpal and tarsal elements. There are several historically debated and controversial topics in craniate research relating to both evolutionary transitions between tetrapod taxa and, more generally, to the evolution of the tetrapod limb (i.e. Wagner and Gauthier 1999; Galis *et al.* 2003; Larsson and Wagner 2003). Besides comparative anatomical work, these questions were most frequently approached by comparative developmental studies within a broad variety of vertebrate taxa (Holmgren 1933, 1952; Shubin and Alberch 1986; Fabrezi 1993; Shubin *et al.* 1995; Fabrezi and Alberch 1996; Blanco *et al.* 1998; Fabrezi and Barg 2001; Prochel *et al.* 2004; Welten *et al.* 2005; Sánchez-Villagra *et al.* 2007; Kundrát 2009; Young *et al.* 2009; Larsson *et al.* 2010; Leal *et al.* 2010).

With the rise of developmental biology, embryological research focused increasingly on exploring universally valid principles and mechanisms of development using model organisms (Mitgutsch 2003). From the perspective of systematics, it has been argued that model organisms should be selected to ensure some kind of comparability across species and allow generalizations about evolutionary systems, or in other words should be representative or typical for a taxon (Kellogg and Shaffer 1993; Bolker 1995). A number of suggestions on how to choose 'proper' model organisms have been made. However, to 'learn about the general by studying the particular' (Kellogg and Shaffer 1993) might not have been the primary objective upon the decision pro or contra a certain species. Thus, comparative developmental biology is currently fashionable, but the question of using and choosing model organisms and the concepts involving comparative developmental research remain debated (Hanken 1993; Bolker 1995; Roush 1996; Metscher and Ahlberg 1999; Mitgutsch 2003; Jenner and Wills 2007; Milinkovitch and Tzika 2007; Duboule 2010). Nevertheless, the large, even intraspecific, diversity in tetrapod autopodial structures reaching from the human hand to flippers (Gegenbaur 1898; Starck 1979) must reflect modifications of developmental programmes.

The renewed interest in comparative embryology (Gilbert 2003; Werneburg 2009) has meant that new non-model species are being established as subjects of study, opening new avenues of research into the developmental genetic bases of the great diversity of morphologies even in fossil taxa (Fröbisch 2008; Delfino and Sánchez-Villagra 2010). Comparative developmental approaches allow not only a better understanding of spatial relationships of skeletal elements prior to possible fusions but also for more robust definition of the temporal sequence of skeletogenic events such as mesenchyme condensation, chondrification or mineralization using a variety of visualization techniques ranging from classic histological staining to the application of molecular markers. These approaches provide important information about the homologies of individual skeletal elements. Temporal information on chondrification has been particularly related to the 'primary axis' (Burke and Alberch 1985; Shubin and Alberch 1986), a certain pattern of chondrification in developing limbs (Richardson *et al.* 2009). Cell condensation patterns, such as those preceding chondrification, were argued to be 'fundamental developmental and selectable units of morphology' (Hall 2003, p. 219); prechondrogenic patterns have been considered a highly informative use in further exploring the issue of digit and autopodial development identity by comparative embryological approaches (Hinchliffe and Griffiths 1983; Oster *et al.* 1983, 1988; Hinchliffe 1989, 2002; Welten *et al.* 2005).

The use of developmental data to establish homologies or to infer phylogenetic relationships has been debated since Haeckel's time and has been renewed with the era of developmental genetics and the discovery of taxonomically widespread developmental control genes. The application of genetics to homology entails comparing gene expression domains to establish the identity of structures, an approach that is problematic given the hierarchical nature of homology statements with respect to structure and processes (for discussion see, e.g. Abouheif 1997; Janies and de Salle 1999; Camardi 2001; Svensson 2004; Scholtz 2005; Müller 2007, among many others). As an example, the precise definition of neural crest exemplifies the difficulties of establishing objective criteria of homologizing structures when different criteria and levels of organization are involved. Molecular signatures alone, even of gene regulatory networks, are not infallible criteria, and 'phenotypic qualifiers prove critical to identify the presence or absence of characters' (Donoghue *et al.* 2008). Consequently, such problems should be approached from various perspectives, acknowledging structural hierarchies, developmental processes and their combination into evolutionary novelties.

For example, a long-standing question relating to evolutionary transformation in reptiles (Romer 1956) that has been revitalized by newly available molecular data is the origin of avian digit identities from theropod ancestors (Wagner and Gauthier 1999; Peterson *et al.* 2007; Delfino and Sánchez-Villagra 2010). On the basis of observations of cartilage condensations in birds, Burke and Feduccia (1997) proposed the digits of the avian hand to be II, III, and IV, not I, II, III as implied by the theropod origin of birds. Larsson and Wagner (2002), observing prechondrogenic condensations using molecular markers, found anlagen of a transient pentadactyl avian hand leading to a wing with the remaining digits II, III, and IV (but see also Welten *et al.* 2005 on the development of the vestigial first digit), a finding generally in agreement with Burke and Feduccia (1997). In support of a variant on the I, II, III hypothesis, Vargas and Fallon (2005a, 2005b) compared expression domains of conserved genes in mice, chicken and polydactylous chicken mutants. They found *Hoxd13* to be expressed in digit I in both mouse and chicken, *Hoxd12* and *Hoxd13* in all other fingers, and concluded that digit I had the same identity in both species. As articulated by Wagner and Gauthier (1999), these findings are consistent with a 'frame shift' in homology, in which digit identity and positions have become uncoupled (Tamura *et al.* 2011). Recent palaeontological data have been interpreted to be consistent with this idea (Xu *et al.* 2009). For these advances in knowledge, a variety of approaches integrating experimental and comparative embryology, palaeontology and morphology have been paramount (Young and Wagner 2011).

A further long-standing question, easily as intriguing as the problems connected to finger reduction, is the rare occurrence of supernumerary digits, or polydactyly, among extant, terrestrial vertebrates. Whereas true polydactyly, both pre- and post-axial, has been demonstrated for several early tetrapods such as *Tulerpeton*, *Acanthostega* or *Ichthyostega* (Coates and Clack 1990; Lebedev and Coates 1995) and might have evolved convergently within amniotes (Wu *et al.* 2003), modern tetrapods varied the number of digits by reduction only, not by addition. Among extant tetrapods, polydactyly is known to be a common, even heritable, malformation (e.g. Lettice *et al.* 2003), and certain breeds of domesticated mammals have a tendency to develop additional digits (Alberch 1985). Otherwise, polydactyly is scarce and only occasionally reported for some cetacean species (Ortega-Ortiz *et al.* 2000). This has often been linked to developmental constraints related to characteristics such as size of embryonic anlagen (e.g. Alberch 1985) or strong pleiotropic genetic interactions related to polydactyly (Galis *et al.* 2001) discussed in connection to the zootype or phylotypic stage concept (Slack *et al.* 1993). Instead of true polydactyly, adding to the considerable diversity of autopodial morphologies within the Tetrapoda, several taxa prominently feature additional skeletal elements, accessory bones, sesamoids and mineralized tendons, in some cases with notable, even individual, variation (Braus 1906; Sarin *et al.* 1999; Fabrezi 2001; Vickaryous and Olson 2007).

Descriptions of radial and tibial autopodial elements can be found throughout the literature, although some of these descriptions, concerning particularly transient, non-ossifying elements, have been questioned (Larsson *et al.* 2010). The terms prepollex and prehallux have been used for such pre-axial elements with inconsistent implications. Whereas some authors used them for equivalents of true digits or their rudiments, others used them for anterior skeletal autopodial elements in a purely positional sense without implications of homology to each other or to true digits. Bardeleben (1885, 1890) for example, found that additional skeletal elements adjacent to fingers or toes ('marginal fingers or toes') were so widely distributed in phylogeny that he questioned the proposed pentadactyl state of mammalian autopodia. As an extreme example, he described a segmented prepollex with 'nail' (but see below) in the springhare, *Pedetes capensis*. Bardeleben (1885) described an embryonic division of the tarsal navicular and an additional toe in lower mammals, and assumed this element, proximal to the hallux, to be a prehallux, in monotremes, edentates, carnivores, rodents, insectivorans and some primates. This rudiment was described as articulating with either tarsal I, the navicular bone, both, or between the tarsal and metatarsal I. Furthermore, Bardeleben (1885) pointed out a complete or incomplete division of the tarsal navicular in several taxa, particularly in rodents. For the hand, likewise, he reported a prepollex (as carpal or metacarpal o) for edentates, rodents, carnivores, 'insectivorans', chiropterans, 'prosimids' and some other primates. As an 'object to demonstrate the simultaneous presence of prepollex and prehallux' in a concluding remark Bardeleben (1885) recommended the mole.

Emery (1890) generally agreed with the view that the carpal navicular (scaphoid) bone is the carpal of a prepollex (principally with reference to anurans) but commented on Bardeleben's description of the prepollex in *Pedetes* referring to Meckel's description of this element as an ossification of the palmar aponeurosis ('Fascia palmaris') and on the distal part as a 'volar bone'. This raises the question whether these elements are simply fascial ossifications or rudiments of an autopodial 'ray'. Answering this question based on ontogenetic studies Emery (1890) found the proximal cartilaginous element in *Pedetes* and rat to develop independently of any tendons, and in a position comparable to the anuran prepollex, and concluded it to be a finger rudiment. Nevertheless, the same author did not agree that the free element of the manus is a true digit but a differentiation of a radial 'Tastballen'. Emery (1890) furthermore reported a cartilaginous, rod-like prepollex in *Sciurus*. He thought this element to be homologous to the radial sesamoid bone of dogs and the os falciforme of the mole. He claimed that all of these developed independently of any tendons. He concluded that these elements might have retained the function of sesamoids but should be considered rudiments of a 'hand ray', the 'free' element not being a true digit, but a secondary new organ deriving from a prepollex-rudiment.

Schmidt-Ehrenberg (1942) likewise described, based on Holmgren (1933), the development of a prepollex and prehallux in *Mus*. Holmgren (1933) noted that the mouse prepollex disappeared at later stages by fusing to other elements. For stem-tetrapods, additional 'digit-like' skeletal elements are – particularly in the light of the fossil record – still somewhat controversial. For amphibians, Westoll (1943) noted the prepollex and prehallux to be truly comparable to digits but not well developed. Galis *et al.* (2001) discussed the condensation of the anuran prehallux and prepollex to be finger-like and these elements possibly to be equal to digits. In contrast, Tokita and Iwai (2010) reported the prepollex ('pseudothumb') in frogs as not being a finger based on its substantially delayed ossification (but see Rieppel 1993b; Hugi *et al.* 2010). For amniote taxa, recent studies generally agree that additional finger-like elements are not true digits but modified wrist bones and – based on their developmental mode – that modified sesamoids cannot be considered true digits either (Tabin 1992; Shubin *et al.* 1997; Galis *et al.* 2001; Mitgutsch *et al.* 2012). Concerning the prehallux, Lewis (1964) pointed out that this tibial sesamoid is frequently confused with the tibiale of rodents and monotremes and that it is enlarged, mimicking an additional pre-axial digit in several taxa. Thus while Reed (1951), for example, criticized Bardeleben's use of the term prepollex in a purely positional sense (although this does not appear to have been Bardeleben's intention), this use is widespread throughout the literature (but see Galis *et al.* 2001).

The mole's thumb

Particularly notable elements of such kind are tibial and radial sesamoids of the talpids of the northern continents (Sánchez-Villagra and Menke 2005; Prochel 2006; Vickaryous and Olson 2007). The massive radial sesamoid of these species, also known as os *radiale externum*, prepollex, 'os falciforme' or 'mole's thumb' (Sánchez-Villagra and Menke 2005) is in all cases accompanied by a distinctive, although smaller, tibial sesamoid (Figure 12.1A; Mitgutsch *et al.* 2012). The members of *Talpa* belong to the monophyletic Talpidae consisting of moles, shrew moles and desmans (Sánchez-Villagra *et al.* 2006); together with shrews, hedgehogs and *Solenodon* they form the clade Eulipotyphla (Lipotyphla *sensu* Asher *et al.* 2009). Talpine moles possess a fossorial mode of locomotion, accompanied by a number of anatomical and physiological characteristics (summarized in great detail by Kley and Kearney 2007; Figure 12.1C). Both the prehallical and the prepollical accessory elements are thought to play important functional roles in burrowing by supporting and widening the autopodia (e.g. Yalden 1966) and thus support their bearer's fossoriality or serve swimming as in desmans.

Whereas little account has been taken concerning the talpid tibial accessory bone (but see Kindahl 1942), several studies discuss forelimb anatomy as well as

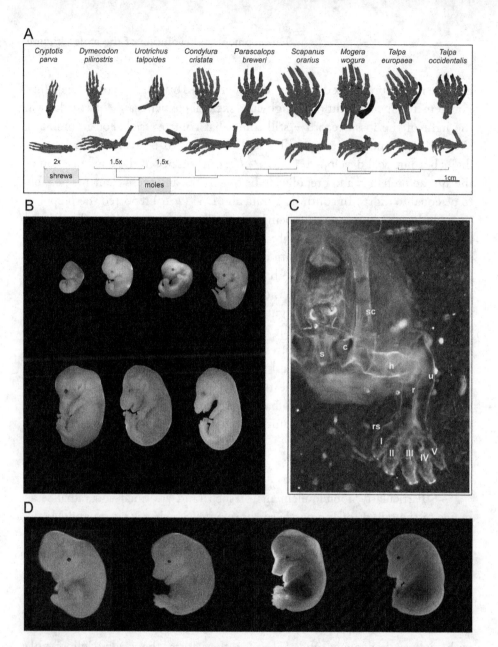

Figure 12.1 (A) Micro CT (computerized tomography) scan-images of autopodia of several talpid species and of *Cryptotis parva*, demonstrating the distribution and proportions of ossified prehallical/prepollical elements (highlighted); phylogenetic relationships based on Sánchez-Villagra *et al.* (2006). *Cryptotis parva*, *Dymecodon pilirostris* and *Urotrichus talpoides* enlarged relative to the other specimens. (From Mitgutsch *et al.* 2012.) (B) Embryonic specimens of *C. parva*, upper row from left to right: 12.5 days post coitum (dpc), 13.5 dpc, 14.5 dpc, 15.5 dpc; lower row left to right: 16.5 dpc, 17.5 dpc, 18.5 dpc. Not to scale. (C) Alcian blue/alizarin red double stained and cleared forelimb of a *Talpa occidentalis* embryo demonstrating the peculiar shoulder girdle and hand anatomy associated with the fossorial life style. (D) Embryonic specimens of *T. occidentalis* from left to right: 17 days post coitum (dpc), early Barrionuevo *et al.* (2004)-stage (BS) e5a, 18 dpc (BS e5b), 19 dpc (BS e5c), 19–21 dpc (BS e6). Not to scale. Roman numerals indicate digits; c, clavicle; h, humerus; r, radius; u, ulna; s, sternum; sc, scapula; rs, radial sesamoid. (See also colour plate.)

the development and skeletal nature of the talpid os falciforme (Edwards 1937; Sánchez-Villagra and Menke 2005; Prochel 2006). Generally, the os falciforme is currently referred to as a sesamoid bone (e.g. Kley and Kearney 2007; Vickaryous and Olson 2007). Throughout the literature, however, the designation 'sesamoid bone' has been applied to a wide variety of skeletal elements that differ considerably in developmental characteristics and, subsequently, in histological features (Prochel 2006; Vickaryous and Olson 2007). Two other terms associated with these elements, prepollex and prehallux, have also been used with different denotations throughout the literature, allowing, together with the finger-like appearance, for confusion of the os falciforme with an actual true digit (e.g. Bardeleben 1885, 1890; see further discussion below). In the more recent literature the prepollical element in *Talpa* is considered to be a true sesamoid (Prochel 2006) with tendinous insertions from the abductor pollicis longus and palmaris longus muscles (Whidden 2000).

Sesamoids are widespread throughout tetrapods (Vickaryous and Olson 2007). In contrast to mineralized tendons, sesamoids are ossicles with a discrete cartilaginous phase of development, developing in tendons around bony prominences. Mechanical factors may play a role during their development and their presence might be intraspecifically quite variable (Sarin *et al.* 1999; Vickaryous and Olson 2007). In addition to the mechanical component, Sarin *et al.* (1999) have reported evidence supporting intrinsic genetic factors playing a role in the formation of sesamoids and discuss a manifestation through Waddington's 'evolutionary canalization' (e.g. Waddington 1953). Le Minor (1994) investigated the distribution of the sesamoid bone of the abductor pollicis longus muscle in primates and found this bone to be a primitive primate characteristic present in all non-human primates (except *Gorilla*), generally articulating with the scaphoid and trapezium. According to this study, this sesamoid is the prepollex of Bardeleben and homologous to the os falciforme present in rodents, chiropterans, 'insectivorans' , *Tupaia*, some marsupials and, in rudimentary form, in fissipeds. Prepollical and prehallical sesamoids can be formed into considerable anatomical structures and involved in complicated functional complexes as seen in the panda (Endo *et al.* 1999) or the elephant (Weissengruber *et al.* 2006). Thus, while sesamoids in association with the abductor pollicis longus muscle seem taxonomically widespread, the independent, finger-like appearance of the structure supported by the talpid os falciforme and the characteristics of the latter are unique specializations.

Polydactyly and digit malformations are parts of several human syndromes and are caused by a variety of mutations affecting limb development (Cohn and Bright 1999). For example, the chromosome 7q36 pre-axial polydactyly, a frequent congenital limb malformation, has been shown to be the result of a point mutation in a *Shh* regulatory element (Lettice *et al.* 2003). Polydactyly can also be induced in various ways experimentally, among others through affecting the ZPA or AER, and a number of polydactylous mutants are known. The chicken talpid³ (ta³) mutant is polydactylous; the fingers have no clear identity, hinting at developmental problems during anterior–posterior patterning (Izpisúa-Belmonte *et al.* 1992). Lallemand *et al.* (2005) observed polydactyly with a pre-axial element longer than the thumb in *Msx1*$^{-/-}$; *Msx2*$^{-/-}$ double null mouse embryos. Generally, *Msx* genes have been discussed to influence the development of species-specific autopodial morphologies (Gañan *et al.* 1998; Hamrick 2001a, 2003). *Msx2* marks the interdigital tissue where, in case the digits get separated, apoptosis will happen, thus influencing further AER development. Apopotosis has been shown to be an important developmental mechanism in species-specific patterning of autopodia (Gañan *et al.* 1998) and in separating digits by interruption of the AER; inhibition of apoptosis in the AER can lead to polydactyly (reviewed by Zuzarte-Luís and Hurlé 2002, 2007). The interdigital tissue has also been shown to have chondrogenic potential (Gañan *et al.* 1994) and to regulate digit identity as shown by transplantation experiments (Dahn and Fallon 2000). *Msx* expression may also repress *Shh* anteriorly, influencing patterning of the acropodium as a *Shh* gradient is involved in determining digit numbers. *Msx*genes are thought to have several influences on limb development, including involvement in AER formation (Lallemand *et al.* 2005, 2009). *Msx* genes have also been shown to be involved in the apoptotic programme; although *Msx* expression seems not sufficient to initiate apoptosis (Davidson 1995; Weatherbee *et al.* 2006; Lallemand *et al.* 2009). Furthermore *Msx* genes have been thought to be involved in the control of bone development and differentiation (Davidson 1995) including the suppression of ectopic cranial neural-crest-derived bones (Roybal *et al.* 2010). Also, defects in potential upstream factors to *Msx1*/*Msx2* such as BMPs are known to cause malformations such as syndactyly and polydactyly (Montero *et al.* 2008). Finally *Msx*-genes could modulate digit number and identity (Bensoussan-Trigano *et al.* 2011).

The aim of a case study by Mitgutsch *et al.* (2012) was to further explore the evolutionary history of the talpid autopodia and particularly the identity of the radial and tibial sesamoids and their developmental relationships to other autopodial skeletal elements. Their early development was studied in the Iberian mole *Talpa occidentalis* and a soricomorph, the North American least shrew *Cryptotis parva*, which lacks these elements (embryonic specimens are shown in Figure 12.1B, D).

An early marker in mesodermal limb chondrification is *Sox9* (Chimal-Monroy *et al.* 2003). The expression of this transcription factor *Sox9* coincides with mesenchymal condensation preceding chondrogenesis during skeletogenesis (e.g. Sahar *et al.* 2005), and plays, among several other functions, essential roles in specifying mesenchymal cells toward a chondrogenic fate (Lefebvre and Smits 2005; Lefebvre *et al.* 2007), possibly throughout the Craniata (Zhang *et al.* 2006). *Sox9* has previously been used as a condensation marker in exploring digit identity in avian embryos (Welten *et al.* 2005). When looking at *Sox9* expression in mole and shrew embryos (Mitgutsch *et al.* 2012), it is clear that *Sox9* marks the prospective domains of chondral autopodial elements (Figure 12.2A, B, D). In *T. occidentalis*, *Sox9* expression becomes apparent in a rod-like manner pre-axial to the region of digit I, after *Sox9* expression in the digits had reached its peak. This pre-axial expression persists in later embryos when *Sox9* transcription has already faded in the phalanges, with the domain extending well into the autopod. Both temporal persistence and spatial situation of *Sox9* expression in these domains do not exactly match with other basipodial, mesopodial or acropodial elements; the expression pattern of this gene in the autopods of the shrew is comparable to those of the moles but there are no signs of condensations in the pre-axial regions of either the hand or the foot comparable to those seen in the mole. Instead, there are proximal, radial and tibial condensations that do not extend into the autopodia. *Fgf8*, by contrast, which marks the apical ectodermal ridge (AER) in the mole hand, shows a pattern comparable to those observed in shrews (Mitgutsch *et al.* 2012) and mice. There is no evidence for an anterior extension of the AER when compared to shrew (Figure 12.2C). Apoptosis is important in patterning species-specific autopodia, as shown, for example, for bird species (Gañan *et al.* 1998) and in separating the fingers. *Msx2* marks the domains between the fingers where apoptosis will happen, thus potentially influencing further AER development. In a mole specimen there is a strong asymmetric anterior *Msx2* expression (Figure 12.2B). This is significant, as it hints that the pre-axial elements may have co-opted a pattern of molecular expression typical for shaping separate digits (Mitgutsch *et al.* 2012).

The mole hand is otherwise webbed in adults; the parts of the hand distal to the webbing consist mainly of claws. The asymmetric anterior expression of *Msx2* in developing mole hands might be a hint at an alteration of anterior autopodial development. Furthermore, *Msx* genes may be involved in anterior repression of *Shh* (Lallemand *et al.* 2005). *Shh* involvement in anterior–posterior appendage patterning is deeply conserved in gnathostomes (Dahn *et al.* 2007). As summarized by Larsson (2007), *Shh* and its downstream targets could be involved both in the regulation of digit numbers and in establishment of digit identity – with the potential exception of digit I – through gradients (e.g. Drossopoulou *et al.* 2000; Wang *et al.* 2000; Litingtung *et al.* 2002; Harfe *et al.*

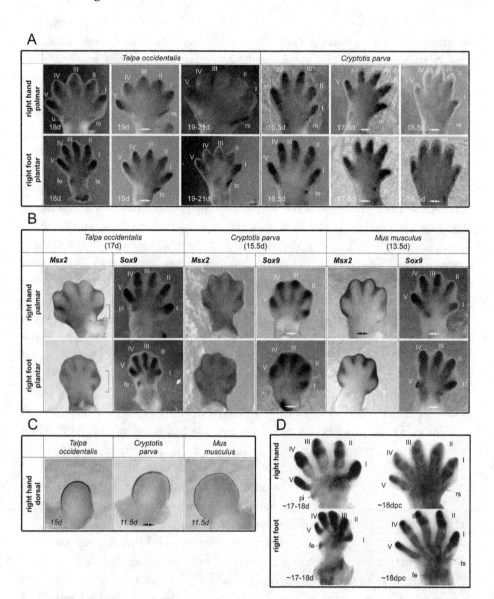

Figure 12.2 (A) *Sox9* expression in autopodia of *Talpa occidentalis* and *Cryptotis parva* embryos. (B) *Msx2* and *Sox9* expression in autopodia of *T. occidentalis*, *C. parva* and *Mus musculus* embryos. (C) *Fgf8* expression in the hands of *T. occidentalis*, *C. parva* and *M. musculus* embryos. (D) Alcian blue stained and cleared hands and feet of embryos of *T. occidentalis* embryos. Ages in days post coitum, centred double arrow indicates mirrored image; square bracket in (B) highlights relative stronger anterior expression in mole autopodia not present in shrew and mouse. Roman numerals label digits; fe, fibulare; pi, pisiforme; re, radial epiphysis; rs, radial sesamoid; te, tibiale; ts, tibial sesamoid. (Selected images from Mitgutsch *et al.* 2011.). (See also colour plate.)

2004; McGlinn *et al.* 2005); both alterations of the zone of polarizing activity (ZPA) or *Shh* expression induced by retinoic acid may lead to polydactyly and changed digit identities, though digit identity is seemingly independent of digit number. The action of such gradients would nicely fit models explaining the patterning of vertebrate limbs based on reaction-diffusion equations (i.e. Hentschel *et al.* 2004).

Given the involvement of gradients and the apparent time-dependent pattern formation through interfering apoptotic regions and the influence of inter-digital regulation on the identity of digits (Dahn and Fallon 2000), one might ask whether the 'pentadactyly constraint' is a result of the individualization and separation of fingers. Supernumerary skeletal elements are often found in paddle- or flipper-like developed autopodia and additional elements are included in the digit rows in webbed autopodia of frogs or in the digging hands of moles that have their digits strongly connected by interdigital soft tissues. Desmans, by contrast, having webbed feet for swimming, feature a prominent tibial sesamoid.

In the light of the presence of the (less pronounced) mole tibial sesamoid, reports on the incidences of sesamoid bones are of interest. In a study by Sarin *et al.* (1999) a significant relationship between the incidence of the fabella (an ossicle associated to the tendon of the gastrocnemius muscle) and os peroneium (an ossicle of the foot) was shown in human individuals. With the results of Pritchett (1984) showing that people with primary osteoarthritis had significant higher incidences of fabellas than the control individuals, it was concluded that certain sesamoids tend to be linked in their appearance. Furthermore, the presence of sesamoid bones may be due to increased tendencies to endochondral ossification (Pritchett 1984; Sarin *et al.* 1999).

It is noteworthy in this context that Alberch (1985) linked polydactyly in dog breeds to relatively large limb buds, a view that has since been supported by the experimental extension of the AER leading to an overgrowth of underlying mesenchyme, as discussed by Lallemand *et al.* (2005). A potential coupling of the appearance of sesamoids, or a gain of a general tendency to acquire ossified elements, might be involved in the evolution of the rod-like tibial sesamoid within the Talpini with their notable proportional differences between fore- and hind limb autopodia (Richardson *et al.* 2009).

An additional point to be discussed in the context of skeletogenesis of the conspicuous mole radial and tibial sesamoids might be the role of gonadal steroids. The presence of ovotestes and masculinized female genitalia in talpid species (Mathews 1935; Jiménez *et al.* 1993; Sánchez 1996; Carmona *et al.* 2008) hint at potential exposure of embryos to elevated levels of testosterone (Jiménez *et al.* 1993; Whitworth *et al.* 1999; Barrionuevo *et al.* 2004; Carmona *et al.* 2009).

Excess testosterone levels have shown to lead to increased rates of ossification in antlers (summarized by Hall 2005). Furthermore, oestrogen has been shown to reversibly mediate the transformation of cartilage into ligamentous tissue in female and castrated male mice. During the reverse process in the latter, the longer that testes are present, the more strongly inhibited is ligament formation (Hall 2005). One might thus speculate that elevated testosterone levels may both mediate extended formation of the pre-axial sesamoid cartilage and support its ossification. Additionally, elevated maternal testosterone levels have been suggested to be a potential cause of polydactyly in human populations by James (1998), arguing that high levels of testosterone are associated with the birth of males. He noted that both pre- and post-axial polydactylous subjects of the study contained high excesses of males and excesses of unaffected brothers. In summary, the relatively large forelimb of the mole, the coupling of the presence of sesamoids and the effects of testosterone are factors that have been discussed in the context of polydactyly and the appearance of additional skeletal elements.

The mole radial sesamoid, although unsegmented, is located within the acropodial (digital) region. It appears to be in series with the digits, both spatial and developmentally. Considering the late development of the anterior elements, it is in agreement with the primary axis model (see also Welten *et al.* 2005 on the presumptive digit I in avian wings). Furthermore, the late distal *Sox9* expression in the os falciforme may hint at an acropodial origin of this element following the establishment of the posterior digits (Mitgutsch *et al.* 2012). This would make the mole pentadactyl, with an accessory digit-like skeletal element, rather than polydactyl. The establishment of the autopodial field was suggested to be one of the first stages in autopodial evolution (Wagner and Chiu 2001). The evolutionary origin of the autopodium itself has been suggested to be coincidental with the origin of the tetrapod limb (Wagner and Chiu 2001). Furthermore the acropodium has been suggested to have appeared before the mesopodium. Early tetrapods have been shown to lack complete complements of wrist bones and posterior digits to articulate with the fibulare in *Acanthostega* and *Ichthyostega* (Johanson *et al.* 2007), digits that ossify relatively early in crown tetrapods (Fröbisch *et al.* 2007). In fact, sarcopterygian pectoral fins have been suggested to possess pre-axial, digit-like skeletal elements (Daeschler and Shubin 1997) and based on the forefin anatomy of *Panderichthys*, Boisvert *et al.* (2008) hypothesized that fingers were derived from distal radials of non-tetrapod sarcopterygians, making true digits and acropodia ancient evolutionary structures (but see Woltering and Duboule 2010).

Recently, increasing evidence has indicated that *HoxD* genes play a major role in establishing digit numbers and identities and for mechanisms with special impact on the digits of the anterior hand ('thumbness'; Deschamps 2008; Montavon *et al.* 2008; Wagner and Vargas 2008; Wang *et al.* 2009; Woltering and Duboule 2010). The model for establishing 'thumbness' (as summarized by Wagner and Vargas 2008; Woltering and Duboule 2010) allows for the anterior digit to have its own developmental programme regulated through *HoxD* genes – reflected in more frequent evolutionary changes compared with other digits – as opposed to the more posterior fingers that appear interdependent in evolutionary phenotype. Wagner and Vargas (2008) additionally pointed out that digital condensations remain undetermined until relatively late stages. Woltering and Duboule (2010) summarized that the identities of the carpus and tarsus, located between the long-shaped bones of the zeugo- and acropodium, are determined by *Hoxa13* expression and low or no *HoxD* gene transcription. The most anterior digit likewise is identified by expression of *Hoxa13* and low expression of *Hoxd13* which links the anterior acropodial region and the wrist. Furthermore, the asymmetric expression of the *Hoxd12* to *Hoxd10* are regulated by *Shh* and thus digits will have distinct morphologies along the autopodium's AP-axis. Investigating late-stage *HoxD* expression in relation to the mole's thumb would thus be a worthwhile subject of investigation.

In conclusion, the late patterning and peculiarities of the anterior acropodial region might allow for the recruitment of non-acropodial structures involving developmental mechanisms patterning the anterior hand. The talpid radial and tibial sesamoids are the skeletal elements supporting the digit-like, abductible structures in mole hands and feet. While such radial sesamoids have been reported to be taxonomically widespread, the evolution of the particular condition in the mole might have been supported by several factors, including the influence of gonadal steroids on the formation of tendon, cartilage and bone, the relatively accelerated growth for the mole forelimb, and the linkage of the incidence of sesamoid bones throughout the skeleton (e.g. hand and foot). The late development of its distal portions in conjunction with the developmental peculiarities of the anterior patterning mechanisms in the acropodial region might have supported an inclusion of this structure into the evolutionarily older acropodial region (see comparative position in non-Talpini, Figure 12.1A), taking the place of an anterior finger and potentially recruiting further developmental mechanisms such as apoptosis that are involved in patterning the autopodium.

Our study serves to illustrate how an understanding of the programmes of limb development contributes to a research programme on homology and morphological diversification. In this enterprise, morphology is the primary source of questions that developmental genetics can address (Wagner 2007).

This understanding of the network of interactions that, ultimately, yields morphological diversification poses new challenges but better enables evolutionary science to understand homology.

Materials and methods

Embryo collection

Embryonic specimens of the Iberian mole *Talpa occidentalis* were collected from pregnant females captured in cultivated areas near Granada (Spain) as described (Barrionuevo *et al.* 2004; Carmona *et al.* 2009). Abdominal palpation of captured female moles permitted detection of pregnancy and an estimate of the age (size) of the embryos before dissection as described by Barrionuevo *et al.* (2004). Once dissected, the gestational age was more accurately determined in days post coitum, according to the crown–rump length of the embryos. The four limbs were fixed overnight in 4% paraformaldehyde at 4 °C, dehydrated in a methanol series and stored at −20 °C in 100% methanol until further use in *in situ* hybridization experiments. Captures of *T. occidentalis* were performed under annual permission granted by the Andalusian Environmental Council and animal handling was performed in accordance with the guidelines of the University of Granada's 'Ethical Committee for Animal Experimentation'. Embryonic specimens of *Cryptotis parva* were obtained from the captive-bred colony at the A. T. Still University (ATSU). A female of reproductive age was placed with two males. Twenty-four hours post-grouping the female was injected with 25 IU of HCG (human chorionic gonadotropin). Thirty-six hours post-injection the female was isolated. The time of ovulation and fertilization was estimated at 14 hours post-injection. The embryos were harvested from 13 to 21 days post-fertilization. All the animal handling was in accordance with ATSU Animal Care Committee guidelines and protocol.

Histology and imaging

Selected embryos were cleared and alizarin red/alcian blue double stained (Dingerkus and Uhler 1977). Specimens were skinned, dehydrated through a graded ethanol series, stained for cartilage with alcian blue, and rehydrated. Subsequently, the embryos were digested in 0.5% trypsin solution in 30% saturated borax solution, stained for mineralized bone with an alizarin red solution in 0.5% KOH and transferred to glycerin through a graded glycerin series. Alcian blue staining of specimens that had previously been stained using *in situ* hybridization was done following the procedures described by Welten *et al.* (2005). Selected *Sox9* whole mount *in situ* hybridized specimens were subsequently dehydrated, stained with alcian blue and cleared through methyl-salicylate.

In situ hybridization

The digoxigenin labelled anti-sense RNA probes were synthesized from plasmids containing PCR products of the major part of the coding sequences of *Sox9* of the Iberian mole (*Talpa occidentalis*) and *Msx2* and *Fgf8* of the mouse (*Mus musculus*), using cDNA retrotranscribed from embryonic mRNA of each species as a template (GenBank accession numbers: HQ260700, HQ260699, HQ260698). Whole mount *in situ* hybridizations were performed according to standard protocols with minor modifications (Mitgutsch *et al.* 2012).

Summary

The limbs of extant tetrapods exhibit a pentadactylous autopod (carpal and tarsal region), which contains the proximal mesopodium (wrist and ankle bones) and the terminal acropodium (the digits), a pattern largely conserved across major groups. The evolution of the pentadactylous autopod seems particularly striking. Digit reduction has occurred in several taxa, but polydactyly is largely unknown in extant species. This is even more remarkable given that extinct early tetrapods, such as *Ichthyostega* and *Acanthostega*, had more than five digits. In addition, polydactyly frequently occurs as a malformation. In a few cases, land vertebrates have more than the typical count of five fingers and toes on each limb, famously represented by the giant panda with its additional 'thumb'. Equally impressive is the remarkable 'os falciforme' of talpid moles. This large, sickle-shaped sesamoid bone is used in concert with the true fingers for digging. All talpids with an os falciforme also show an extra bone in the foot, the tibial sesamoid. Here, we discuss how talpid moles circumvent the pentadactyl constraint by recruiting wrist sesamoids into their digital region using a novel developmental pathway. Our investigation involved studying the expression of three genes (*Sox9*, *Msx2* and *Fgf8*) during limb development, in the Iberian mole *Talpa occidentalis*, the North American shrew *Cryptotis parva*, and the mouse.

Acknowledgements

We thank the Paläontologisches Institut und Museum, Universität Zürich and the Institute of Biology, Leiden University, for support. This study was supported by the Swiss National Fund to MRSV (31003A_133032/1) and by grants of the Spanish ministry of Science and Innovation (CGL2004–00863/BOS and CGL2008–00828/BOS) and of the Andalusian Government (CVI2057) to RJ. The authors gratefully acknowledge the support of the Smart Mix programme of the Netherlands Ministry of Economic Affairs, and the

Netherlands Scientific Organisation. We would also like to thank J. Hugi, L. B. Wilson, C. Kolb, D. Koyabu, J. Neenan, T. Scheyer and I. Werneburg for comments and discussion and D. G. Lupiañez, F. M. Real and R. K. Dadhich for valuable help in mole captures. C. Zollikofer and N. Morimoto, University of Zürich, kindly made available the micro-CT facilities and supervised their use. Rob Asher and two anonymous reviewers provided valuable comments on earlier versions of this manuscript.

REFERENCES

Abouheif, E. (1997). Developmental genetics and homology: a hierarchical approach. *Trends in Ecology and Evolution*, **12**, 405–8.

Alberch, P. (1985). Developmental constraints: why St. Bernards often have an extra digit and poodles never do. *American Naturalist*, **126**, 430–3.

Asher, R., Bennett, N. and Lehmann, T. (2009). The new framework for understanding placental mammal evolution. *BioEssays*, **31**, 853–64.

Bardeleben, K. (1885). Zur Morphologie des Hand-und Fussskelets. *Jenaische Zeitschrift für Naturwissenschaften*, **19** Suppl., 84–8.

Bardeleben, K. (1890). Über die Hand-und Fuß-Muskeln der Säugetiere, besonders die des Praepollex (Praehallux) und Postminimus. *Anatomischer Anzeiger*, **5**, 435–44.

Barrionuevo, F., Zurita, F., Burgos, M. and Jiménez, R. (2004). Developmental stages and growth rate of the mole *Talpa occidentalis* (Insectivora, Mammalia). *Journal of Mammalogy*, **85**, 120–5.

Bensoussan-Trigano, V., Lallemand, Y., Saint Cloment, C. and Robert, B. (2011). *Msx1* and *Msx2* in limb mesenchyme modulate digit number and identity. *Developmental Dynamics*, **240**, 1190–202.

Blanco, M., Misof, B. and Wagner, G. (1998). Heterochronic differences of Hoxa-11 expression in *Xenopus* fore- and hind limb development: evidence for lower limb identity of the anuran ankle bones. *Development Genes and Evolution*, **208**, 175–87.

Boisvert, C., Mark-Kurik, E. and Ahlberg, P., (2008). The pectoral fin of *Panderichthys* and the origin of digits. *Nature*, **456**, 636–8.

Bolker, J. A. (1995). Model systems in developmental biology. *BioEssays*, **17**, 451–5.

Braus, H. (1906). Die Entwickelung der Form der Extremitäten und des Extremitäten-skeletts. In *Handbuch der vergleichenden und experimentellen Entwickelungslehre der Wirbeltiere, Dritter Band*, ed. O. Hertwig. Jena, Germany: Verlag von Gustav Fischer, pp. 167–338.

Burke, A. C. and Alberch, P. (1985). The development and homology of the chelonian carpus and tarsus. *Journal of Morphology*, **186**, 119–31.

Burke, A. C. and Feduccia, A. (1997). Developmental patterns and the identification of homologies in the avian hand. *Science*, **278**, 666–86.

Buscalioni, A. D., Ortega, F., Rasskin-Gutman, D. and Pérez-Moreno, B. P. (1997). Loss of carpal elements in crocodilian limb evolution: morphogenetic model corroborated by palaeobiological data. *Biological Journal of the Linnean Society*, **62**, 133–44.

Camardi, G. (2001). Richard Owen, morphology and evolution. *Journal of the History of Biology*, **34**, 481–515.

Carmona, F. D., Motokawa, M., Tokita, M., *et al.* (2008). The evolution of female mole ovotestes evidences high plasticity of mammalian gonad development. *Journal of Experimental Zoology–Molecular and Developmental Evolution*, **310B**, 259–66.

Carmona, F. D., Lupiáñez, D. G., Real, F. M., *et al.* (2009). Sox9 is not required for the cellular events of testicular organogenesis in XX mole ovotestes. *Journal of Experimental Zoology–Molecular and Developmental Evolution)*, **312B**, 734–48.

Chimal-Monroy, J., Rodriguez-Leon, J., Montero, J. A., *et al.* (2003). Analysis of the molecular cascade responsible for mesodermal limb chondrogenesis: Sox genes and BMP signaling. *Developmental Biology*, **257**, 292–301.

Coates, M. I. (1995). Fish fins or tetrapod limbs – a simple twist of fate? *Current Biology*, **5**, 844–8.

Coates, M. I. and Clack, J. A. (1990). Polydactyly in the earliest known tetrapod limbs. *Nature*, **347**, 66–9.

Coates, M., Jeffery, J. and Ruta, M. (2002). Fins to limbs: what the fossils say. *Evolution and Development*, **4**, 390–401.

Cohn, M. J. and Bright, P. E. (1999). Molecular control of vertebrate limb development, evolution and congenital malformations. *Cell and Tissue Research*, **296**, 3–17.

Cohn, M. J. and Tickle, C. (1999). Developmental basis of limblessness and axial patterning in snakes. *Nature*, **399**, 474–9.

Cooper, L. N. and Dawson, S. D. (2009). The trouble with flippers: a report on the prevalence of digital anomalies in Cetacea. *Zoological Journal of the Linnean Society*, **155**, 722–35.

Daeschler, E. and Shubin, N. (1997). Fish with fingers? *Nature*, **391**, 133.

Dahn, R. D. and Fallon, J. F. (2000). Interdigital regulation of digit identity and homeotic transformation by modulated BMP signaling. *Science*, **289**, 438–41.

Dahn, R. D., Davis, M. C., Pappano, W. N. and Shubin, N. H. (2007). Sonic hedgehog function in chondrichthyan fins and the evolution of appendage patterning. *Nature*, **445**, 311–14.

Davidson, D. (1995). The function and evolution of Msx genes: pointers and paradoxes. *Trends in Genetics*, **11**, 405–11.

Delfino, M. and Sánchez-Villagra, M. R. (2010). A survey of the rock record of reptilian ontogeny. *Seminars in Cell and Developmental Biology*, **21**, 432–40.

Delfino, M., Fritz, U. and Sánchez-Villagra, M. R. (2010). Evolutionary and developmental aspects of phalangeal formula variation in pig-nose and soft-shelled turtles (Carettochelyidae and Trionychidae). *Organisms Diversity and Evolution*, **10**, 69–79.

Deschamps, J. (2008). Tailored Hox gene transcription and the making of the thumb. *Genes and Development*, **22**, 293–396.

Dingerkus, G. and Uhler, L. D. (1977). Enzyme clearing of alcian blue stained whole small vertebrates for demonstration of cartilage. *Stain Technology*, **52**(4), 229–32.

Donoghue, P. C. J., Graham, A. and Kelsh, R. N. (2008). The origin and evolution of the neural crest. *BioEssays*, **30**, 530–41.

Drossopoulou, G., Lewis, K. E., Sanz-Ezquerro, J. J., *et al.* (2000). A model for anteroposterior patterning of the vertebrate limb based on sequential long- and short-range Shh signaling and Bmp signaling. *Development*, **127**, 1337–48.

Duboule, D. (2010). The evo-devo comet. *EMBO reports*, **11**, 489.

Edwards, L. F. (1937). Morphology of the forelimb of the mole (*Scalops aquaticus*, L.) in relation to its fossorial habits. *Ohio Journal of Science*, **37**, 20–41.

Emery, C. (1890). Zur Morphologie des Hand-und Fußskeletts. *Anatomischer Anzeiger*, **5**, 283–94.

Endo, H., Yamagiwa, D., Hayashi, Y., *et al.* (1999). Role of the giant panda's 'pseudo-thumb'. *Nature*, **397**, 309–10.

Fabrezi, M. (1993). The anuran tarsus. *Alytes*, **11**, 47–63.

Fabrezi, M. (2001). A survey of prepollex and prehallux variation in anuran limbs. *Zoological Journal of the Linnean Society*, **131**, 227–48.

Fabrezi, M. and Alberch, P. (1996). The carpal elements of anurans. *Herpetologica*, **52**, 188–204.

Fabrezi, M. and Barg, M. (2001). Patterns of carpal development among anuran amphibians. *Journal of Morphology*, **249**, 210–20.

Fedak, T. J. and Hall, B. K. (2004). Perspectives on hyperphalangy: patterns and processes. *Journal of Anatomy*, **204**, 151–63.

Fröbisch, N. B. (2008). Ossification patterns in the tetrapod limb – conservation and divergence from morphogenetic events. *Biological Reviews of the Cambridge Philosophical Society*, **83**, 571–600.

Fröbisch, N. B., Carroll, R. L. and Schoch, R. R. (2007). Limb ossification in the Paleozoic branchiosaurid *Apateon* (Temnospondyli) and the early evolution of preaxial dominance in tetrapod limb development. *Evolution and Development*, **9**, 69–75.

Galis, F., Alphen, J. J. M. van and Metz, J. A. J. (2001). Why five fingers? Evolutionary constraints on digit numbers. *Trends in Ecology and Evolution*, **16**, 637–46.

Galis, F., Alphen, J. J. M. van and Metz, J. A. J. (2002). Digit reduction: via repatterning or developmental arrest? *Evolution and Development*, **4**, 249–51.

Galis, F., Kundrát, M. and Sinervo, B. (2003). An old controversy solved: bird embryos have five fingers. *Trends in Ecology and Evolution*, **18**, 7–9.

Gañan, Y., Macias, D. and Hurle, J. M. (1994). Pattern regulation in the chick autopodium at advanced stages of embryonic development. *Developmental Dynamics*, **199**, 64–72.

Gañan, Y., Macias, D., Basco, R. D., Merino, R. and Hurle, J. M. (1998). Morphological diversity of the avian foot is related with the pattern of Msx gene expression in the developing autopod. *Developmental Biology*, **196**, 33–41.

Gegenbaur, C., (1898). *Vergleichende Anatomie der Wirbelthiere mit Berücksichtigung der Wirbellosen. Band 1: Einleitung, Integument, Skeletsystem, Muskelsystem, Nervensystem und Sinnesorgane*, Leipzig, Germany: Verlag von Wilhelm Engelmann.

Gilbert, S. F. (2003). The morphogenesis of evolutionary developmental biology. *International Journal of Developmental Biology*, **47**, 467–77.

Goodwin, B. C. and Trainor, L. E. H. (1983). The ontogeny and phylogeny of the pentadactyl limb. In *Development and Evolution*, ed. B. Goodwin, N. Holder and C. C. Wylie. Cambridge, UK: Cambridge University Press, pp. 75–98.

Hall, B. (2003). Unlocking the black box between genotype and phenotype: cell condensations as morphogenetic (modular) units. *Biology and Philosophy*, 18, 219–47.

Hall, B, (2005). *Bones and Cartilage: Developmental and Evolutionary Skeletal Biology.* Amsterdam, Netherlands: Elsevier Academic Press.

Hall, B. (2007). *Fins into Limbs.* Chicago, IL: University of Chicago Press.

Hamrick, M. W. (2001a). Development and evolution of the mammalian limb: adaptive diversification of nails, hooves, and claws. *Evolution and Development*, 3, 3555–63.

Hamrick, M. W. (2001b). Primate origins: evolutionary change in digital ray patterning and segmentation. *Journal of Human Evolution*, 40, 339–51.

Hamrick, M. W. (2002). Developmental mechanisms of digit reduction. *Evolution and Development*, 4, 247–8.

Hamrick, M. W. (2003). Evolution and development of mammalian limb integumentary structures. *Journal of Experimental Zoology–Molecular and Developmental Evolution*, 298B, 152–63.

Hanken, J. (1993). Model systems versus outgroups: alternative approaches to the study of head development and evolution. *American Zoologist*, 33, 448–56.

Harfe, B. D., Scherz, P. J., Nissim, S., *et al.* (2004). Evidence for an expansion-based temporal Shh gradient in specifying vertebrate digit identities. *Cell*, 118, 517–28.

Hentschel, H. G. E., Glimm, T., Glazier, J. A. and Newman, S. A. (2004). Dynamical mechanisms for skeletal pattern formation in the vertebrate limb. *Proceedings of the Royal Society of London B*, 271, 1713–22.

Hinchliffe, J. (1989). Evolutionary aspects of the developmental mechanisms underlying the patterning of the pentadactyl limb skeleton in birds and other tetrapods. *Fortschritte der Zoologie*, 35, 226–9.

Hinchliffe, J. R, (2002). Developmental basis of limb evolution. *International Journal of Developmental Biology*, 46, 835–45.

Hinchliffe, J. and Griffiths, P. (1983). The prechondrogenic patterns in tetrapod limb development and their phylogenetic significance. In *Development and Evolution*, ed. B. Goodwin, N. Holder and C. C. Wylie. Cambridge, UK: Cambridge University Press, pp. 99–121.

Holder, N. (1983). The vertebrate limb: patterns and constraints in development and evolution. In *Development and Evolution*, ed. B. Goodwin, N. Holder and C. C. Wylie. Cambridge, UK: Cambridge University Press, pp. 399–425.

Holmgren, N. (1933). On the origin of the tetrapod limb. *Acta Zoologica*, 14, 185–295.

Holmgren, N. (1952). An embryological analysis of the mammalian carpus and its bearing upon the question of the origin of the tetrapod limb. *Acta Zoologica*, 33, 1–115.

Hugi, J., Mitgutsch, C. and Sánchez-Villagra, M. R. (2010). Chondrogenic and ossification patterns and sequences in White's skink *Liopholis whitii* (Scincidae, Reptilia). *Zoosystematics and Evolution*, 86, 21–32.

Izpisúa-Belmonte, J. C., Ede, D. A., Tickle, C. and Duboule, D. (1992). The misexpression of posterior Hox-4 genes in talpid (ta3) mutant wings correlates with the absence of anteroposterior polarity. *Development*, **114**, 959–63.

James, W. H. (1998). Hypothesis: one cause of polydactyly. *Journal of Theoretical Biology*, **192**, 1–2.

Janies, D. and de Salle, R. (1999). Development, evolution, and corroboration. *Anatomical Record (New Anatomist)*, **257**, 6–14.

Jenner, R. A. and Wills, M. A. (2007). The choice of model organisms in evo-devo. *Nature Reviews Genetics*, **8**, 311–19.

Jiménez, R., Burgos, M., Sánchez, A., *et al.* (1993). Fertile females of the mole *Talpa occidentalis* are phenotypic intersexes with ovotestes. *Development*, **118**, 1303–11.

Johanson, Z., Joss, J. M. P., Sutija, M., Boisvert, C. and Ahlberg, P. E. (2007). Fish fingers: digit homologues in sarcopterygian fish fins. *Journal of Experimental Zoology–Molecular and Developmental Evolution*, **308B**, 757–68.

Kellogg, E. A. and Shaffer, H. B. (1993). Model organisms in evolutionary studies. *Systematic Biology*, **42**, 409–14.

Kindahl, M. (1942). Einige Mitteilungen über die Entwicklung der Hand und des Fußes bei *Talpa europaea* L. *Zeitschrift für mikroskopisch-anatomische Forschung*, **52**, 267–73.

Kley, N. and Kearney, M. (2007). Adaptations for digging and burrowing. In *Fins into Limbs: Evolution, Development, and Transformation*, ed. B. Hall. Chicago, IL: University of Chicago Press, pp. 284–309.

Kundrát, M. (2009). Primary chondrification foci in the wing basipodium of *Struthio camelus* with comments on interpretation of autopodial elements in Crocodilia and Aves. *Journal of Experimental Zoology–Molecular and Developmental Evolution*, **312B**, 30–41.

Lallemand, Y., Nicola, M., Ramos, C., *et al.* (2005). Analysis of Msx1; Msx2 double mutants reveals multiple roles for Msx genes in limb development. *Development*, **132**, 3003–14.

Lallemand, Y., Bensoussan, V., Saint Cloment, C. and Robert, B. (2009). Msx genes are important apoptosis effectors downstream of the Shh/Gli3 pathway in the limb. *Developmental Biology*, **331**, 189–98.

Larsson, H. (2007). MODEs of developmental evolution: an example with the origin and definition of the autopodium. In *Major Transitions in Vertebrate Evolution*, ed. J. Anderson and H. Sues. Bloomington, IN: Indiana University Press, pp. 150–81.

Larsson, H. C. E. and Wagner, G. P. (2002). Pentadactyl ground state of the avian wing. *Journal of Experimental Zoology–Molecular and Developmental Evolution*, **294**, 146–51.

Larsson, H. C. E. and Wagner, G. P. (2003). Old morphologies misinterpreted. *Trends in Ecology and Evolution*, **18**, 10.

Larsson, H. C. E., Heppleston, A. C. and Elsey, R. M. (2010). Pentadactyl ground state of the manus of *Alligator mississippiensis* and insights into the evolution of digital reduction in Archosauria. *Journal of Experimental Zoology–Molecular and Developmental Evolution*, **314B**, 571–9.

Le Minor, J. (1994). The sesamoid bone of musculus abductor pollicis longus (os radiale externum or prepollex) in Primates. *Acta Anatomica*, **150**, 227–31.

Leal, F., Tarazona, O. A. and Ramírez-Pinilla, M. P. (2010). Limb development in the gekkonid lizard *Gonatodes albogularis*: a reconsideration of homology in the lizard carpus and tarsus. *Journal of Morphology*, **271**, 1328–41.

Lebedev, O. and Coates, M. (1995). The postcranial skeleton of the Devonian tetrapod *Tulerpeton curtum* Lebedev. *Zoological Journal of the Linnean Society*, **114**, 307–48.

Lefebvre, V. and Smits, P. (2005). Transcriptional control of chondrocyte fate and differentiation. *Birth Defects Research (Part C)*, **75**, 200–12.

Lefebvre, V., Dumitriu, B., Penzo-Méndez, A., Han, Y. and Pallavi, B. (2007). Control of cell fate and differentiation by Sry-related high-mobility-group box (Sox) transcription factors. *International Journal of Biochemistry and Cell Biology*, **39**, 2195–214.

Lettice, L. A., Heaney, S. J., Purdie, L. A., *et al.* (2003). A long-range Shh enhancer regulates expression in the developing limb and fin and is associated with preaxial polydactyly. *Human Molecular Genetics*, **12**, 1725–35.

Lewis, O., (1964). The homologies of the mammalian tarsal bones. *Journal of Anatomy*, **98**, 195–208.

Litingtung, Y., Dahn, R. D., Li, Y., Fallon, J. F. and Chiang, C. (2002). Shh and Gli3 are dispensable for limb skeleton formation but regulate digit number and identity. *Nature*, **418**, 979–83.

Lunde, D. P. and Schutt, W. A., Jr. (1999). The peculiar carpal tubercles of male *Marmosops parvidens* and *Marmosa robinsoni* (Didelphidae: Didelphinae). *Mammalia*, **63**, 495–504.

Mathews, L. (1935). The oestrous cycle and intersexuality in the female mole (*Talpa europaea* Linn). *Proceedings of the Royal Society of London*, **2**, 347–83.

McGlinn, E., van Bueren, K. L., Fiorenza, S., *et al.* (2005). *Pax9* and *Jagged1* act downstream of Gli3 in vertebrate limb development. *Mechanisms of Development*, **122**, 1218–33.

Metscher, B. D. and Ahlberg, P. E. (1999). Zebrafish in context: uses of a laboratory model in comparative studies. *Developmental Biology*, **210**, 1–14.

Milinkovitch, M. and Tzika, A. (2007). Escaping the mouse trap: the selection of new evo-devo model species. *Journal of Experimental Zoology–Molecular and Developmental Evolution*, **308B**, 337–46.

Mitgutsch, C. (2003). On Carl Gegenbaur's theory on head metamerism and the selection of taxa for comparisons. *Theory in Biosciences*, **122**, 204–29.

Mitgutsch, C., Richardson, M. K., Jiménez, R., *et al.* (2012). Circumventing the polydactyly 'constraint': the mole's 'thumb'. *Biology Letters*, **8**, 74–7.

Montavon, T., Le Garrec, J. F., Kerszberg, M. and Duboule, D. (2008). Modeling Hox gene regulation in digits: reverse collinearity and the molecular origin of thumbness. *Genes and Development*, **22**, 346–59.

Montero, J. A., Lorda-Diez, C. I., Gañan, Y., Macias, D., Hurle, J. M. (2008). Activin/TGFβ and BMP crosstalk determines digit chondrogenesis. *Developmental Biology*, **321**, 343–56.

Motani, R. (1999). On the evolution and homologies of ichthyopterygian forefins. *Journal of Vertebrate Paleontology*, **19**, 28–41.

Müller, G. B. (2007). Evo-devo: extending the evolutionary synthesis. *Nature Reviews Genetics*, **8**, 943–9.

Ortega-Ortiz, J. G., Villa-Ramírez, B. and Gersenowies, J. R. (2000). Polydactyly and other features of the manus of the vaquita, *Phocoena sinus. Marine Mammal Science*, **16**, 277–86.

Oster, G. F., Murray, J. D. and Harris, A. K. (1983). Mechanical aspects of mesenchymal morphogenesis. *Journal of Embryology and Experimental Morphology*, **78**, 83–125.

Oster, G. F., Shubin, N., Murray, J. D. and Alberch, P. (1988). Evolution and morphogenetic rules: the shape of the vertebrate limb in ontogeny and phylogeny. *Evolution*, **42**, 862–84.

Peterson, K. J., Summons, R. E. and Donoghue, P. C. J. (2007). Molecular palaeobiology. *Palaeontology*, **50**, 775–809.

Pritchett, J. (1984). The incidence of fabellae in osteoarthrosis of the knee. *Journal of Bone and Joint Surgery*, **66**, 1379–80.

Prochel, J. (2006). Early skeletal development in *Talpa europaea*, the common European mole. *Zoological Science*, **23**, 427–34.

Prochel, J. and Sánchez-Villagra, M. (2003). Carpal ontogeny in *Monodelphis domestica* and *Caluromys philander* (Marsupialia). *Zoology*, **106**(1), 73–84.

Prochel, J., Vogel, P. and Sánchez-Villagra, M. (2004). Hand development and sequence of ossification in the forelimb of the European shrew *Crocidura russula* (Soricidae) and comparisons across therian mammals. *Journal of Anatomy*, **205**, 99–111.

Reed, C. A. (1951). Locomotion and appendicular anatomy in three soricoid insectivores. *American Midland Naturalist*, **45**, 513–671.

Richardson, M., Jeffery, J. and Tabin, C J. (2004). Proximodistal patterning of the limb: insights from evolutionary morphology. *Evolution and Development*, **6**, 1–5.

Richardson, M. K. and Oelschläger, H. H. A. (2002). Time, pattern, and heterochrony: a study of hyperphalangy in the dolphin embryo flipper. *Evolution and Development*, **4**, 435–44.

Richardson, M. K., Gobes, S. M., van Leeuwen, A. C., *et al.* (2009). Heterochrony in limb evolution: developmental mechanisms and natural selection. *Journal of Experimental Zoology–Molecular Development and Evolutio*, **312B**, 639–64.

Rieppel, O. (1993a). Studies on skeleton formation in reptiles. II. *Chamaeleo hoehnelii* (Squamata: Chamaeleoninae), with comments on the homology of carpal and tarsal bones. *Herpetologica*, **49**, 66–78.

Rieppel, O. (1993b). Studies on skeleton formation in reptiles: patterns of ossification in the skeleton of *Chelydra serpentina* (Reptilia, Testudines). *Journal of Zoology*, **231**, 487–509.

Rolian, C. (2008). Developmental basis of limb length in rodents: evidence for multiple divisions of labor in mechanisms of endochondral bone growth. *Evolution and Development*, **10**, 15–28.

Romer, A. S. (1956). *Osteology of the Reptiles*. Chicago, IL: Chicago University Press.

Roush, W. (1996). Zebrafish embryology builds better model vertebrate. *Science*, **272**, 1103.

Roybal, P. G., Wu, N. L., Sun, J., *et al.* (2010). Inactivation of Msx1 and Msx2 in neural crest reveals an unexpected role in suppressing heterotopic bone formation in the head. *Developmental Biology*, **343**, 28–39.

Sahar, D., Longaker, M. and Quarto, N. (2005). Sox9 neural crest determinant gene controls patterning and closure of the posterior frontal cranial suture. *Developmental Biology*, **280**, 344–61.

Sánchez, A., Bullejos, M., Burgos, M., *et al.* (1996). Females of four mole species of genus *Talpa* (Insectivora, Mammalia) are true hermaphrodites with ovotestes. *Molecular Reproduction and Development*, **44**, 289–94.

Sánchez-Villagra, M. R. and Menke, P. R. (2005). The mole's thumb – evolution of the hand skeleton in talpids (Mammalia). *Zoology*, **108**, 3–12.

Sánchez-Villagra, M. R., Horovitz, I. and Motokawa, M. (2006). A comprehensive morphological analysis of talpid moles (Mammalia) phylogenetic relationships. *Cladistics*, **22**, 59–88.

Sánchez-Villagra, M. R., Mitgutsch, C., Nagashima, H. and Kuratani, S. (2007). Autopodial development in the sea turtles *Chelonia mydas* and *Caretta caretta*. *Zoological Science*, **24**, 257–63.

Sarin, V., Erickson, G., Giori, N., Bergman, A. and Carter, D. (1999). Coincident development of sesamoid bones and clues to their evolution. *Anatomical Record (New Anatomist)*, **257**, 174–80.

Schmidt-Ehrenberg, E. (1942). Die Embryogenese des Extremitätenskelettes der Säugetiere. *Revue Suisse de Zoologie*, **49**, 33–131.

Scholtz, G. (2005). Homology and ontogeny: pattern and process in comparative developmental biology. *Theory in Biosciences*, **124**, 121–43.

Sears, K. (2008). Molecular determinants of bat wing development. *Cells, Tissues, Organs*, **187**, 6–12.

Sears, K. E., Behringer, R. R., Rasweiler, J. J. T. and Niswander, L. A. (2006). Development of bat flight: morphologic and molecular evolution of bat wing digits. *Proceedings of the National Academy of Sciences of the United States of America*, **103**, 6581–6.

Sears, K. E., Behringer, R. R., Rasweiler, J. J., 4th, Niswander, L. A. (2007). The evolutionary and developmental basis of parallel reduction in mammalian zeugopod elements. *American Naturalist*, **169**, 105–17.

Shubin, N. (1995). The evolution of paired fins and the origin of tetrapod limbs. *Evolutionary Biology*, **28**, 39–86.

Shubin, N. H. and Alberch, P. (1986). A morphogenetic approach to the origin and basic organization of the tetrapod limb. *Evolutionary Biology*, **20**, 319–87.

Shubin, N., Wake, D. B. and Crawford, A. J. (1995). Morphological variation in the limbs of *Taricha granulosa* (Caudata: Salamandridae): evolutionary and phylogenetic implications. *Evolution*, **49**, 874–84.

Shubin, N., Tabin, C. and Carroll, S. (1997). Fossils, genes and the evolution of animal limbs. *Nature*, **388**, 639–48.

Shubin, N. H., Daeschler, E. B. and Coates, M. I. (2004). The early evolution of the tetrapod humerus. *Science*, **304**, 90–3.

Shubin, N. H., Daeschler, E. B. and Jenkins, F. A., Jr. (2006). The pectoral fin of *Tiktaalik roseae* and the origin of the tetrapod limb. *Nature*, **440**, 764–71.

Shubin, N., Tabin, C and Carroll, S. (2009). Deep homology and the origins of evolutionary novelty. *Nature*, **457**, 818–23.

Slack, J. M. W., Holland, P. W. H. and Graham, C. F. (1993). The zootype and the phylotypic stage. *Nature*, **361**, 490–2.

Stafford, B. J. and Thorington, R. W. (1998). Carpal development and morphology in archontan mammals. *Journal of Morphology*, **235**, 135–55.

Starck, D. (1979). *Vergleichende Anatomie der Wirbeltiere auf evolutionsbiologischer Grundlage. Band 2: Das Skeletsystem. Allgemeines, Skeletsubstanzen, Skelet der Wirbeltiere einschließlich Lokomotionstypen.* Berlin: Springer-Verlag.

Svensson, M. E. (2004). Homology and homocracy revisited: gene expression patterns and hypotheses of homology. *Development Genes and Evolution*, **214**, 418–21.

Tabin, C. J. (1992). Why we have (only) five fingers per hand: Hox genes and the evolution of paired limbs. *Development*, **116**, 289–96.

Tamura, K., Nomura, N., Seki, R., Yonei-Tamura, S. and Yokoyama, H. (2011). Embryological evidence identifies wing digits in birds as digits 1, 2, and 3. *Science*, **331**, 753–7.

Thewissen, J. G. M., Cohn, M. J., Stevens, L. S., *et al.* (2006). Developmental basis for hind-limb loss in dolphins and origin of the cetacean bodyplan. *Proceedings of the National Academy of Sciences of the United States of America*, **103**, 8414–18.

Thorington, R. W., Jr., and Stafford, B. J. (2001). Homologies of the carpal bones in flying squirrels (Pteromyinae): a review. *Mammal Study*, **26**, 61–8.

Tokita, M. and Iwai, N. (2010). Development of the pseudothumb in frogs. *Biology Letters*, **6**, 517–20.

Vargas, A. O. and Fallon, J. F. (2005a). Birds have dinosaur wings: the molecular evidence. *Journal of Experimental Zoology–Molecular and Developmental Evolution*, **304B**, 86–90.

Vargas, A. O. and Fallon, J. F. (2005b). The digits of the wing of birds are 1, 2, and 3. a review. *Journal of Experimental Zoology–Molecular and Developmental Evolution*, **304B**, 206–19.

Vickaryous, M. and Olson, W. (2007). Sesamoids and ossicles in the appendicular skeleton. In *Fins Into Limbs: Evolution, Development, and Transformation*, ed. B. Hall. Chicago, IL: University of Chicago Press, pp. 323–41.

Waddington, C. H. (1953). Genetic assimilation of an acquired character. *Evolution*, **7**, 118–26.

Wagner, G. P. (2007). The current state and the future of developmental evolution. In *From Embryology to Evo-Devo: A History of Developmental Evolution*, ed. M. D. Laubichler and J. Maienschein. Cambridge, MA: MIT Press, pp. 525–45.

Wagner, G. P. and Chiu, C. H. (2001). The tetrapod limb: a hypothesis on its origin. *Journal of Experimental Zoology–Molecular and Developmental Evolution*, **291**, 226–40.

Wagner, G. P. and Gauthier, J. A. (1999). 1, 2, 3= 2, 3, 4: a solution to the problem of the homology of the digits in the avian hand. *Proceedings of the National Academy of Sciences of the United States of America*, **96**, 5111–16.

Wagner, G. P. and Vargas, A. O. (2008). On the nature of thumbs. *Genome Biology*, **9**, 213.

Wang, B., Fallon J. F. and Beachy, P. A. (2000). Hedgehog-regulated processing of Gli3 produces and anterior/posterior repressor gradient in the developing vertebrate limb. *Cell*, **100**, 423–34.

Wang, Z., Yuan, L. H., Rossiter, S. J., *et al.* (2009). Adaptive evolution of 5'HoxD genes in the origin and diversification of the cetacean flipper. *Molecular Biology and Evolution*, **26**, 613–22.

Weatherbee, S. D., Behringer, R. R., Rasweiler IV, J. J. and Niswander, L. A. (2006). Interdigital webbing retention in bat wings illustrates genetic changes underlying amniote limb diversification. *Proceedings of the National Academy of Sciences of the United States of America*, **103**, 15 103.

Weisbecker, V. and Nilsson, M. (2008). Integration, heterochrony, and adaptation in pedal digits of syndactylous marsupials. *BMC Evolutionary Biology*, **8**, 160.

Weissengruber, G. E., Egger, G. F., Hutchinson, J. R., *et al.* (2006). The structure of the cushions in the feet of African elephants (*Loxodonta africana*). *Journal of Anatomy*, **209**, 781–92.

Welten, M. C. M., Verbeek, F. J., Meijer, A. H. and Richardson, M. K. (2005). Gene expression and digit homology in the chicken embryo wing. *Evolution and Development*, **7**, 18–28.

Werneburg, I. (2009). A standard system to study vertebrate embryos. *PloS One*, **4**, p.e5887.

Westoll, T. S. (1943). The origin of the primitive tetrapod limb. *Proceedings of the Royal Society of London B–Biological Sciences*, **131**, 373–93.

Whidden, H. P. (2000). Comparative myology of moles and the phylogeny of the Talpidae (Mammalia, Lipotyphla). *American Museum Novitates*, **3294**, 1–53.

Whitworth, D. J., Licht, P., Racey, P. A. and Glickman, S. E. (1999). Testis-like steroidogenesis in the ovotestis of the European mole, *Talpa europaea*. *Biology of Reproduction*, **60**, 413–18.

Woltering, J. M. and Duboule, D. (2010). The origin of digits: expression patterns versus regulatory mechanisms. *Developmental Cell*, **18**, 526–32.

Wu, X. C., Li, Z., Zhou, B.-C. and Dong, Z.-M. (2003). A polydactylous amniote from the Triassic period. *Nature*, **426**, 516.

Xu, X., Clark, J. M., Mo, J., *et al.* (2009). A Jurassic ceratosaur from China helps clarify avian digital homologies. *Nature*, **459**, 940–4.

Yalden, D. W. (1966). The anatomy of mole locomotion. *Journal of Zoology*, **149**, 55–64.

Young, R. L. and Wagner, G. P. (2011). Why ontogenetic homology criteria can be misleading: lessons from digit identity transformations. *Journal of Experimental Zoology–Molecular Development and Evolution*, **314B**, 165–70.

Young, R. L., Caputo, V., Giovannotti, M., *et al.* (2009). Evolution of digit identity in the three-toed Italian skink *Chalcides chalcides*: a new case of digit identity frame shift. *Evolution and Development*, **11**, 647–58.

Zhang, G., Miyamoto, M. M. and Cohn, M. J. (2006). Lamprey type II collagen and Sox9 reveal an ancient origin of the vertebrate collagenous skeleton. *Proceedings of the National Academy of Sciences of the United States of America*, **103**, 3180–5.

Zuzarte-Luís, V. and Hurlé, J. M. (2002). Programmed cell death in the developing limb. *International Journal of Developmental Biology*, **46**, 871–6.

Zuzarte-Luís, V. and Hurlé, J. M. (2007). Apoptosis in fin and limb development. In *Fins Into Limbs: Evolution, Development, and Transformation*, ed. B. Hall. Chicago, IL: University of Chicago Press, pp. 103–8.

13

Manus horribilis: the chicken wing skeleton

MICHAEL K. RICHARDSON

Introduction

Evolution is a natural experiment that has been running for millions of years (Shubin 1991). Developmental biologists and palaeontologists can learn from this experiment (Zákány and Duboule 2007; see also chapters in this volume by Anthwal and Tucker; Buchholtz; Kuratani and Nagashima; Mitgutsch *et al.*; Schmid; Sears *et al.*; Smith and Johanson). A particular challenge is the homology of elements distal to the ulna and radius in modern birds. Developmental biologists differ on how many skeletal primordia actually develop in the embryonic wing bud, and the homologies of the permanent bones of the avian wrist are also uncertain.

Study of the avian wing is important for several reasons. First, the wing skeleton is an important issue in discussions about avian origins (Ostrom 1975; Müller 1991; Feduccia 2002; Prum 2002; Vargas and Fallon 2005b; Feduccia *et al.* 2007). Furthermore, the chicken is a key model species in developmental biology, and the development of its wing has been intensively studied in the context of pattern formation theory (Tickle 2004). Finally, the fact that the avian wing is studied by developmental biologists, palaeontologists, morphologists and others makes it a suitable subject of enquiry for the integrated discipline of evo devo (Galis *et al.* 2003).

Here, I review the developmental anatomy and homologies of the avian carpus and digits. I consider the types of data provided by morphology, developmental biology and molecular biology. Much of the work that I cite relates to the domestic chicken *Gallus gallus* (Linnaeus 1758, p. 158; for nomenclature of *Gallus gallus* see Donegan 2008). The chicken is a basal member of the Neognathae (Livezey and Zusi 2007) and belongs, along with ducks and

From Clone to Bone: The Synergy of Morphological and Molecular Tools in Palaeobiology, ed. Robert J. Asher and Johannes Müller. Published by Cambridge University Press.

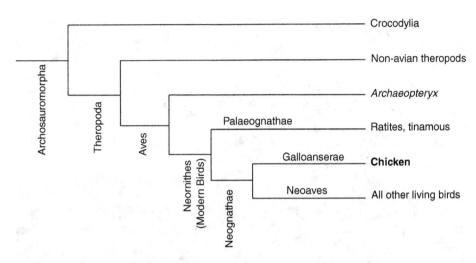

Figure 13.1 Phylogeny of the chicken, based on Livezey and Zusi (2007).

geese, to the Galloanserae. This latter clade is probably the sister group to all other Neoaves (modern birds excluding the Palaeognathae; Figure 13.1).

Throughout this chapter I shall use the II-III-IV convention for naming the structures of the first, second and third adult wing digits, respectively, counted from the radial (preaxial) border. These three digits (Figures 13.2, 13.3) are the digitus alulae manus, digitus majoris manus and digitus minoris manus, respectively (Baumel and Witmer 1993). When citing an author who numbers the avian wing digits as I-II-III, I have translated their numbering into II-III-IV. To make this clear, all named structures conforming to the II-III-IV terminology are written here in SMALL CAPITALS. *In summary therefore, the three permanent adult digits of the avian wing, and their associated distal carpal elements, are rendered in this chapter as II (digitus alulae manus), III (digitus majoris manus) and IV (digitus minoris manus) regardless of the numbering scheme used in the original article under discussion.* Anatomical abbreviations used in the figures are given in Table 13.1.

The carpus

The carpus (wrist or mesopodium) forms a relatively large part of the early wing bud, and grows relatively slowly (Lewis 1977); most of its elements, other than those on the primary axis, develop comparatively late (Richardson *et al.* 2009). Few or no transcripts of hox genes other than *hoxa13* and *hoxd13* are expressed in the carpus, and this may explain why carpal elements do not develop into long bones and lack epiphyses (Woltering and Duboule 2010). Development of the carpus is illustrated in Figures 13.4 and 13.5.

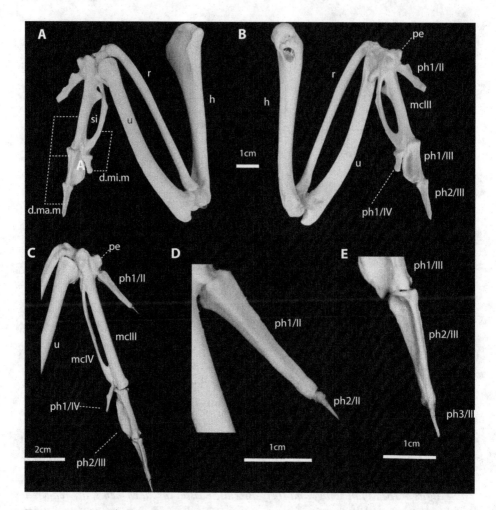

Figure 13.2 (A, B) Articulated left wing skeleton of adult female chicken, *Gallus gallus*. (A) Dorsal aspect. (B) Ventral aspect. (C) Articulated right wing skeleton of an adult male duck, *Anas platyrhynchos*, showing the styliform phalanges on the tips of DIGIT II (D, detail) and DIGIT III (E, detail). (Both specimens in this figure are from the collections of the University of Leiden Zoology Department, and were prepared by Mr. Schelte Zijlstra.) Abbreviations: d.ma.m, digitus majoris manus; d.mi.m, digitus minoris manus. For other abbreviations, see Table 13.1.

Gegenbaur's nomenclature

It is customary to write about the 'emerging' discipline of evo devo. However, evo devo of the wing has been emerging since at least 1864. Carl Gegenbaur regarded the carpus in urodeles as being the least specialized among tetrapods, and used it as the basis of a system of osteological nomenclature (Gegenbaur 1864) still used in modified form today. He divided the carpus into

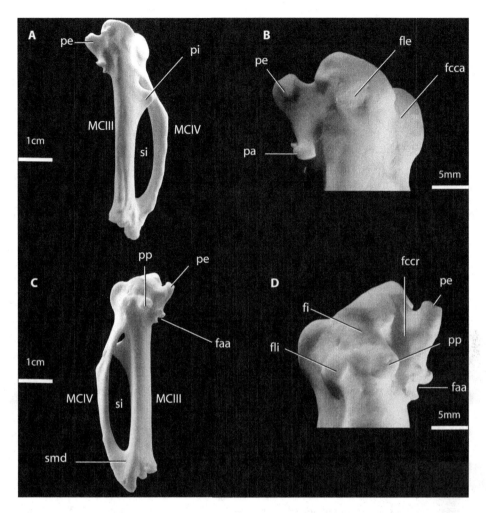

Figure 13.3 Adult chicken carpometacarpus from the left wing in Figure 13.2A, B. (A, B) Dorsal aspect. (C, D) Ventral aspect. For abbreviations, see Table 13.1.

proximal and distal rows. Lying between these, and encircled by them, is the **centrale**. He coined the tems **radiale** and **ulnare** for the two elements flanking the **intermedium** in the proximal row. His distal row comprised a series of carpalia (**distal carpals**), which articulate with the proximal ends of the metacarpals.

Unfortunately, Gegenbaur's assumption that extant urodeles show a primitive pattern is not correct (Shubin 2002). It is also quite difficult to identify homologies between some of Gegenbaur's elements and those in birds. Indeed, Holmgren (1933) claimed that the avian proximal carpus was almost too complex to be interpreted, and Schestakowa (1927) said that nearly every scientist who studies the avian carpus reaches a different conclusion about its

Table 13.1 *Anatomical terms and abbreviations used in this chapter*

Annotation	Anatomical term
I-II-IV	Example of a hypothesis about the homology of digits in a tridactyl hand, with the value on the left of the formula referring to the pre-axial adult digit. The Roman numerals indicate the digit in a pentadactyl hand with which each is assumed to be homologous
faa	facies articularis alularis
fcca	fovea carpalis caudalis
fccr	fovea carpalis cranialis
fle	facies ligamentalis externa; this, and the fli, are for the dorsal and ventral parts, respectively, of the ligament joining the ulnare to the carpometacarpus (Ballmann 1969)
fli	facies ligamentalis interna (Ballmann 1969)
h	humerus
mc	metacarpal
ocr	os carpi radiale
ocu	os carpi ulnare
om	os carpi magnum
pa	processus alularis
pi	processus intermetacarpalis
pe	processus extensorius (Livezey and Zusy 2006) for extensor carpi radialis, or e. metacarpi radialis (Montagna 1945)
ph$_1$/II	proximal phalanx of digit II
ph$_2$/II	second phalanx of digit II
pp	processus pisiformis
r	radius
re	radiale
si	spatium intermetacarpale
smd	symphysis (synostosis) metacarpalis distalis
u	ulna
ue	ulnare
unc	unciform (Shufeldt 1881)

Sources: (unless otherwise stated): Baumel and Witmer (1993); Livezey and Zusy (2006).

development and homologies. In histological preparations, carpal condensations are indistinct, and the boundaries may be further obscured by fusions (Norsa, 1895; Braus 1906; Holmgren 1933; Hinchliffe and Hecht 1984; Hinchliffe 1989, 2002). Because of these technical limitations, and because of the influence of ideologies such as recapitulation and typology (Richardson *et al.* 1999; Richardson and Keuck 2002), the number of pre-cartilaginous condensations described by different authors ranges from 4 to 13 (Hinchliffe 1977).

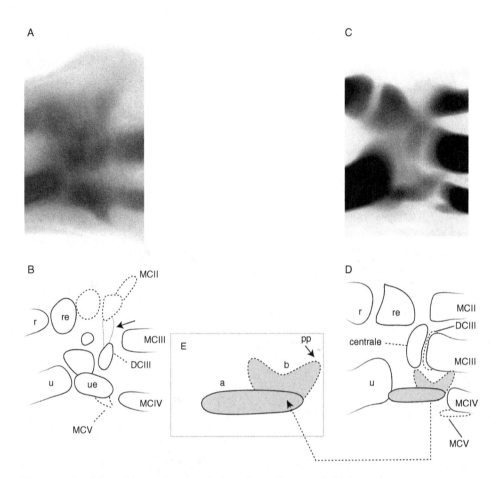

Figure 13.4 Alcian blue stained and cleared specimens of chicken wings
(M. K. Richardson, unpublished data). (A) First specimen, with (B) interpretative
drawing. Arrow indicates the condensation between DISTAL CARPALS II and III (see
Müller and Alberch 1990). (C) Second specimen, with interpretative drawing (D).
(E) Detail of (D), showing the pisiform and distal carpal IV or [IV+V]. There are
various ways of interpreting this complex (Hinchliffe 1977; Montagna 1945).
Abbreviations: a, pisiform; b, distal carpal IV or [IV+V]; pp, processus pisiformis. For
other abbreviations, see Table 13.1. The interpretation shown here in (E) is based on
preliminary observations from my lab by T. Hoppenbrouwers and V. de Winter and
shows a pisiform lying superficial to element (grey shading, dotted outline), which we
think may be derived from DISTAL CARPAL IV or DISTAL CARPALS IV+V; it has one
limb connected to the nascent semilunar complex, and the other limb projecting as
the 'processus pisiformis'.

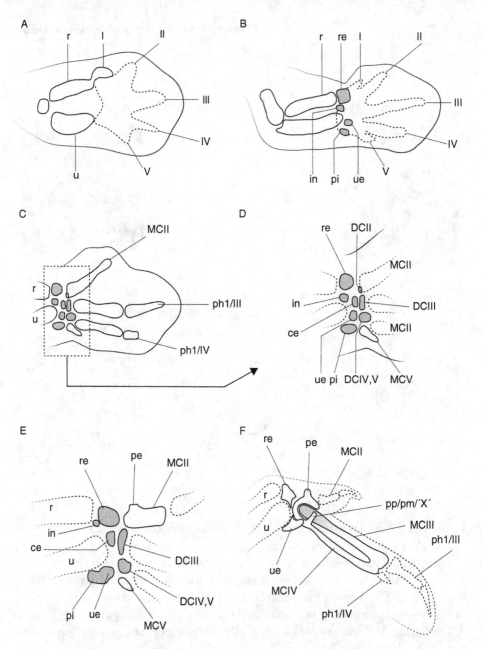

Figure 13.5 Prein's study of condensation patterns in the chicken wing (Prein 1914). Grey shading in (B–E) indicates carpal elements. (A) Right wing, after 5.5 days of incubation (Prein's text fig. 2 inverted). (B) Left forelimb after 6 days of incubation (Prein's text fig. 4 rotated 180°). (C) Left forelimb after 7 days of incubation. (D) Boxed area in (C). (E) Left forelimb after 8 days of incubation (after Prein 1914: text fig. 7). (F) Right wing after 14 days of incubation; radioventral view (Prein 1914: text fig. 11 inverted). Abbreviations: ce, centrale; DC, distal carpal; in, intermedium; pm, processus muscularis; for 'X' see below. Element in (F) labelled 'pp/pm/X' indicates possible homology with processus pisiformis (pp), process muscularis (pm), or unknown element (X). For other abbreviations, see Table 13.1.

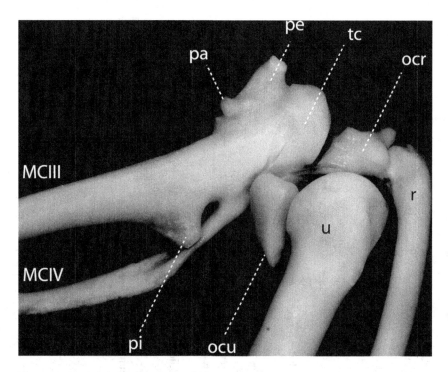

Figure 13.6 Adult chicken wing (the same specimen as the wing in Figure 13.2A, B, after partial digestion with trypsin to loosen the ligamentous attachments). The phalanx of DIGIT II has been removed to show the articular surface of METACARPAL II. Abbreviation: tc, trochlea carpalis. For other abbreviations, see Table 13.1.

Proximal carpus

In the adult, there are two free bones in the carpus (Figure 13.6), the os carpi radiale ('radiale') and os carpi ulnare ('ulnare'). These names do not necessarily indicate a developmental origin from the embryonic radiale and ulnare. In fact, the consensus is that the os carpi ulnare develops from the pisiform, while the embryonic ulnare itself disappears during development. The os carpi radiale is probably derived from the embryonic radiale, with some contributions from other elements. Differing views on the developmental of this region are summarized in Table 13.2.

The disappearance of the ulnare and its replacement by the pisiform is possibly an autapomorphic feature in Neornithes (James and Pourtless 2009). The pisiform, which develops late, is initially distal and post-axial to the ulna, in the vicinity of DISTAL CARPAL IV; it then moves to the ventral surface of the carpus to take the place of the disappearing ulnare (Sieglbauer 1911; Prein 1914; Holmgren 1955; Hinchliffe 1977). METACARPAL V also becomes displaced ventrally in birds (Prein 1914). The migration of skeletal primordia from the limb margin onto the

Table 13.2 Composition of the proximal carpus according to selected authors. Key: [x + y], elements x and y fuse; x → y, x develops into y.

Species	Developmental anatomy of proximal carpus	Reference
Ostrich (*Struthio camelus*)	[radiale + intermedium] → intermedioradiale → os carpi radiale; other proximal elements poorly described; only four specimens studied	Nassonov 1896
Ostrich (*Struthio camelus*)	[radiale + centrale] → radiocentrale → os carpi radiale; pisiform → os carpi ulnare	Norsa 1895
Ostrich (*Struthio camelus*)	There is a radiale, intermedium, ulnare, pseudoulnare, pisiform and a centrale. [intermedium + radiale] → os carpi radiale; pseudoulnare → os carpi ulnare	Kundrát 2009
Kiwis (various species of *Apteryx* including *A. australis*)	[radiale + intermedium] → os carpi radiale; ulnare → os carpi ulnare	Parker 1891; Parker 1892
Chicken (*Gallus gallus*)	radiale → os carpi radiale; ulnare → os carpi ulnare	Gegenbaur 1864
Chicken (*Gallus gallus*)	intermedio-radiale → os carpi radiale; centralo-ulnare → os carpi ulnare (these two elements form *in situ*, never having consisted first of separate elements)	Parker 1888
Chicken (*Gallus gallus*)	[intermedium + radiale] → os intermedio-radiale → os carpi radiale; the ulnare regresses and is replaced by (or possibly fuses with), the pisiform which → os carpi ulnare. A centrale becomes connective tissue	Prein 1914
Chicken (*Gallus gallus*)	[radiale + centrale I] → os carpi radiale; intermedium regresses to a strand; [ulnare + centrale III + pisiform] → carpi ulnare (four centralia are described)	Montagna 1945
Chicken (*Gallus gallus*)	A radiale, ulnare, intermedium and pisiform are present	Hinchliffe 1977; Hinchliffe and Hecht 1984
Chicken (*Gallus gallus*) + sparrow?	radiale → os carpi radiale; [ulnare + intermedium] → os carpi ulnare	Rosenberg 1873

Anser, Anas	The intermedium disappears; [centrale radiale + radiale] → os carpi radiale; ulnare disappears completely; proximal part of 'digit V' (presumably the pisiform) → os carpi ulnare	Steiner 1922
Duck (*Anas platyrhynchos*)	[rudimentary intermedium + radiale] → intermedioradiale → os carpi radiale; ulnare initially contains within its distal part the rudiments of DISTAL CARPALS IV and V; it later regresses; pisiform → os carpi ulnare	Sieglbauer 1911
Hoatzin (*Opisthocomus hoatzin*)	A free radiale, intermedium and ulnare, and a biparite centrale (only relatively late stages studied) later forming the elements as he described for the chick (see entry above)	Parker 1895
African darter (*Anhinga rufa = Plotus levaillantii*)	A radiale, intermedium and centrale radiale are present, together with an ulnare and a centrale ulnare which later completely disappear; [radiale + intermedium + centrale radiale] → os carpi radiale; 'digit V' (presumably the pisiform) → os carpi ulnare	Schestakowa 1927
Tern (*Sterna hirundo* [= *Sterna wilsonii*])	[intermedium + radiale] → intermedio-radiale → os carpi radiale; ulnare fuses with the another element, possibly a centrale or distal carpal IV, → ulnare–centrale → os carpi ulnare	Leighton 1894
Gull (*Larus canus*)	[ulnare + pisiform], or possibly the ulnare alone, → os carpi ulnare; the intermedium forms only a tendon; [tiny radiale + centrale II] → 'radiale' of all other authors → os carpi radiale	Holmgren 1933; Holmgren 1955

Note: Holmgren (1955) finds in the ostrich an independent, chondrified intermedium.

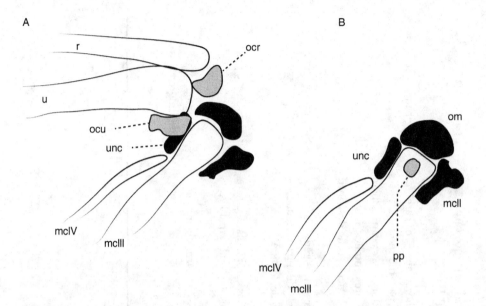

Figure 13.7 The wing of a two-month-old grouse (*Centrocercus* sp.: Galliformes), redrawn and schematized after Shufeldt (1881), and illustrating his view of the crescentic series of distal carpal elements (black fill) contributing to the proximal head of the carpometacarpus. (A) Dorsal aspect of the right wing. (B) Ventral aspect of the left wing (radius and ulna not shown). For abbreviations, see Table 13.1.

ventral surface of the carpus is seen in other species with digit loss: see the ventral migration of DIGITS II and IV in the horse, for example (Ewart 1894a, 1894b).

The embryonic distal carpus

The distal carpal condenstations contribute to the trochlear surface of the adult carpometacarpus (Figure 13.3). In the chicken, the distal carpus is made up of DISTAL CARPALS II–IV (Parker 1888). These, together with other elements, form a transient, crescentic condensation (Shufeldt 1881) found in juvenile birds (Figure 13.7A, B, in black). The concavity of this element embraces the proximal heads of the metacarpal bones and eventually fuses with them. It is also found in the ostrich *Struthio camelus* (Nassonov 1896), and may be homologous with the **semilunate** of adult maniraptorans including *Archaeopteryx lithographica*. The semilunate of maniraptorans, however, does not fuse with the metacarpals (Rosenberg 1873; Prum 2002). Kundrát (2009) considers the semilunate to be homologous to DISTAL CARPALS II+III of the ostrich. Others argue that, given the uncertainties about carpal development and digital homologies in birds, the status of this element is unclear (James and Pourtless 2009).

The pre-axial component of the semilunate element in birds is usually considered to be METACARPAL II (Rosenberg 1873; Shufeldt 1881; Leighton 1894; Prein 1914; Sieglbauer 1911; Steiner 1922; Schestakowa 1927; Kundrát 2009). The central component is DISTAL CARPAL III (Parker 1862; Shufeldt 1881; Prein 1914; Hinchliffe and Hecht 1984) or fused DISTAL CARPALS II+III (Rosenberg 1873; Leighton 1894; Steiner 1922; Schestakowa 1927; Kundrát 2009), and is closely related to the proximal end of METACARPAL III. The post-axial component of the semilunate in birds is most likely to be fused DISTAL CARPALS IV and V (Rosenberg 1873; Sieglbauer 1911; Prein 1914; Steiner 1922; Schestakowa 1927). In the ostrich it may be formed by DISTAL CARPAL IV, DISTAL CARPAL V having disappeared (Kundrát 2009).

The post-axial component of the semilunate complex is intimately associated with the ulnare-pisiform (Figures 13.4, 13.5, 13.7) and is described by some authors (Rosenberg 1873; Shufeldt 1881; Sieglbauer 1911; Schestakowa 1927) as extending a process preaxiad on the ventral side of the manus. The process, according to Sieglbauer (1911), extends towards METACARPAL II across the volar face of META-CARPAL III and ultimately fuses to the latter forming the processus pisiformis (Figures 13.3C, D; 13.5F). This process is also known as the processus muscularis of METACARPAL III (Sieglbauer 1911; Lambrecht 1914; Montagna 1945) or least helpfully of all, as 'element X' (Montagna 1945). Kundrát (2009) interprets element X of the chicken as being analogous with the DISTAL CARPAL IV/pseudoulnare complex of the ostrich. Element X may be a synapomorphy of birds which has no counterpart in the saurian pentadactyl manus (Hinchliffe and Hecht 1984).

Digit homology

The issue addressed in this section is the homology of the avian digits with those of some pentadactyl ancestor. Digit position and digit phenotype are two different levels at which the homology of digits can be considered. For discussion of 'levels' of homology see Hall (1999); and for further discussions of avian digit homology in particular, see Hinchliffe and Hecht (1984); Hecht and Hecht (1994); Burke and Feduccia (1997); Chatterjee et al. (1998); Feduccia (1999); Wagner and Gauthier (1999); Feduccia and Nowicki (2002); Larsson and Wagner (2002); Welten et al. (2005). The term 'digit identity' is often used ambiguously in the literature. I shall therefore use instead the terms 'digit position' and 'digit phenotype'.

Digit position

The position of a digit is defined by its topographical relationships to other structures. The usual way of recording digit position is to number the digits in series. These positions are referred to as condensations CI–V by Wagner and Gauthier (1999). The avian forelimb is tridactyl, and so it is not clear which two digit positions have been lost. One therefore needs to identify landmarks that are stable across phylogeny, and which can be used to calibrate the digit numbering in different species. Below, and in Table 13.3, I summarize some of the ways of calibrating the numbering of digits in relation to DIGITS I–V in the pentadactyl archosaurian manus.

I–II–III on the basis of phenotype

The simplest argument for a I–II–III positional homology of the avian digits is that they have the same phenotype (see 'Digit phenotype', below) as those three digits in pentadactyl archosaurs (Gegenbaur 1864). At least, they do if we define phenotype on the basis of phalangeal count and, as Gegenbaur (1864) pointed out, only if we limit the analysis to those birds that have claws on the first two digits. The latter proviso ensures that we are dealing with a complete digit and not a truncated one. An additional phenotypic character supporting a I–II–III hypothesis is the relative narrowness of both METACARPAL III in the crocodilian manus and the third permanent metacarpal in the avian hand (Gegenbaur 1864).

II–III–IV on the basis of forelimb-hind limb comparison

Owen (1866) noted that the tarsometatarsus of the avian foot includes fused METATARSALS II–III–IV; he therefore thought it likely that the carpometacarpus, by analogy, was made up of the the homologous metacarpals. Holmgren (1933) concluded that the bird wing digits are II–III–IV because of the similarity between the rudimentary DIGIT V in the foot and wing, and their angle of insertion on the fibulare and ulnare, respectively.

I–II–III on the basis of a trend in archosaurs towards the reduction of digits IV and V

A pattern of digit reduction identified in archosaurs affects DIGITS IV and V; these two digits become smaller and lose claws and phalanges during evolution (Rabl 1910; Steiner 1934; Romer 1956; Stephan 1992). The fact that the first two adult digits are clawed in some birds is consistent with this pattern, because it suggests that the radial digits are not truncated (Rosenberg 1911; Heilmann 1926). However, Holmgren (1955) argued that *Archaeopteryx lithographica* also had DIGITS II, III and IV, and that its ancestors showed a reduction of DIGITS I and V, not IV and V as is generally supposed. Holmgren's challenge is in turn refuted by Vargas and Fallon (2005b) who argue that it is unlikely that the three digits in birds and *Archaeopteryx lithographica* somehow converged on the same phalangeal formula. Which brings us to the frame shift.

Table 13.3 Some views on the homology of avian manal digits. Meckel (1825) and Cuvier (1835) did not explicitly make statements about the numbering of avian manal digits with respect to those in the pentadactyl limb; a, alula, bastard wing, ala spuria (Reichenow 1913); digitus alulae manus; b, digitus majoris manus; c, digitus minoris manus (Baumel and Witmer 1993)

	Transient radial digit in embryo	Radial permanent digit[a]	Middle permanent digit[b]	Ulnar permanent digit[c]	Ulnar transient digit in embryo
This chapter		II	III	IV	
Nitzsch 1811		'thumb'	large finger	little finger	
Meckel 1825		'thumb'	first finger	second finger	
Cuvier 1835		'thumb'			
Rosenberg 1873		I	II	III	IV (ossified rudiment)
Morse 1874		II	III	IV	V
Wray 1887		I	II	III	
Parker 1888	prepollex rudiment	I	II	III	IV rudiment
Leighton 1894		II	III	IV	
Norsa 1895	I	II	III	IV	V
Nassonov 1896		I	II	III	
Mehnert 1897	I (transient)	II	III	IV	(transient)
Barfurth 1911		II	III	IV	not found
Rosenberg 1911	not found	I	II	III	IV (transient)
Sieglbauer 1911	I (rudiment)	II	III	IV	V (rudiment)
Prein 1914	I (pre-cartilaginous ray only)	II	III	IV	V

Table 13.3 *(cont.)*

	Transient radial digit in embryo	Radial permanent digit[a]	Middle permanent digit[b]	Ulnar permanent digit[c]	Ulnar transient digit in embryo	
Steiner 1922	prepollex	I	II	III	IV	V
Heilmann 1926	prepollex	I	II	III	IV	
Schestakowa 1927	prepollex	I	II	III	IV	V (moves proximal)
Montagna 1945	distal carpal I	II	III	IV		
Holmgren 1955	I in ostrich (transient) / prepoll in ostrich (transient)	II	III	IV	V (transient)	pisiform part of ulnare–pisiform?
Müller and Alberch 1990	not found	II	III	IV		
Hinchliffe 1977	rudiment	II	III	IV	V (rudiment)	
Burke and Feduccia 1997	rudiment	II	III	IV		
Feduccia and Nowicki 2002	rudiment	II	III	IV		
Kundrát et al. 2002	rudiment	II	III	IV		
Larsson and Wagner 2002	rudiment	II	III	IV		
Welten et al. 2005	rudiment	II	III	IV		

II–III–IV on the basis of a frame shift

In theropods, the sequence of digit reduction seen in the fossil record is V→IV→III→II→I and this pattern is unique among amniotes, according to Hecht and Hecht (1994). If it is agreed that birds posess DIGITS II–III–IV, then they do not share this pattern of digit reduction. Various explanations for this apparent mismatch have been proposed (reviewed by Hecht and Hecht 1994). One influential hypothesis is that birds show a developmental 'frame shift' that has altered the phenotype of the digits (Wagner and Gauthier 1999). The frame shift theory has been criticized as an ad hoc auxiliary hypothesis used to explain away the incongruence of palaeontological and developmental data (James and Pourtless 2009). Conversely, morphological changes interpreted as frame shifts have been described in cyclopamine-treated chicken embryos, as well as in the tridactyl skink *Chalcides chalcides* (Vargas and Wagner 2009; Young *et al.* 2009).

II–III–IV or I–II–III on the basis of the primary axis

The primary axis identifies DIGIT IV in the forelimb and hind limb of turtles, lizards and crocodilians (Rabl 1910; Holmgren 1933; Burke and Alberch 1985; Richardson *et al.* 2009). In the chick hindlimb the primary axis passes through DIGIT IV and in the chick wing it passes through the digitus minoris manus. On these grounds, one could argue that the digitus minoris manus should be regarded as DIGIT IV. One problem with using the primary axis as a marker of position is that we cannot explain its development in terms of the underlying mechanisms. However, one clue is that the sonic hedgehog pathway shows a spatiotemporal gradient that is functionally centred on DIGITS IV and V in the mouse (Harfe *et al.* 2004). The primary axis could therefore be an emergent pattern resulting from anteroposterior and proximodistal gradients of developmental timing in the limb.

Sieglbauer (1911) confirmed that DIGIT IV was the first to develop in the foot of the duck *Anas platyrhynchos* and the ulnar permanent digit (digitus minoris manus) was the first to develop in the wing. He therefore concluded that the digitus minoris manus of the wing is homologous with DIGIT IV of the foot. Müller and Alberch (1990) reached similar conclusions on the basis of a study of limb chondrogenesis in *Alligator mississippiensis*. They noticed a band of condensed mesenchyme connecting distal mesopodials II–III in both the alligator and chick, consistent with the II–III–IV model (Figure 13.4A, C).

Some authors have found that the primary axis in the avian wing is distorted by ulnar deviation, so that it passes through DIGIT V, at least in early stages (Schestakowa 1927). Until this issue is re-examined, it seems sensible to follow the caution of Vargas and Fallon (2005b) against assuming that the primary axis is 'universal'.

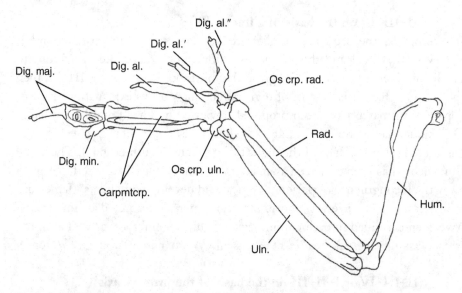

Figure 13.8 The wing of a kestrel, *Falco tinnunculus* (left wing, dorsal aspect), showing five digits. This specimen is interpreted as showing a duplication of DIGIT II, rather than the reappearance of a lost DIGIT I. (From fig.7 in Frey *et al.* 2001, courtesy of R. Frey, and with permission from Springer.) Abbreviations: Carpmtcrp., carpometacarpus; Dig. al., DIGIT II; Dig. al.', first accessory digit; Dig. al.'', secondary accessory digit; Dig. maj., DIGIT III; Dig. min., DIGIT IV; Os crp. uln., os carpi radiale; Rad., radius; Uln., ulna; Os crp. rad., os carpi radiale.

II-III-IV on the basis of functional 'rescue' of the missing digit I

Several types of manipulation lead to the experimental formation of cartilaginous structures, or even digits, anterior to DIGIT II. Development of such an ectopic digit can be triggered in the chicken wing by misexpression of *pitx1* (Szeto *et al.* 1999). This is probably due to the partial transformation of the wing into a hind limb phenotype rather than a rescue of DIGIT I. A cartilage nodule pre-axial to DIGIT II also appears in some cyclopamine-treated limbs (Vargas and Wagner 2009).

Misexpression of *hoxd11* in the developing chick foot (Morgan *et al.* 1992) transforms DIGIT I into a DIGIT II phenotype, while in the wing it results in the formation of an extra digit anterior to DIGIT II, and having a DIGIT II phenotype. One possible interpretation of these findings (Welten *et al.* 2005) is that the 'lost' DIGIT I in the wing was rescued. Alternatively, the extra digit may simply be a duplication caused by an ectopic source of polarizing activity at the radial margin of the wing (Tickle *et al.* 1982). Indeed, duplication of DIGIT II (Figure 13.8), rather than rescue of DIGIT I, seems to be the most likely explanation for some examples of pre-axial polydactyly in birds (Barfurth 1911; Frey *et al.* 2001).

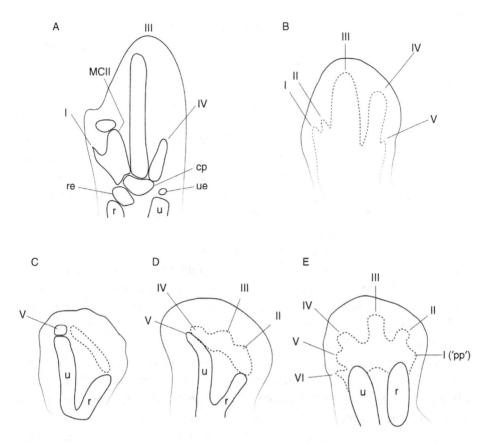

Figure 13.9 Examples of missing digits found in the avian wing bud. (A) Ostrich embryo manus, histological section (from Mehnert 1897: fig. 63); Mehnert sees a prominent ray for DIGIT I. (B) Peanut agglutinin staining (from Larsson and Wagner 2002). (C, D, E) *Plotus* forelimb (from Schestakowa 1927: Plate I, fig. 3). Abbreviations: 'pp', the so-called prepollex; other abbreviations are as in Table 13.1.

II-III-IV on the basis of pentadactyly in the embryo

A transient stage of embryonic pentadactyly has been reported in ungulates, whose adult skeleton shows a reduced number of digits (Mehnert 1897). Transient pentadactyly or hexadactyly (Figure 13.9) has even been reported in the embryonic avian wing (Norsa 1895; Mehnert 1897; Sieglbauer 1911; Schestakowa 1927) and these reports are consistent with models in which DIGITS I and V are the two lost in birds (Barfurth 1911). More recently, pentadactyly has been confirmed by alcian blue staining (Feduccia and Nowicki 2002), Indian-ink mapping of vascular networks (Kundrát *et al.* 2002), peanut agglutinin labelling (Larsson and Wagner 2002) and *Sox9* expression (Welten *et al.* 2005).

Hinchliffe (1977) used $^{35}SO_4$ autoradiography, a technique which visualizes early chondrogenic elements in the chick wing. He found no evidence that any 'missing' digits make a transient appearance during development; only those detectable by routine histology (II–V) were seen. Nonetheless, 'lost' elements may form transiently as prechondrogenic mesenchymal condensations. Thus Sieglbauer (1911), Prein (1914), Steiner (1922), Schestakowa (1927), Welten *et al.* (2005) and Kundrát (2009) all find a condensation radial to the digitus alulae manus which they see as a rudimentary DIGIT I or prepollex. The processus extensorius of METACARPAL II (Figure 13.2B, C and Figure 13.5E, F) may seem like a tempting candidate for a lost DIGIT I – at least it tempted Holmgren (1955). However, this structure appears to develop quite independently of any digit, lost or otherwise (Prein 1914).

Rosenberg (1873) was probably the first to describe what we would now call METACARPAL V in the wing. He noted a spindle-shaped rudiment, which ossified and then fused with the proximal shaft of METACARPAL IV (on its volar face, towards the post-axial side) to form the *crista* of the latter. Many other authors have confirmed the presence of a rudimentary METACARPAL V (Parker 1888; Sieglbauer 1911; Steiner 1922; Hinchliffe 1977; Kundrát 2009)

The pisiform is often considered to be – or is confused with – a rudimentary DIGIT VI or 'postminimus' (Mehnert 1897; Steiner 1922; Schestakowa 1927; Holmgren 1952). However, Hinchliffe (1977) argues that the pisiform is too proximal to be a digit.

II-III-IV on the basis of 'laws' of digit reduction

Some authors (Barfurth 1911; Sieglbauer 1911) assign the identities II-III-IV to the avian digits on the basis that digit reduction in mammals occurs commonly in the pattern I→V→II→IV→III. This pattern was identified by Flower who said:

If one [digit] is absent [in mammals], it is most commonly the first. (Flower 1870, p. 255)

This, in turn, influenced Morse in the framing of his statement, sometimes referred to as **Morse's law** (Shubin *et al.* 1997; Wagner *et al.* 1999):

... when the number of fingers or toes is reduced in Mammalia and Reptilia, they are always taken away from the sides of the member, the thumb first disappearing and then the little finger. (Morse, 1874, p. 153)

Morse's statement seems to constitute *two* laws: (1) digit reduction is bilateral, (2) DIGIT I is reduced before DIGIT V; perhaps unwise to conflate these two. Young *et al.* (2009) attribute these laws to developmental constraints.

Personally, I think Morse's law is simply an observation that something happens often. Digits are indeed often lost bilaterally, and DIGIT I often disappears before DIGIT V; the skink genus *Hemiergis* provides a good example of both (Shapiro 2002). However, there are some very significant exceptions which suggest that these phenomena are lineage specific.

Ironically, one of the exceptions to Morse's 'law' are the birds. The sequence of digit loss in the wing is a Morse-like I→V→II→IV→III, but in the foot it is a law-breaking V→I→II→IV→III (Montagna 1945). There is a similar forelimb/hind limb discrepancy in some tortoises. In *Testudo graeca*, for example, DIGIT I is reduced in the manus while DIGIT V is reduced in the pes (Rabl 1910). In the skink genus *Calchides*, DIGIT V alone may be reduced with no reduction in DIGIT I (reviewed by Young *et al.* 2009). Some tortoises show a trend towards the reduction of the post-axial digits in the hind limb rather than bilateral reduction. In *Testudo marginata* [= *T. campanulata*] and *Psammobates geometricus* [= *T. gemoetrica*], TOE IV is very rudimentary and TOE V missing (Rabl 1910). Further exceptions to Morse's law are the loss of DIGIT V first on the manus of all living Caviidae, the same also in the capybara *Hydrochoerus hydrochaeris* (Lande 1977) and the Grant's golden mole *Eremitalpa granti* (Kindahl 1949); the loss of DIGIT V, while a rudimentary DIGIT I is still present in the manus of the two-toed sloths *Choloepus hoffmanni* and *C. didactylus* (Mendel 1981); and the reduction of DIGIT V, and the loss of its claw, while DIGIT I remains better developed and clawed, in the manus of the anteater *Tamandua tetradactyla* (Taylor 1978).

II-III-IV on the basis of the relations of the pisiform

This argument (Hinchliffe and Hecht 1984; Müller and Alberch 1990) is as follows. The most posterior element in the chick wing is the pisiform, and it lies on the postaxial side of the transient metacarpal (which in turn lies on the postaxial side of the digitus minoris manus). In pentadactyl archosaurs and turtles, the pisiform occupies a similar boundary position and is post-axial to DIGIT V (Gegenbaur 1876; Sánchez-Villagra *et al.* 2009). On the grounds of these topographic relations, the rudimentary metacarpal in the chick wing should be numbered METACARPAL V.

II-III-IV on the basis of anterior cell death

Hinchliffe and Hecht (1984) argue that massive cell death on the anterior border of the chick wing, in a region known as the anterior necrotic zone (Hinchliffe and Ede 1973), is consistent with the elimination of an anterior digit. In *wingless* chicken mutants, the anterior necrotic zone is precociously developed and abnormally extended (Hinchliffe 1976). The chicken wing also possesses a posterior necrotic zone, proximal to the ZPA and partly overlapping it, which may represent former sonic hedgehog-expressing cells from the ZPA (Harfe *et al.* 2004). The posterior necrotic zone remains proximal to the handplate (Harfe *et al.* 2004). There is some disagreement about whether necrotic zones of the chick limb are comparable to those seen in rodents (Martin 1990; Milaire 1992).

An aborted middle finger on the basis of an 'intercalary' element

A rudimentary or 'intercalary' structure has been described as lying between the metacarpals of DIGITS III and IV in the avian wing, and this has been interpreted as the transiently developed metacarpal of a 'lost' middle digit (Heusinger 1820; Parker 1887; Tschan 1889). Parker later retracted this opinion (Parker 1888), on the grounds that the intecalary ray did not show a digit-like developmental pattern and was not present in ratites.

Digit phenotype

The phenotype of a digit reflects its intrinsic properties, not its relations to external landmarks or its number in a series. Intrinsic properties include the number and shape of phalanges (Table 13.4), the presence or absence of claws (Table 13.5), and characteristic patterns of gene expression (Table 13.6).

Mechanisms for specifying digit phenotype

Great progress has been made in understanding the specification of digit phenotype, and in how anteroposterior positional information is translated into differences in phalanx number. Earlier models (Tabin 1992), in which individual digits were encoded by *hox* genes acting as homeotic selector genes, have been modified (Tabin, interviewed by Richardson 2009). *Hox* genes appear to be more involved in the recruitment of cells into mesenchymal condensations, and in their subsequent growth (Dollé *et al.* 1993; Favier and Dollé 1997). They seem to act in a dose-dependent manner rather than in the manner of a combinatorial code (Zákány *et al.* 1997).

Dose is also important in the case of sonic hedgehog signalling. Different concentrations and duration of exposure influence the development of limb mesenchyme along different digit phenotypes (Harfe *et al.* 2004; Scherz *et al.* 2007). Interestingly, manual morphology in the *Hemiergis* lizards was found to correlate with differences in the spatial or temporal pattern of sonic hedgehog expression (Shapiro *et al.* 2003). Levels of Gli3 repressor (Ahn and Joyner 2004) and bone morphogenetic proteins (Drossopoulou *et al.* 2000) also play a role in specifying digit phenotype.

Table 13.4 Selected examples where both the phalangeal formula *and* the presence or absence of claws has been noted

Most common in birds	1-2-1	Baumel and Witmer 1993
Huoshanornis huji	2c-3c-1	Wang *et al.* 2010
Gracilitarsus mirabilis	2c-2-1	Mayr 2003
Hoatzin	2c-3c-1	Banzhaf 1929
(*Opisthocomus hoatzin*) nestling[a]		
Hoatzin adult	2c-2-1	Baumel and Witmer 1993
Ostrich embryo	2c-3c-2	Kundrát 2009
(*Struthio camelus*)		
Young ostrich	2c-3-2	Owen 1866
Ostrich adult	2c-3c-1	Kundrát 2009
Some anatids	2-3-1	Baumel and Witmer 1993;
		Shufeldt 1909
Chicken embryo (*Gallus gallus*)	2c-3-2	Prein 1914
Chicken embryo (*Gallus gallus*)	2c-3c-2	Parker 1888
Chicken adult	2c-2-1	Prein 1914
Ducks and swans	2-3-1	Meckel 1825
Anas sp.	2c-3c-1	Sieglbauer 1911

[a] In the hoatzin, *Opisthocomus hoatzin*, DIGIT IV may have a claw in the embryo, but it has disappeared by the nestling stage (Banzhaf 1929).

Sanz-Ezquerro and Tickle (2003) found that extra phalanges could be induced by implanting Shh-impregnated beads. All toes in the chick foot could be induced to form extra phalanges, but the same was only true of DIGIT II in the wing. The authors suggest that when *fgf8* switches off in the ridge, a programme of **tip formation** is implemented which involves *wnt14* and leads to the development of the ungual. They also found that formation of an additional phalanx was associated with extended activity of the apical ridge. This supports the hypothesis (Richardson and Oelschlager 2002) that hyperphalangy in animals such as dolphins has a heterochronic basis.

The cells which translate AP signals into phalangeal formula may reside in the **phalanx-forming region** that caps each digit and is derived from the mesenchyme under the AER (Suzuki *et al.* 2008). The phalanx-forming regions of each digit appear to have a different profile of phosphorylated SMADs. This could explain something observed in the skink genus *Hemiergis* (Shapiro *et al.* 2003): metacarpals that form phalanges have more uncondensed mesenchyme at their tips (possibly the phalanx-forming region) than vestigial metacarpals that do not.

Table 13.5 Taxonomic distribution of claws in Neornithes. There is evidence of progressive distal truncation of DIGITS III and IV in the evolution of modern birds.

II	III	IV	Taxonomic distribution
claw	**claw**	**claw**	Struthioniformes
claw	**claw**	no claw	Tinamiformes, Opisthocomiformes
claw	**claw**	no claw	These formulae variably present within the following orders:
claw	**claw**	no claw	Podicipediformes, Ciconiiformes, Anseriformes, Cathartiformes,
claw	no claw	no claw	Falconiformes, Gruiformes, Charadriiformes, Psittaciformes,
no claw	no claw	no claw	Strigiformes
claw	no claw	no claw	Gaviiformes, Pelecaniformes (Phaëthontidae), Phoenicopteriformes
claw	no claw	no claw	These formulae variably present within the following orders: Galliformes,
no claw	no claw	no claw	Cuculiformes, Apodiformes
no claw	no claw	no claw	Procellariiformes (or exceptionally a rudimentary first claw), Sphenisciformes, Pelecaniformes, (excluding Phaëthontidae), Pterocliformes, Columbiformes, Caprimulgiformes (or exceptionally a rudimenary first claw), Trochiliformes, Coliiformes, Trogoniformes, Coraciiformes, Piciformes, Passeriformes

Source: Adapted from Stephan (1992).

Table 13.6 Some molecular markers for different anatomical structures or regions

Anatomical structure or region (context)	Marker/expression profile	Reference
DIGIT I (the pre-axial margin of the autopod in the mouse forelimb, and chick hind limb, during late-phase *hox* gene expression)	*hoxd10⁻*, *hoxd11⁻*, *hoxd12⁻*, *hoxd13⁺*, *hoxa13⁺*	Chiang *et al.* 2001; Vargas *et al.* 2008; Vargas and Fallon 2005a
DIGIT II (the region of the distal tip of this digit in the chick wing)	*mario⁺*	Amano and Tamura 2005
DIGIT I (mouse forelimbs and hind limbs)	*Pax9⁺*	Peters *et al.* 1998
DIGIT I (*Alligator mississippiensis* forelimb and hind limb, chicken hind limb);	*hoxd11⁻*	Vargas *et al.* 2008
Tips of DIGITS I–IV in hind limb but only DIGIT II of the forelimb (wing DIGITS III and IV lack expression)	BAMBI	Casanova and Sanz-Ezquerro 2007

Digit reduction and morphoclines

Digits III and IV in the wing are unique among chicken digits in being refractory to the induction of extra phalanges (Sanz-Ezquerro and Tickle 2003). These two digits also lack claws in many species of bird (Tables 13.4, 13.5). An obvious hypothesis to explain such observations is that DIGITS III and IV are affected by distal truncation, and that this in turn reflects the pattern of post-axial digit reduction seen in theropods. This pattern of digit reduction has continued in modern birds (Stephan 1992).

Three patterns of digit reduction are seen in tetrapods: (1) reduction in size of elements, (2) distal truncation – the loss of terminal phalanges and/or claws; and (3) loss of non-terminal phalanges.

Size reduction, with no loss of elements, is seen in the ostrich wing. Distal truncation involves loss of the claw first, then the terminal phalanx, and so on for each successive proximal element. This distal-to-proximal rolling back of the normal developmental sequence is seen in avian wing DIGITS IV and V and in teiid lizards (Presch 1975). Interestingly, the gene BAMBI, an inhibitor of

BMP signalling, is expressed at the tips of digits in mice when the terminal phalanges are developing (Grotewold *et al.* 2001). In the chicken wing, BAMBI is expressed in the four pedal digits, and in the wing at the tip of DIGIT II; but it is not expressed in wing DIGITS III or IV (Casanova and Sanz-Ezquerro 2007). This is consistent with distal truncation.

Digit reduction via the loss of non-terminal phalanges, without any loss of claws, was noted by Sieglbauer (1911, p. 283 footnote 1) and Young *et al.* (2009) who comment on digit reduction without claw-loss as seen in the skink *Chalcides chalcides*. The loss of non-terminal phalanges, with retention of the claws, is seen in *Hemiergis* skinks (Shapiro 2002). In *Hox* mutant mice, digits lose non-terminal phalanges and become smaller as *Hoxd* genes are inactivated, without losing their claws (Zákány *et al.* 1997). Ultimately, though, a digit in these mutants may be reduced to an unsegmented, unclawed rod of cartilage (Zákány *et al.* 1997) rather like the METACARPAL V of the chicken wing. Casanova and Sanz-Ezquerro (2007) give further examples of the loss of phalanges without the loss of claws or nails.

The reduction of digits IV and V in crocodilians is accompanied by the loss of claws on those two digits (Gegenbaur 1864; Müller and Alberch 1990). The last two phalanges on DIGITS IV and V may remain cartilaginous in the adult giving a phalangeal formula of 2c-3c-4c-5-4 (Müller and Alberch 1990; Rabl 1910). Note that 'c' indicates that the terminal phalanx forms the bony core of an epidermal claw. The very interesting fossil bird *Huoshanornis huji* has a phalangeal formula 2c-3c-1; the claw on DIGIT II is well developed, although digit II itself is short (Wang *et al.* 2010). In *Archaeopteryx lithographica*, DIGIT II is clawed and long, while in modern birds it is short and often unclawed. This suggests a morphocline with progressive reduction in the proximal phalanx of DIGIT II (Zhang and Zhou 2000) and, ultimately, loss of the claw.

Phalangeal formula of truncated wing digits

A common way of defining digit phenotype, in developmental studies and in palaeontology, is by indicating the number of phalanges on each digit, from pre-axial to post-axial. For *Archaeopteryx lithographica* the wing has the phalangeal formula 2c-3c-4c. Whether this formula corresponds to that of modern birds is not entirely obvious because Neornithes have lost certain claws and phalanges during evolution (and may recapitulate this loss by forming and then losing them during development).

As if to make the phalangeal formula of birds more difficult to count, there is the problem of claws dropping off the dead birds collected for museums (Stephan 1992), and of the terminal phalanges being lost during the maceration or storage of skeletons (Meckel 1825). Despite these annoyances, we can hardly improve on the 200-year-old summary of Nitzsch (1811): excluding the ungual phalanges, the phalangeal formula of birds is typically 1:2:1. This number is sometimes increased by the addition of a true ungual phalanx, or a styliform (pointed) phalanx with no real epidermal claw covering it. These two types of addition are seen quite often on DIGIT II, occasionally on DIGIT III and very rarely on DIGIT IV.

The few cases where DIGIT IV has more than one phalanx include the description by Baur (1885) of a second cartilaginous phalanx on DIGIT IV in *Anas domestica*. Further, *Numenius* embryos show a small extra phalanx on DIGIT IV according to Parker (1888). Wray (1887) found in the ostrich wing a distinct ossified proximal phalanx on DIGIT IV, and also a distal cartilaginous rod that fades out distally into connective tissue. He finds that this distal rod contains at its proximal end an ossified second phalanx, while the region of the rod distal to this, he suggests, may represent unsegmented phalanges 3 and possibly 4. If Wray's speculation could be proved correct, then it would of course make the avian digit IV consistent in phalangeal formula with the middle digit of pentadactyl archosaurs. Nassonov (1896) also studied the ostrich and found that while digit IV does indeed develop two phalanges, the distal one regresses and fails to develop a claw.

Parker shares Wray's hope of finding the reptilian phalangeal formula. He says of the hoatzin (*Opisthcomos hoatzin*) embryo:

The phalanges of the [digit IV] are in a curiously-aborted condition; there is an attempt at segmentation into what should be, according to the Reptilian norma, four phalanges. (Parker 1895, p. 72)

DIGITS II and III in the *Opisthocomus hoatzin* embryo have large, well developed claws (Parker, 1895). In *Opisthocomus hoatzin* nestlings, the claws on DIGITS II and III persist, but in the adult hoatzin all claws are reduced to an epidermal tubercle (Banzhaf 1929).

An ossified phalangeal formula of 2-3-1 is found in *Anas platyrhynchos*, *Cairina moschata* and *Somateria mollissima* (Maxwell 2008), and I illustrate this formula with the duck wing in Figure 13.2. Hogg (1980) studied chicken ossification patterns in the chicken wing with radiography and alizarin red staining. He found a phalangeal formula of 2-2-1 in adults, and this was occasionally (in 3.8% of cases) reduced to 1-2-1. Sieglbauer (1911) sees a transient claw on DIGIT IV in *Anas platyrhnchos* that later regresses. Schestakowa (1927) found that the number of phalanges in the digits is higher in embryos and then regresses due to loss of the small distal phalanges. In *Plotus* he found a formula of 2-3-2, with DIGIT III having two cartilaginous phalanges; and in *Anser*, 2-3-3, with DIGIT III having two cartilaginous phalanges and a mesenchymal terminal phalanx. In the penguins of the Spheniscidae, the phalangeal formula is 0-2-1 (Watson 1883).

Parker (1888) finds a formula 2c-3c-2 in the embryonic chick. The claw on DIGIT III is never fully developed and the distal phalanx disappears by fusion with the middle phalanx. The claw of DIGIT II develops more fully, and is covered with horny epidermis. The second phalanx on DIGIT IV also disappears. In the wing bud of *Apteryx*, DIGIT III has 2 or 3 phalanges and forms the sole digit of the wing (Parker, 1891). The ostrich has claws on wing DIGITS II and III; *Rhea americana* and *R. pennata* on DIGIT II only; *Cassuarius cassuarius*, *Dromaius novae-hollandae*, *Apteryx australis* and *A. owenii*, on DIGIT III only (Stephan 1992).

Future challenges

Developmental anatomy is an interdisciplinary endeavour that grows with contributions across multiple disciplines. Success in identifying avian wing homologies will hinge on increased understanding of embryology, developmental genetics and palaeontology. For example, simple lineage-marking studies could help solve some major unresolved questions, such as the developmental composition of the os carpi ulnare and the os carpi radiale.

It will also be useful to follow up on reports that cells may 'remember' at least part of their developmental transcription profile into adulthood. Whether one could ever rewrite the anatomy textbooks, with molecular signatures standing next to the anatomical names, remains to be seen. But one could at least study homology simply by taking tiny tissue samples from different regions of an adult animal. In this way, the embryo could be bypassed altogether for some purposes.

This is likely to be necessary in view of the increasing difficulty, at least for birds and mammals, of obtaining embryos from a large phylogenetic sample. It is either expensive or impossible to establish breeding colonies of many of the most interesting species; some are so rare that one has little chance of obtaining wild material, and in any case there are increasingly severe (but essential) restrictions on collecting material in the wild. Finally, the study of embryos always means extra work because multiple stages have to be analysed. Nevertheless, information on the development and embryology of non-model species – even rare ones – does exist (e.g. Thewissen *et al.* 2006; Mitgutsch *et al.* 2012, this volume; Sears *et al.* this volume), and it is reasonable to expect additional insight on limb development to be forthcoming in vertebrates beyond chicken and mouse. Such studies could help place developmental data in a phylogenetic context and, ultimately, to better understand the chicken wing.

Summary

There has long been disagreement about the developmental homologies of skeletal elements in the avian manus. This is because digits, claws and carpal elements have been lost or reduced during avian evolution, making it difficult to identify homologies with the bones of the pentadactyl limb. Here, I review some of the morphological, developmental and molecular evidence relating to the carpal and digital homology of birds. I find that developmental anatomy may be at the limits of its ability to address some of the homology issues. Fortunately, recent years have seen major advances in our understanding of the avian wing. There are now molecular markers for early chondrogenic foci, and a few anatomical structures have been defined in terms of characteristic gene expression profiles. An important theoretical advance has been the 'frame shift' theory, which is supported by an overwhelming body of circumstantial evidence. Another recent step forward has been the characterization of a 'phalanx-forming region' at the tip of each chick digit. This region translates positional identity into phalangeal formula. As to the future of wing homology studies, there is some promise of being able to characterize the different anatomical parts of an adult in terms of unique expression profiles, some of them remembered since developmental stages. It is conceivable that one day it may be possible to take a series of tiny tissue samples from different parts of an adult animal, bypassing the need for embryos altogether.

Acknowledgements

I thank D. Duboule, M. Shapiro and J. Zákány for brief but helpful communications about various topics related to this chapter; and G. Wagner, and H. Larsson for permission to copy figures. R. Frey very kindly supplied the figure of the pentadactyly kestrel wing. Two of my students, T. Hoppenbrouwers and V. de Winter, kindly agreed to let me use their unpublished observations on the development of the skeleton of the chicken wing for this chapter.

REFERENCES

Ahn, S. and Joyner, A. L. (2004). Dynamic changes in the response of cells to positive hedgehog signaling during mouse limb patterning. *Cell*, **118**, 505–16.

Amano, T. and Tamura, K. (2005). Region-specific expression of mario reveals pivotal function of the anterior nondigit region on digit formation in chick wing bud. *Development Dynamics*, **233**, 326–36.

Ballmann, P. (1969). Die Vögel aus der altburdigalen Spaltenfüllung von Wintershof (West) bei Eichstätt in Bayern. *Zitteliana*, **1**, 5–60.

Banzhaf, W. (1929). Die vorderextremitat von opisthocomus cristatus (vieillot). *Zoomorphology*, **16**, 113–233.

Barfurth, D. (1911). Experimentelle Untersuchung Über die Vererbung der Hyperdactylie bei Hühnern. *Archiv für Entwicklungsmechanik*, **33**, 255–73.

Baumel, J. J. and Witmer, L. M. (1993). Osteologia. In *Handbook of Avian Anatomy: Nomina Anatomica Avium*, 2nd edn., ed. J. J. Baumel, A. S. King, J. E. Breazile, H. E. Evans and J. C. Vanden Berge. Cambridge, MA: Nuttall Ornithological Club, pp. 45–122.

Baur, G. (1885). A second phalanx in the third digit of a carinate-bird's wing. *Science*, **5**, 355.

Braus, H. (1906). Die Entwickelung der Form der Extremit 'ten und des Extremit' tenskeletts. In *Handbuch der vergleichenden und experimentellen Entwickelungslehre der Wirbeltiere*, ed. O. Hertwig. Jena, Germany: Gustav Fischer, pp. 167–338.

Burke, A. C. and Alberch, P. (1985). The development and homology of the chelonian carpus. *Journal of Morphology*, **186**, 119–31.

Burke, A. C. and Feduccia, A. (1997). Developmental patterns and the identification of homologies in the avian hand. *Science*, **278**, 666–8.

Casanova, J. C. and Sanz-Ezquerro, J. J. (2007). Digit morphogenesis: is the tip different? *Development, Growth and Differentiation*, **49**, 479–91.

Chatterjee, S., Garner, J. P., Thomas, A. L. R., Burke, A. C. and Feduccia, A. (1998). Counting the fingers of birds and dinosaurs. *Science*, **280**, 355.

Chiang, C., Litingtung, Y., Harris, M. P. S., Li, Y., Beachy, P. A. and Fallon, J. F. (2001). Manifestation of the limb prepattern: limb development in the absence of sonic hedgehog function. *Developmental Biology*, **236**, 421–35.

Cuvier, G. (1835). *Leçons d'Anatomie Comparée de Georges Cuvier*. Paris: Crochard.

Dollé, P., Dierich, A., LeMeur, M., *et al.* (1993). Disruption of the Hoxd-13 gene induces localized heterochrony leading to mice with neotenic limbs. *Cell*, **75**, 431–41.

Donegan, T. M. (2008). Comment on the proposed conservation of *Columba roseogrisea* Sundevall, 1857 (currently *Streptopelia roseogrisea*; Aves, Columbidae) (Case 3380). *Bulletin of Zoological Nomenclature*, **65**, 63.

Drossopoulou, G., Lewis, K. E., Sanz-Ezquerro, J. J., *et al.* (2000). A model for anteroposterior patterning of the vertebrate limb based on sequential long- and short-range Shh signalling and Bmp signalling. *Development*, **127**, 1337–48.

Ewart, J. C. (1894a). The development of the skeleton of the limbs of the horse, with observations on polydactyly. *Journal of Anatomy and Physiology*, **28**, 236–56.

Ewart, J. C. (1894b). The Development of the Skeleton of the Limbs of the Horse, with Observations on Polydactyly: Part II. *Journal of Anatomy and Physiology*, **28**, 342–369.

Favier, B. and Dollé, P. (1997). Developmental functions of mammalian Hox genes. *Molecular Human Reproduction*, **3**, 115–31.

Feduccia, A. (1999). 1,2,3 = 2,3,4: accommodating the cladogram. *Proceedings of the National Acadamy of Sciences of the United States of America*, **96**, 4740–2.

Feduccia, A. (2002). Birds are dinosaurs: simple answer to a complex problem. *The Auk*, **119**, 1187–201.

Feduccia, A. and Nowicki, J. (2002). The hand of birds revealed by early ostrich embryos. *Naturwissenschaften*, **89**, 391–3.

Feduccia, A., Martin, L. D. and Tarsitanoc, S. (2007). *Archaeopteryx* 2007: quo vadis? *The Auk*, **124**, 373–80.

Flower, W. H. (1870). *Osteology of the Mammalia*. London: Macmillan.

Frey, R., Albert, R., Krone, O. and Lierz, M. (2001). Osteopathy of the pectoral and pelvic limbs including pentadactyly in a young kestrel (*Falco t. tinnunculus*). *Journal of Ornithology*, **142**, 335–66.

Galis, F., Kundrát, M. and Sinervo, B. (2003). An old controversy solved: bird embryos have five fingers. *Trends in Ecoology and Evolution*, **18**, 7–9.

Gegenbaur, C. (1864). *Untersuchungen zur vergleichenden Anatomie der Wirbelthiere: Erstes heft. Carpus und Tarsus*. Leipzig, Germany: Engelmann, 127 pp.

Gegenbaur, C. (1876). Zur Morphologie der Gliedmassen de Wirbletiere. *Morphologie Jahrbuch*, **2**.

Grotewold, L., Plum, M., Dildrop, R., Peters, T. and Ruther, U. (2001). Bambi is coexpressed with Bmp-4 during mouse embryogenesis. *Mechanisms of Development*, **100**, 327–30.

Hall, B. K. (1999). *Homology*. Novartis Foundation Symposium 222. Chichester, UK: Wiley, 256 pp.

Harfe, B. D., Scherz, P. J., Nissim, S., *et al.* (2004). Evidence for an expansion-based temporal shh gradient in specifying vertebrate digit identities. *Cell*, **118**, 517–28.

Hecht, M. K. and Hecht, B. M. (1994). Conflicting developmental and paleontological data: the case of the bird manus. *Acta Palaeontologica Polonica*, **38**, 329–38.

Heilmann, G. (1926). *The Origin of Birds*. London: Witherby, 208 pp.

Heusinger, C. F. (1820). Zootomische Analekten. *Deutsches Archiv für die Physiologie*, **6**, 544–52.

Hinchliffe, J. R. (1976). The development of winglessness (ws) in the chick embryo. *Colloque International du C.N.R.S. Paris*, **266**, 175–82.

Hinchliffe, J. R. (1977). The chondrogenic pattern in chick limb morphogenesis: a problem of development *and* evolution. In *Vertebrate Limb and Somite Morphogenesis*, ed. D. A. Ede, J. R. Hinchliffe and M. Balls. Cambridge, UK: Cambridge University Press, pp. 293–309.

Hinchliffe, J. R. (1989). Reconstructing the archetype: innovation and conservatism in the evolution and development of the pentadactyl limb. In *Complex Organismal Functions: Integration and Evolution in Vertebrates*, ed. D. B. Wake and G. Roth. Dahlem Konferenzen. Chichester, UK: John Wiley, pp. 171–89.

Hinchliffe, J. R. (2002). Developmental basis of limb evolution. *International Journal of Developmental Biology*, **46**, 835–45.

Hinchliffe, J. R. and Ede, D. A. (1973). Cell death and the development of limb form and skeletal pattern in normal and wingless (ws) chick embryos. *Journal of Embryology and Experimental Morphology*, **30**, 753–72.

Hinchliffe, J. R. and Hecht, M. K. (1984). Homology of the bird wing skeleton: embryological versus paleontological evidence. *Evolutionary Biology*, **18**, 21–39.

Hogg, D. A. (1980). A re-investigation of the centres of ossification in the avian skeleton at and after hatching. *Journal of Anatomy*, **130**, 725–43.

Holmgren, N. (1933). On the origin of the origin of the tetrapod limb. *Acta Zoologica*, **14**, 185–295.

Holmgren, N. (1952). An embryological analysis of the mammalian carpus and its bearing upon the question of the origin of the tetrapod limb. *Acta Zoologica*, **33**, 1–115.

Holmgren, N. (1955). Studies on the phylogeny of birds. *Acta Zoologica*, **36**, 243–328.

James, F. C. and Pourtless, J. A. (2009). Cladistics and the origin of birds: a review and two new analyses. *Ornithological Monographs*, **66**, 1–78.

Kindahl, M. (1949). The embryonic development of the hand and foot of *Eremitalpa (Chrysochloris) granti* (Broom). *Acta Zoologica*, **30**, 1–20.

Kundrát, M. (2009). Primary chondrification foci in the wing basipodium of *Struthio camelus* with comments on interpretation of autopodial elements in Crocodilia and Aves. *Journal of Experimental Zoology B–Molecular and Developmental Evolution*, **312**, 30–41.

Kundrát, M., Seichert, V., Russell, A. P. and Smetana, K. (2002). Pentadactyl pattern of the avian wing autopodium and pyramid reduction hypothesis. *Journal of Experimental Zoology B–Molecular and Developmental Evolution*, **294**, 152–9.

Lambrecht, K. (1914). Morphologie des Mittelhandknochens – Os metacarpi – der Vögel. *Aquila*, **21**, 53–84.

Lande, R. (1977). Evolutionary mechanisms of limb loss in tetrapods. *Evolution*, **32**, 73–92.

Larsson, H. C. and Wagner, G. P. (2002). Pentadactyl ground state of the avian wing. *Journal of Experimental Zoology*, **294**, 146–51.

Leighton, V. L. (1894). The development of the wing of *Sterna wilsonii*. *American Naturalist*, **28**, 761–74.

Lewis, J. (1977). Growth and determination in the developing limb. In *Vertebrate Limb and Somite Morphogenesis*, ed. D. A. Ede, J. R. Hinchliffe and M. Balls. Cambridge, UK: Cambridge University Press, pp. 215–28.

Linnaeus, C. (1758). *Systema Naturæ*, 10th edn. Stockholm: Laurentii Salvii, 824 pp.

Livezey, B. C. and Zusy, R. L. (2006). Phylogeny of Neornithes. *Bulletin of Carnegie Museum of Natural History*, 1–544.

Livezey, B. C. and Zusi, R. L. (2007). Higher-order phylogeny of modern birds (Theropoda, Aves: Neornithes) based on comparative anatomy. II. Analysis and discussion. *Zoological Journal of the Linnean Society*, **149**, 1–95.

Martin, P. (1990). Tissue patterning in the developing mouse limb. *International Journal of Developmental Biology*, **34**, 323–36.

Maxwell, E. E. (2008). Ossification sequence of the avian order anseriformes, with comparison to other precocial birds. *Journal of Morphology*, **269**, 1095–113.

Mayr, G. (2003). A new specimen of the tiny Middle Eocene bird *Gracilitarsus mirabilis* (new family: Gracilitarsidae). *Condor*, **103**, 78–84.

Meckel, J. F. (1825). *System der vergleichenden Anatomie: Skelet der Vögel*. Halle, Germany: Renger, 638 pp.

Mehnert, E. (1897). *Kainogenesis als Ausdruk differenter phylogenetischer Energien*. Jena, Germany: Verlag von Gustav Fischer, 165 pp.

Mendel, F. C. (1981). The hand of two-toed sloths (*Choloepus*): its anatomy and potential uses relative to size of support. *Journal of Morphology*, **169**, 1–19.

Milaire, J. (1992). A new interpretation of the necrotic changes occurring in the developing limb bud paddle of mouse embryos based upon recent observations in four different phenotypes. *International Journal of Developmental Biology*, **36**, 169–78.

Mitgutsch, C., Richardson, M. K., Jiménez, R., *et al.* (2012). Circumventing the pentadactyly 'constraint': autopodial recruitment of pre-axial structures in true moles. *Biology Letters*, **8**, 74–7.

Montagna, W. (1945). A re-investigation of the development of the wing of the fowl. *Journal of Morphology*, **76**, 87–113.

Morgan, B. A., Izpisua-Belmonte, J. C., Duboule, D. and Tabin, C. J. (1992). Targeted misexpression of *Hox-4.6* in the avian limb bud causes apparent homeotic transformations. *Nature*, **358**, 236–9.

Morse, E. S. (1874). On the tarsus and carpus of birds. *Annals of the Lyceum of Natural History of New York*, **10**, 141–58.

Müller, G. B. (1991). Evolutionary transformation of limb pattern: heterochrony and secondary fusion. In *Developmental Patterning of the Vertebrate Limb*, ed. J. R. Hinchliffe, J. M. Hurle and D. Summerbell. New York: Plenum Press, pp. 395–405.

Müller, G. B. and Alberch, P. (1990). Ontogeny of the limb skeleton in *Alligator mississippiensis*: developmental invariance and change in the evolution of archosaur limbs. *Journal of Morphology*, **203**, 151–64.

Nassonov, N. (1896). Sur le développement du squelette des extrémités de l'autruche. *Bibliographie Anatomique*, **4**, 160–7.

Nitzsch, C. L. (1811). *Osteografische Beiträge zur Naturgeschichte der Vögel*. Leipzig, Germany: Reclam, pp. 1–122.

Norsa, E. (1895). Recherches sur la morphologie des membres antérieures des oiseaux. *Archives Italiennes de Biologie*, **22**, 232–41.

Ostrom, J. H. (1975). The origin of birds. *Annual Review of Earth and Planetary Sciences*, **3**, 55–77.

Owen, R. (1866). *On the Anatomy of Vertebrates*, Vol. 2 *Birds and Mammals*. London: Longmans, Green.

Parker, T. J. (1891). Observations on the anatomy and development of *Apteryx*. *Philosophical Transactions of the Royal Society of London B*, **182**, 25–134.

Parker, T. J. (1892). Additional observations on the development of *Apteryx*. *Philosophical Transactions of the Royal Society of London B*, **183**, 73–84.

Parker, W. K. (1862). On the osteology of *Balæniceps rex* (Gould). *Transactions of the Zoological Society of London*, **4**, 269–351.

Parker, W. K. (1887). On the morphology of birds. *Proceedings of the Royal Society of London*, **42**, 52–8.

Parker, W. K. (1888). On the structure and development of the wing in the common fowl. *Philosophical Transactions of the Royal Society of London B*, **179**, 385–98.

Parker, W. K. (1895). On the morphology of a reptilian bird, *Opisthocomus cristatus*. *Transactions of the Zoological Society of London*, **13**, 43–85.

Peters, H., Neubuser, A., Kratochwil, K. and Balling, R. (1998). Pax9-deficient mice lack pharyngeal pouch derivatives and teeth and exhibit craniofacial and limb abnormalities. *Genes and Development*, **12**, 2735–47.

Prein, F. (1914). Die Entwicklung des Vorderen Extremitätenskelettes Beim Haushuhn. *Anatomische Hefte*, **51**, 643–90.

Presch, W. (1975). The evolution of limb reduction in the teiid lizard genus *Bachia*. *Bulletin of Southern California Academy of Science*, **74**, 113–21.

Prum, R. O. (2002). Why ornithologists should care about the theropod origin of birds. *The Auk*, **119**, 1–17.

Rabl, C. (1910). *Bausteine zu einer Theorie der Extremitäten der Wirbeltiere*. Leipzig, Germany: Engelmann, 290 pp.

Reichenow, A. (1913). *Die Vögel: Handbuch der systematischen Ornithologie*. Stuttgart, Germany: Ferdinand Enke.

Richardson, M. K. (2009). Molecular tools, classic questions – an interview with Clifford Tabin. *International Journal of Developmental Biology*, **53**, 725–31.

Richardson, M. K. and Keuck, G. (2002). Haeckel's ABC of evolution and development. *Biological Reviews of the Cambridge Philosophical Society*, **77**, 495–528.

Richardson, M. K. and Oelschlager, H. H. (2002). Time, pattern, and heterochrony: a study of hyperphalangy in the dolphin embryo flipper. *Evolution and Development*, **4**, 435–44.

Richardson, M. K., Minelli, A. and Coates, M. I. (1999). Some problems with typological thinking in evolution and development. *Evolution and Development*, **1**, 5–7.

Richardson, M. K., Gobes, S. M., van Leeuwen, A. C., *et al.* (2009). Heterochrony in limb evolution: developmental mechanisms and natural selection. *Journal of Experimental Zoology B–Molecular and Developmental Evolution*, **312**, 639–64.

Romer, A. S. (1956). *Osteology of the Reptiles*. Chicago, IL: University of Chicago Press, 772 pp.

Rosenberg, A. (1873). Über die Entwickelung des Extremitätenskelets bei einigen durch Reduktion ihrer Gliedmassen charakterisierten Wirbeltieren. *Zeitschrift für wissenschaft Zoologie*, **23**, 116–71.

Rosenberg, F. T. (1911). Beiträge zur Entwicklungsgeschichte und Biologie der Colymbidae. *Zeitschrift für wissenschaft Zoologie*, **97**, 199–217.

Sánchez-Villagra, M. R., Muller, H., Sheil, C. A., *et al.* (2009). Skeletal development in the Chinese soft-shelled turtle *Pelodiscus sinensis* (Testudines: Trionychidae). *Journal of Morphology*, **270**, 1381–99.

Sanz-Ezquerro, J. J. and Tickle, C. (2003). Fgf signaling controls the number of phalanges and tip formation in developing digits. *Current Biology*, **13**, 1830–6.

Scherz, P. J., McGlinn, E., Nissim, S. and Tabin, C. J. (2007). Extended exposure to sonic hedgehog is required for patterning the posterior digits of the vertebrate limb. *Developmental Biology*, **308**, 343–54.

Schestakowa, G. S. (1927). Die Entwicklung des Vlogelfügels. Bulletin de la Société des Naturalistes de Moscou *(Biologie)*, **36**, 163–210.

Shapiro, M. D. (2002). Developmental morphology of limb reduction in *Hemiergis* (Squamata: Scincidae): chondrogenesis, osteogenesis, and heterochrony. *Journal of Morphology*, **254**, 211–31.

Shapiro, M. D., Hanken, J. and Rosenthal, N. (2003). Developmental basis of evolutionary digit loss in the Australian lizard *Hemiergis*. *Journal of Experimental Zoology B–Molecular and Developmental Evolution*, **297**, 48–56.

Shubin, N. (1991). The implications of 'the Bauplan' for develoment and evolution of the tetrapod limb. In *Developmental Patterning of the Vertebrate Limb*, ed. J. R. Hinchliffe. New York: Plenum Press, pp. 411–42.

Shubin, N., Tabin, C. and Carroll, S. (1997). Fossils, genes and the evolution of animal limbs. *Nature*, **388**, 639–48.

Shubin, N. H. (2002). Origin of evolutionary novelty: examples from limbs. *Journal of Morphology*, **252**, 15–28.

Shufeldt, R. W. (1881). Osteology of the North American Tetraonidae. *Bulletin of the United States Geological Survey* **6**, 309–50.

Shufeldt, R. W. (1909). *Osteology of Birds*. Education Department Bulletin of the New York State Museum. Albany, NY: University of the State of New York.

Sieglbauer, F. (1911). Zur Entwicklung der Vogelextremität. *Zeitschrift für wissenschaft Zoologie*, **97**, 262–313.

Steiner, H. (1922). Die ontogenetische und phylogenetiche Entwicklung des Vogelflügelskelettes. *Acta Zoologica*, **3**.

Steiner, H. (1934). Ueber die embryonale Hand- und Fuss-Skelett-Anlage bei den Crocodiliern, sowie über ihre Beziehungen zur Flügelanlage und zur ursprunglichen Tetrapoden-Extremität. *Revue Suisse De Zoologie*, **41**, 383–96.

Stephan, B. (1992). Vorkommen und Ausbildung der Fingerkrallen bei rezenten Vögeln. *Journal of Ornithology*, **133**, 251–77.

Suzuki, T., Hasso, S. M. and Fallon, J. F. (2008). Unique SMAD1/5/8 activity at the phalanx-forming region determines digit identity. *Proceedings of the National Academy of Sciences of the United States of America*, **105**, 4185–90.

Szeto, D. P., Rodriguez-Esteban, C., Ryan, A. K., et al. (1999). Role of the bicoid-related homeodomain factor Pitx1 in specifying hindlimb morphogenesis and pituitary development. *Genes and Development*, **13**, 484–94.

Tabin, C. J. (1992). Why we have (only) five fingers per hand: hox genes and the evolution of paired limbs. *Development*, **116**, 289–96.

Taylor, B. K. (1978). Anatomy of forelimb in anteater (*Tamandua*) and its functional implications. *Journal of Morphology*, **157**, 347–67.

Thewissen, J. G. M., Cohn, M. J, Stevens, L. S., et al. (2006). Developmental basis for hind-limb loss in dolphins and origin of the cetacean bodyplan. *Proceedings of the National Academy of Sciences of the United States of America*, **103**, 8414–18.

Tickle, C. (2004). The contribution of chicken embryology to the understanding of vertebrate limb development. *Mechanisms of Development*, **121**, 1019–29.

Tickle, C., Alberts, B., Wolpert, L. and Lee, J. (1982). Local application of retinoic acid to the limb bond mimics the action of the polarizing region. *Nature*, **296**, 564–6.

Tschan, A. (1889). *Recherches sur l'extremité antérieure des oiseaux et des reptiles*. Geneva, Switzerland: Ch. Pfeffer, 63 pp.

Vargas, A. O. and Fallon, J. F. (2005a). Birds have dinosaur wings: the molecular evidence. *Journal of Experimental Zoology B–Molecular and Developmental Evolution*, **304**, 86–90.

Vargas, A. O. and Fallon, J. F. (2005b). The digits of the wing of birds are 1, 2, and 3. A review. *Journal of Experimental Zoology B–Molecular and Developmental Evolution*, **304**, 206–19.

Vargas, A. O. and Wagner, G. P. (2009). Frame-shifts of digit identity in bird evolution and cyclopamine-treated wings. *Evolution and Development*, **11**, 163–9.

Vargas, A. O., Kohlsdorf, T., Fallon, J. F., Vandenbrooks, J. and Wagner, G. P. (2008). The evolution of HoxD-11 expression in the bird wing: insights from *Alligator mississippiensis*. *PLoS One*, **3**, e3325.

Wagner, G. P. and Gauthier, J. A. (1999). 1,2,3 = 2,3,4: a solution to the problem of the homology of the digits in the avian hand. *Proceedings of the National Academy of Sciences of the United States of America*, **96**, 5111–16.

Wagner, G. P., Khan, P. A., Blanco, M. J., Misof, B. E. R. N. and Liversage, R. A. (1999). Evolution of Hoxa-11 expression in amphibians: is the urodele autopodium an innovation? *American Zoologist*, **39**, 686–94.

Wang, X., Zhang, Z., Gao, C., *et al.* (2010). A new enantiornithine bird from the Early Cretaceous of Western Liaoning, China. *Condor*, **112**, 432–7.

Watson, M. (1883). Report on the anatomy of the Spheniscidæ collected during the voyage of H.M.S. Challenger. In *Report of the Scientific Results of the Voyage of H.M.S. Challenger during the years 1873–1876, Zoology*, Vol. 7, ed. J. Murray. Edinburgh, UK: Neill, pp. 1–244.

Welten, M. C., Verbeek, F. J., Meijer, A. H. and Richardson, M. K. (2005). Gene expression and digit homology in the chicken embryo wing. *Evolution and Development*, **7**, 18–28.

Woltering, J. M. and Duboule, D. (2010). The origin of digits: expression patterns versus regulatory mechanisms. *Developmental Cell*, **18**, 526–32.

Wray, R. S. (1887). Note on a vestigial structure in the adult ostrich representing the distal phalanges of digit III. *Proceedings of the Scientific Meetings of the Zoological Society of London*, **1887**, 283–4.

Young, R. L., Caputo, V., Giovannotti, M., *et al.* (2009). Evolution of digit identity in the three-toed Italian skink *Chalcides chalcides*: a new case of digit identity frame shift. *Evolution and Development*, **11**, 647–58.

Zákány, J. and Duboule, D. (2007). The role of Hox genes during vertebrate limb development. *Current Opinion in Genetics and Development*, **17**, 359.

Zákány, J., Fromental-Ramain, C., Warot, X. and Duboule, D. (1997). Regulation of number and size of digits by posterior Hox genes: a dose-dependent mechanism with potential evolutionary implications. *Proceedings of the National Academy of Sciences of the United States of America*, **94**, 13 695–700.

Zhang, F. and Zhou, Z. (2000). A primitive enantiornithine bird and the origin of feathers. *Science*, **290**, 1955–9.

Index

Printed in the United States
by Baker & Taylor Publisher Services